Manufacturing Engineering and Materials Science

This book, which is part of a two-volume handbook set, gives a comprehensive description of recent developments in materials science and manufacturing technology, aiming primarily at its applications in biomedical science, advanced engineering materials, conventional/non-conventional manufacturing techniques, sustainable engineering design, and related domains.

Manufacturing Engineering and Materials Science: Tools and Applications provides state-of-the-art research conducted in the fields of technological advancements in surface engineering, tribology, additive manufacturing, precision manufacturing, electromechanical systems, and computer-assisted design and manufacturing.

The book captures emerging areas of materials science and advanced manufacturing engineering and presents the most recent trends in research for emerging researchers, field engineers, and academic professionals.

Sustainable Manufacturing Technologies: Additive, Subtractive, and Hybrid

Series Editors: Chander Prakash, Sunpreet Singh, Seeram Ramakrishna, and Linda Yongling Wu

This book series offers the reader comprehensive insights of recent research breakthroughs in additive, subtractive, and hybrid technologies while emphasizing their sustainability aspects. Sustainability has become an integral part of all manufacturing enterprises to provide various techno-social pathways toward developing environmental friendly manufacturing practices. It has also been found that numerous manufacturing firms are still reluctant to upgrade their conventional practices to sophisticated sustainable approaches. Therefore this new book series is aimed to provide a globalized platform to share innovative manufacturing mythologies and technologies. The books will encourage the eminent issues of the conventional and non-conventual manufacturing technologies and cover recent innovations.

Sustainable Advanced Manufacturing and Materials Processing
Methods and Technologies
Edited by Sarbjeet Kaushal, Ishbir Singh, Satnam Singh, and Ankit Gupta

3D Printing of Sensors, Actuators, and Antennas for Low-Cost Product Manufacturing
Edited by Rupinder Singh, Balwinder Singh Dhaliwal, and Shyam Sundar Pattnaik

Lean Six Sigma 4.0 for Operational Excellence Under the Industry 4.0 Transformation
Edited by Rajeev Rathi, Jose Arturo Garza-Reyes, Mahender Singh Kaswan, and Mahipal Singh

Handbook of Post-Processing in Additive Manufacturing
Requirements, Theories, and Methods
Edited by Gurminder Singh, Ranvijay Kumar, Kamalpreet Sandhu, Eujin Pei, and Sunpreet Singh

Manufacturing Engineering and Materials Science
Tools and Applications
Edited by Abhineet Saini, B. S. Pabla, Chander Prakash, Gurmohan Singh, Alokesh Pramanik

Manufacturing Technologies and Production Systems
Principles and Practices
Edited by Abhineet Saini, B. S. Pabla, Chander Prakash, Gurmohan Singh, Alokesh Pramanik

For more information on this series, please visit: www.routledge.com/Sustainable-Manufacturing-Technologies-Additive-Subtractive-and-Hybrid/book-series/CRCSMTASH

Manufacturing Engineering and Materials Science

Tools and Applications

Edited by: Abhineet Saini, B. S. Pabla, Chander Prakash, Gurmohan Singh, and Alokesh Pramanik

CRC Press
Taylor & Francis Group
Boca Raton London New York

CRC Press is an imprint of the
Taylor & Francis Group, an **informa** business

Designed cover image: Unsplash

First edition published 2024
by CRC Press
2385 Executive Center Drive, Suite 320, Boca Raton FL 33431

and by CRC Press
4 Park Square, Milton Park, Abingdon, Oxon, OX14 4RN

CRC Press is an imprint of Taylor & Francis Group, LLC

ISBN: 978-1-032-42964-9 (hbk)
ISBN: 978-1-032-43408-7 (pbk)
ISBN: 978-1-003-36715-4 (ebk)

DOI: 10.1201/9781003367154

Typeset in Times
by Apex CoVantage, LLC

Contents

Editor Biographies

Dr Abhineet Saini is working as a Professor in the Department of Mechanical Engineering, Chitkara University, Punjab. He received his PhD degree in mechanical engineering from Panjab University, Chandigarh, India, and has more than 11 years of teaching and research experience in the field of manufacturing technology, which includes conventional machining, biomaterials, composite materials, additive manufacturing, CAD/CAM, and engineering optimization. He has been associated with a number of international and national conferences in the role of session chair for technical sessions. He has been the reviewer of many peer-reviewed journals and has authored/co-authored 20+ articles in peer-reviewed international journals of repute, conferences, and book chapters.

Dr B. S. Pabla is an experienced academician and researcher with 36 years of experience in academics and five years in industry. He earned his PhD in mechanical engineering from Panjab University, Chandigarh, India, and has served at various administrative positions and is presently working as a professor in the Mechanical Engineering Department at NITTTR Chandigarh. He has authored/co-authored books and research articles in various national and international journals of repute. He has also filed four patents, of which one has been granted and three are under process. Dr Pabla was awarded the Eminent Engineering Personality of the Year in 2014 by the Institution of Engineers (India) Haryana State Centre, India, and has visited a number of national and international universities/organizations for academic and research purposes.

Dr Chander Prakash is a Professor at the School of Mechanical Engineering, Lovely Professional University, Jalandhar, India. He has received a PhD in mechanical engineering from Panjab University, Chandigarh, India. His area of research is biomaterials, rapid prototyping, 3D printing, advanced manufacturing, modelling, simulation, and optimization. He has more than 15 years of teaching experience and six years of research experience. He has authored 100+ research papers and 20+ book chapters. He is also a guest editor of two journals, as well as a book series editor for the *Sustainable Manufacturing Technologies: Additive, Subtractive, and Hybrid* for CRC Press/Taylor and Francis.

Dr Gurmohan Singh has over 12 years of teaching, research, and industry experience. His areas of interest include additive manufacturing, biomedical engineering, materials science, and integrative research areas. He received his PhD degree in mechanical engineering from Chitkara University, Punjab, India, and has been associated with a number of international and national conferences as a session chair for technical sessions. He has also been a reviewer for many peer-reviewed journals and has authored/co-authored more than ten research articles/book chapters. He has also been part of organizing committees for three international conferences.

Dr Alokesh Pramanik is currently a senior lecturer in the Mechanical Engineering Department, School of Civil and Mechanical Engineering at Curtin University. He earned his PhD degree in mechanical engineering from the University of Sydney and has more than 15 years of research experience in the fields of manufacturing and composite materials at different universities. He has published more than 110 research articles, which include several books, many book chapters, and many reputed journal articles. His areas of research are synthesis/development, surface modification, and advanced/precision machining of metallic and non-metallic biomaterials.

Contributors

Anjali
Chitkara University Institute of Engineering and Technology
Chitkara University
Punjab, India

Indu Aggarwal
Chitkara School of Planning and Architecture
Chitkara University
Punjab, India

Naveen Aggarwal
University Institute of Engineering and Technology
Panjab University
Chandigarh, India

Priyanka Agarwal
Chitkara School of Planning and Architecture
Chitkara University
Punjab, India

Md. Sofiul Alom
Department of Chemistry
University of Louisville
Louisville, KY, USA

Ravish Arora
Department of Mechanical Engineering
Maharishi Markandeshwar Engineering College (deemed to be University)
Mullana, India

Aarti Bansal
Chitkara University Institute of Engineering and Technology
Chitkara University
Punjab, India

Mohit Bansal
ABES Engineering College
Ghaziabad, Uttar Pradesh, India

Puneet Bansal (Research Scholar)
Department of Electronics & Communication Engineering
I. K. G. Punjab Technical University
Kapurthala, Punjab, India
and
Department of Electronics & Communication Engineering
University Institute of Engineering & Technology
Kurukshetra University
Kurukshetra, Haryana-136119, India

Payal Bassi
Associate Professor
Chitkara Business School
Chitkara University
Rajpura, Punjab

Nerswn Basumatary
Department of Civil engineering
Dr. B R Ambedkar National Institute of Technology
Jalandhar, India

Parishrut Singh Charak:
UIET, Panjab University
Chandigarh, India

Himani Chugh
Chandigarh Group of Colleges
Landran, Mohali, Punjab, India

Rajeev Kumar Dang
Assistant Professor, UIET, PUSSGRC
Hoshiarpur, Punjab, India

Gurjot S. Dhaliwal
Intertribal Research and Resource Center
United Tribes Technical College
Bismarck, ND, USA

Atul Dutta
Chitkara School of Planning and Architecture
Chitkara University
Punjab, India

Atul Garg
Chitkara University Institute of Engineering and Technology
Chitkara University
Punjab, India

Nidhi Bansal Garg
Chitkara University Institute of Engineering and Technology
Chitkara University
Punjab, India

Kamal Deep Garg
Chitkara University Institute of Engineering and Technology
Chitkara University
Punjab, India

Meenu Garg
Chitkara University Institute of Engineering and Technology
Chitkara University
Punjab, India

Parul Gaur
UIET, Panjab University
Chandigarh, India

Paul Gaur
UIET, Panjab University
Chandigarh, India

Cheenu Goel
Associate Professor
Chitkara Business School
Chitkara University
Rajpura, Punjab, India

Rakesh Goyal
Chitkara University Institute of Engineering and Technology
Chitkara University
Punjab, India

Mandy Guinn
Environmental Science Department
United Tribes Technical College
Bismarck, ND, USA

Isha Gupta
Chitkara University Institute of Engineering and Technology
Chitkara University
Punjab, India

Prashant Gupta
Chitkara University Institute of Engineering and Technology
Chitkara University
Punjab, India

Preeti Gupta
University Institute of Engineering and Technology
Panjab University
Chandigarh, India

Shaman Gupta
Department of Mechanical Engineering
Maharishi Markandeshwar Engineering College (deemed to be University)
Mullana, India

Vandana Gupta
Indira Gandhi National College
Ladwa, Haryana

Ram Krishna Hona
Environmental Science Department
United Tribes Technical College
Bismarck, ND, USA

T. Jambhulkar
National Institute of Technology Raipur
Chhattisgarh, India

Shinnu Jangra
Chitkara University Institute of Engineering and Technology
Chitkara University
Punjab, India

Poonam Jindal
Chitkara University Institute of Engineering and Technology
Chitkara University
Punjab, India

B. J. Kalaiselvi
Puducherry Technological University
Puducherry, India

Sachin Kalsi
Department of Mechanical Engineering
Chandigarh University
Mohali, India

Kanish Kapoor
Dr B. R. Ambedkar National Institute of Technology
Jalandhar, India

Mohit Kakkar
Chitkara University Institute of Engineering and Technology
Chitkara University
Punjab, India

Amandeep Kaur
Chitkara University Institute of Engineering and Technology
Chitkara University
Punjab, India

Mehardeep Kaur
Chitkara School of Planning and Architecture
Chitkara University
Punjab, India

Rajesh Kumar
Chitkara University Institute of Engineering and Technology
Chitkara University
Punjab, India

Rakesh Kumar
Department of Mechanical Engineering
Chandigarh University
Mohali, Punjab, India
and
Department of Regulatory & Quality Assurance
Auxein Medical, Sonipat, Haryana, India

Santosh Kumar
Department of Mechanical Engineering
Chandigarh Group of Colleges
Landran Mohali, Punjab, India

Satish Kumar
Department of Mechanical Engineering
Chandigarh Group of Colleges
Landran, Punjab, India

Neeru Mago
PUSSGRC
Hoshiarpur, Punjab, India

Reetu Malhotra
Chitkara University Institute of Engineering and Technology
Chitkara University
Punjab, India

Archana Mantri
Chitkara University Institute of Engineering and Technology
Chitkara University
Punjab, India

Kapil Mehta
Chandigarh Group of Colleges
Mohali, Punjab, India

Sahil Mehta
Chitkara University Institute of Engineering and Technology
Chitkara University
Punjab, India

Shivani Malhotra
Chitkara University Institute of Engineering and Technology
Chitkara University
Punjab, India

Amruthamol N. A.
PG Student, Dr. B. R. Ambedkar National Institute of Technology
Jalandhar, India

Harpreet Kaur
Chitkara University Institute of Engineering and Technology
Chitkara University
Punjab, India

Sushil Kumar Narang
Chitkara University Institute of Engineering and Technology
Chitkara University
Punjab, India

Shahbaz P.
Dr B. R. Ambedkar National Institute of Technology
Jalandhar, India

Pankaj
Chandigarh University
Punjab, India

Uttam S. Phuyal
University of Mt Olive
Mount Olive, NC, USA

Punam
University College Miranpur Punjabi University
Patiala, Punjab, India

R. K. Sahu
National Institute of Technology Raipur
Chhattisgarh, India

Abhineet Saini
Chitkara University Institute of Engineering and Technology
Chitkara University
Punjab, India

Jasminder Kaur Sandhu
Chandigarh University
Mohali, Punjab, India

Mamatha Sandhu
Chitkara University Institute of Engineering and Technology
Chitkara University
Punjab, India

Atipriya Sharma
Chitkara University Institute of Engineering and Technology
Chitkara University
Punjab, India

K. D. Sharma
Chitkara University Institute of Engineering and Technology
Chitkara University
Punjab, India

Neeraj Sharma
Department of Mechanical Engineering
Maharishi Markandeshwar Engineering College (deemed to be University)
Mullana, India

N. K. Sharma
Department of Mechanical Engineering
Chandigarh University
Mohali, India

Vijay Kumar Sharma
Chitkara University Institute of Engineering and Technology
Chitkara University
Punjab, India

Shubham Sharma
Department of Mechanical Engineering
Chandigarh University
Mohali, India

Sumit Sharma
Dr B. R. Ambedkar National Institute of Technology
Jalandhar, India

Shaveta
Guru Nanak College
Ferozepur, Punjab, India

Gurjinder Singh
Chitkara University Institute of Engineering and Technology
Punjab, India

Harvinder Singh
Chandigarh Group of Colleges
Landran Mohali, Punjab, India

Jagjit Singh
Department of Mechanical Engineering
Chandigarh University
Mohali, India

Paramveer Singh
Dr B. R. Ambedkar National Institute of Technology
Jalandhar, India

S. P. Singh
Dr B. R. Ambedkar National Institute of Technology
Jalandhar, India

Hitesh Singla
Chitkara University Institute of Engineering and Technology
Chitkara University
Punjab, India

Yogesh Kumar Singla
University of Idaho
Moscow, Idaho, USA

Vandana Mohindru Sood
Chitkara University Institute of Engineering and Technology
Chitkara University
Punjab, India

S. Sukandhiya
Puducherry Technological University
Puducherry, India

R. Suruthy
Puducherry Technological University
Puducherry, India

Tilak Thakur
Punjab Engineering College, (Deemed to be University)
Chandigarh, India

B. Uthayakumar
Puducherry Technological University
Puducherry, India

Renu Vig
Panjab University
Chandigarh, India

Tanner Vio
United Tribes Technical College
Bismarck, ND, USA

Hemender Yadav
Chitkara University Institute of Engineering and Technology
Chitkara University
Punjab, India

Harvinder Singh
Chitkara University Institute of Engineering and Technology
Chitkara University
Punjab, India

1 Nanomaterials for Cyber-Physical Systems

A Review

*Vandana Mohindru Sood, Kapil Mehta,
Sushil Kumar Narang, Kamal Deep Garg*

1.1 INTRODUCTION

Nowadays various current computing systems employ a hybrid of independently developed cyber and physical systems. Increased performance and complex usage patterns have significantly altered the dynamics and connections of cyber and physical elements. These changes necessitate fundamentally new design approaches that integrate cyber and physical components at all levels [1]. As a result of this requirement, progressions in the field of cyber-physical system (CPS) research have occurred. CPSs are the systems that provide integrations of networking, computation, and physical processes, or the systems that deeply intertwined physical and software components, separately operating on diverse spatial and temporal scales. These systems exhibit numerous and different behavioural modalities and interact with each other in a variety of varying ways. CPSs integrate tangible process dynamics with software and communication dynamics, providing abstractions as well as modelling, analysis, and design practices for the unified whole. The primary goal of CPS research programme is to integrate physical and cyber design as deeply as possible.

In cyber systems, computations are performed with the goal of maximizing renewable resource utilization, and a suitable decision is made, based on which the physical resources are further controlled. CPS design necessitates extensive reasoning due to exclusive challenges and multifaceted functionality, reliability, and performance requirements. As shown in Figure 1.1, the CPS interfaces are made up of a transmission network and other transitional elements, such as organized sensors, analogue-to-digital converters (ADCs), actuators, and digital-to-analogue converters (DACs), that are accountable for connecting cyber systems to the physical world [2].

Sensors and actuators are in charge of changing other types of energy to electricity (analogue signal) and vice versa. CPS is an intellectual challenge that focuses on the intersection, rather than the unification, of the physical and the cyber. It joins environmental, mechanical, civil, biomedical, electrical, aeronautical, chemical, and industrial engineering models and approaches with computer science models and methods [3].

DOI: 10.1201/9781003367154-1

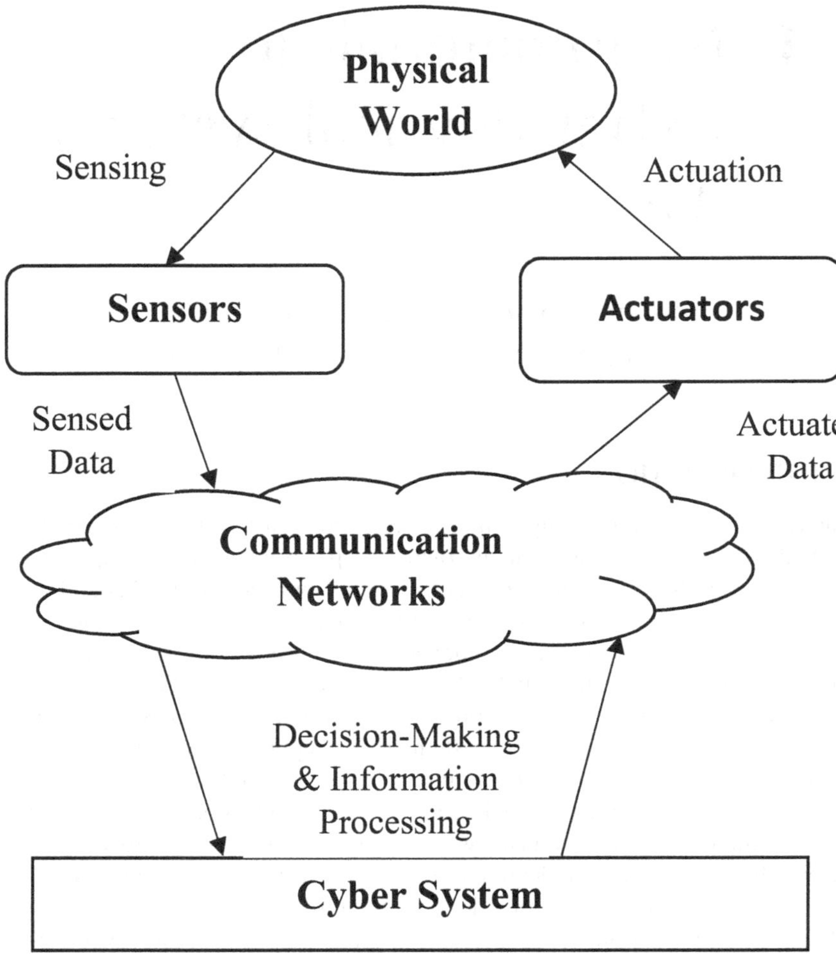

FIGURE 1.1 Architecture of the cyber-physical system.

This figure here shows the architecture of the cyber-physical system, which include sensors, analogue-to-digital converters (ADCs), actuators, and digital-to-analogue converters (DACs), that are accountable for connecting cyber systems to the physical world.

Because CPSs syndicate physical dynamics with the process of computation processes, it appears that they can benefit from the authority of deterministic models. Inappropriately, combining these deterministic models for physical dynamics and computation is almost always nondeterministic [4]. The main motive for this is that programmes lack temporal semantics. The goals of the CPS are to integrate knowledge from various domains into a consistent body of knowledge that is supported by the fundamental values of technical, natural, social, formal, and human sciences, as well as to develop a system-level consideration and abstract frameworks for this family of structures [5].

Overall, the discipline appears to be rather immature, with some unresolved problems related to computation and communication. Mobility also infers limited size and energy consumption constraints. All-in-one connectivity with different devices and immovable networks is a critical enabler for ambient intelligence models, necessitating higher data rates for wireless links. Sensing, intelligence, and increased data rates necessitate extra memory and computing power, which, when combined with size constraints, pose significant issues in thermal management [6].

All of these necessities, when combined, result in a situation that cannot be resolved using current technologies. The execution of the physical components of CPSs, such as sensors, actuators, transducers, and transponders, necessitates the use of cutting-edge materials [7]. While functionally enhanced materials and multifunctional materials were the true technological breakthroughs a decade ago, carbon nanotubes, quantum dots, and other materials are now the concentration of research. The use of nanotechnologies enables material scientists to define material properties at the atomic and molecular levels by creating nanostructures [8].

Over the last decade, much effort has been directed toward improving conventional interrelated expertise for chip-to-chip, on-chip, and board-to-board transmission by lowering the resistivity of conductors such as copper (Cu) and dropping the dielectric constant of interlayer dielectric constituents through the use of low-k polymers. Because of this, interconnects are regarded as the hardest challenges that giga-scale system incorporation faces [9]. Another problem with copper-wire interconnect resonant digital information is that it is inadequate to provide the connections needed by a considerably increasing transistor count, stifling the advancement of forthcoming ultra-large-scale computing.

This chapter will provide an overview of various nanomaterials used in the design of a variety of CPS components and applications that lead to the accomplishment of forthcoming transmission technologies, computation, and smart sensor networks.

1.2 MATERIALS AND METHODS

CPS must incorporate concepts of communication and computing systems, sensing and control of physical systems, and human-computer interface. CPS composition science needs to adapt new architecture designs, classify system configuration from materials and subsystems, and use QoS (quality of service) theories, protocol composition, and new modelling languages and tools to specify, analyse, synthesize, and simulate various compositions [10]. We will be able to do the following with the proposed scientific and engineering foundation:

1. Make use of novel and efficient programming concepts and hardware functions.
2. Capture the physical and material constraints that cooperate with the cyber system constraints, such as complexity, sturdiness, protection, and security.

3. Iteratively create both system structure and system behaviour models, as well as mapping behaviours to structures and vice versa.
4. Perform a quantifiable trade-off evaluation that take into account the existing technologies and the constraints imposed by physical components, cyber components, and human operators.
5. Guarantee the safety, constancy, robustness, and safety of the CPS in the face of genuine models of environmental uncertainties, operator errors, flaws in physical or cyber components, and security attacks.

Nanomaterials, or more precisely nanotechnology, are rapidly gaining traction. Over the past two years, significant growth has been done in the monitoring and production of innovative materials on the nanoscale at the level of the atom, molecules, and supramolecular assemblies [11]. Several nanostructured constituents have revealed great ability as building blocks for future small, scalable, and energy-efficient electronics and electromechanical, magnetic, and photonic systems that will alter communication and computing.

1.3 NANOMATERIALS FOR CYBER-PHYSICAL SYSTEMS

Nanomaterial advancements are becoming critical to progress in all of these areas, including CPSs. Nanomaterials will realistically transform communications connections to enable quicker data transmission rates for networks with universal intelligent ambient schemes at home, public, and workplaces [12]. At the moment, nanodevices for memory, miniaturized processors, interconnects, circuits, and perhaps self-powered computing systems of the future with extraordinary intelligence, scalability, and energy efficiency are posing engineering and scientific challenges rather than an outstanding field of fundamental research. Nanoparticles are categorized according to their morphology, size, and chemical and physical possessions. Carbon-created nanoparticles, metal nanoparticles, ceramic nanoparticles, polymeric nanoparticles, semiconductor nanoparticles, and lipid-based nanoparticles are rare specimens [13]. The various components of cyber-physical systems designed using nanomaterials are listed in Table 1.1.

TABLE 1.1
Nanomaterials Used in the Design of Various Components of Cyber-Physical Systems

Components of a Cyber-Physical Systems	Nanomaterials
Sensor nodes	Photovoltaics, quantum dots, plasmonic nanotube transistors, photonic crystals
Network routers	Quantum dots, photonic crystals
Interface gadgets	Carbon nanotubes (CNTs), fullerenes, graphene
Gateway	CNTs
Material for communication devices	CNT-based in-chip connections, thermoelectric material
Nano optical interconnects	Photonic interconnects
Nano storage devices	Spintronic, CNT memory, ferroelectric RAMs

The following nanomaterials can be used to design CPS components for improved performance and computation:

1. **Carbon nanotubes (CNTs):** CNTs are one-dimensional nanostructures, and as such, they should be regarded as a novel material with distinct physicochemical qualities that hold the great ability for a varied variety of applications, including additives to polymers and catalysts, gas discharge tubes in wired networks, electromagnetic wave fascination and screening, lithium battery anodes, energy alteration, hydrogen storing, composite materials (fillers and coatings), and nanoparticles. CNTs have a wide range of new unusual electrical, mechanical, and magnetic properties that could pave the way for a breakthrough in nanoelectronics [14, 15].
2. **Photovoltaics:** Photovoltaic (PV) technology is used to convert the energy absorbed from sunlight into electricity. These are used to improve the efficiency and performance of various sensing devices. Photovoltaic work produces no carbon dioxide, which benefits environmental safety.
3. **Plasmonic nanotube transistors:** Over the last half-decade, the field of plasmonics—which uses electron waves generated when photons strike a metal structure to carry out optoelectronic processes—has gained traction in the research community [16]. These nano transistors are widely used in various communication devices, allowing for faster data transmission over networks.
4. **Quantum dots:** Quantum dots have been used as operational materials in a diversity of applications, such as memory components, optical sensors, and quantum lasers. The nanostructure of quantum dots is zero-dimensional [17].
5. **Photonic crystals:** Photonics is a future technology that integrates all optical components and devices into a photonic chip. These chips will meet the growing demand for signal dispensation and high-speed data communication between users. These will greatly improve light wave communications due to their high reliability, low error rate, and high speed of operation. This eventually leads to signal routing between diverse ports in an integrated circuit with wide-ranging connectivity and enhanced performance [18].
6. **Supercapacitors:** The supercapacitor is a type of energy-storing device that endures regular charge and liberation cycles at high current and short duration. They have a higher transient response, a higher power density, a lower weight, a lower volume, and a lower internal resistance, making them suitable for a wide range of applications.
7. **Nanoelectromechanical systems (NEMS):** NEMS are nanoscale devices that combine electrical and mechanical functionality. Because of NEMS's most appealing features, such as smaller mass, high electrical efficiency, and high surface area to volume ratio, it is chosen for highly effective applications, such as high-frequency resonators and ultrasensitive sensors.

8. **Thermoelectric materials:** The thermoelectric effect describes a phenomenon in which either a temperature difference or an electric current creates a temperature difference. These materials are used in energy harvesting applications that harvest various forms of energy from the environment and convert it to electricity on a continuous basis.
9. **Photonics:** Photonics is playing an increasingly significant role in driving invention across a wide range of fields. Photonic applications range from optical data communications to imaging, lighting, and displays, as well as the industrial sector, life sciences, healthcare, safety, and security [19].
10. **Spintronics:** Spintronic devices are widely used in large storage devices. It is used to adjust large volumes of data to a smaller area; for example, one trillion bits per square inch (1.5 Gbit/mm^2) or coarsely one terabyte of data can be deposited on a single-sided 3.5-inch diameter disc.

1.4 CYBER-PHYSICAL SYSTEM—APPLICATIONS

As the demand for CPS grows, so does the range of applications. The incorporation of nanomaterial technologies into CPS results in improved performance that can be seen in a wide range of applications, including agriculture, energy, healthcare, transportation, and manufacturing [20, 21, 22]. Some applications are discussed as follows:

1. **Healthcare and medicine:** National health information networks, operating rooms, electronic patient record initiatives, home care, and other healthcare and medicine domains are progressively managed by computational systems having hardware and software components attached and are real-time models with protection and timing necessities [23]. Health CPS (HCPS) will substitute traditional health devices that work independently in the future. Various health devices collaborate with sensors and networks to detect patients' physical conditions in real time, particularly for critical patients, like those with heart disease [24, 25].
2. **Vehicular systems and transportation:** Modern vehicles depend upon CPSs, which provide improved displays, entertainment, and information while also managing the vehicle's motion and energy utilization. CPS benefits have been used in a variety of vehicular systems [26, 27]. In order to improve safety, advanced sensing and computing capabilities will be extensively used in next-generation transportation systems, such as railway, air traffic, and car control [28].
3. **Smart homes and buildings:** Smart homes include sensors and actuators that are set up so that they can be handled remotely via the internet [29, 30]. The activities of the users can thus be tracked. The concept of smart communities expands on the notion of smart homes by utilizing networking between a cluster of smart homes. Specific homes are designed to include multi-efficient sensors, and physical response is provided to advance community safety, better healthcare quality, and home safety as needed [31].

4. **Electric power grid:** A CPS is made up of power electronics, a power grid, and embedded control software, and its design is deeply inclined by security, fault tolerance, decentralized control, and economic considerations [32, 33].
5. **Data centres:** A data centre comprises online applications and services to ensure computation and communication, as well as physical components to ensure proper operation; thus, a data centre can be modelled as a CPS. The multifaceted interaction of cyber and physical components in a large-scale data centre necessitates using CPS design principles to address data centre management issues [34, 35].

1.5 CONCLUSION AND FUTURE SCOPE

CPS is an emerging standard for designing present and future engineered schemes, and it is expected to have a significant influence on our connections with the physical world. The concept of CPS focuses on the unified system strategy rather than the cyber or physical system separately. Nanomaterials, both raw and refined, are critical to future technological expansion and industrial revolutions. The current era will be referred to as the nanomaterials age, as it will deliver our humanity with the tools it needs to address several technological fences. Several nanoscale device-based innovations can play a unique and significant role in increasing the data communication rapidity of forthcoming communication systems. While it is incredible to forecast the final timeline of distribution and applications of novel nanodevices in climbable profitable systems at this time, the scientific community is likely to be kept busy in the next decade by resolute efforts to overcome barriers to large-scale and low-cost integration of functional nanomaterials in CPS, as well as understanding their environmental impact. Work would be done in the near future on developing CPSs that use nano components for industry or agriculture for various monitoring activities. In the future, CPS will include not only physical but also chemical and biological subsystems in which information manifests in a variety of substances and forms and exists on multiple spatial and time scales.

REFERENCES

[1] Bonci, Andrea, Alessandro Carbonari, Alessandro Cucchiarelli, Leonardo Messi, Massimiliano Pirani, and Massimo Vaccarini. "A cyber-physical system approach for building efficiency monitoring." *Automation in Construction* 102 (2019): 68–85.

[2] Ahmed, Syed Hassan, Gwanghyeon Kim, and Dongkyun Kim. "Cyber physical system: Architecture, applications and research challenges." *2013 IFIP Wireless Days (WD)* (2013): 1–5.

[3] Chen, Hong. "Applications of cyber-physical system: A literature review." *Journal of Industrial Integration and Management* 2, no. 3 (2017): 1750012.

[4] Ding, Derui, Qing-Long Han, Xiaohua Ge, and Jun Wang. "Secure state estimation and control of cyber-physical systems: A survey." *IEEE Transactions on Systems, Man, and Cybernetics: Systems* 51, no. 1 (2020): 176–190.

[5] Yaacoub, Jean-Paul A., Ola Salman, Hassan N. Noura, Nesrine Kaaniche, Ali Chehab, and Mohamad Malli. "Cyber-physical systems security: Limitations, issues and future trends." *Microprocessors and Microsystems* 77 (2020): 103201.

[6] Shi, Jianhua, Jiafu Wan, Hehua Yan, and Hui Suo. "A survey of cyber-physical systems." In *2011 International Conference on Wireless Communications and Signal Processing (WCSP)*, pp. 1–6. IEEE, 2011.
[7] Lee, Edward A. "The past, present and future of cyber-physical systems: A focus on models." *Sensors* 15, no. 3 (2015): 4837–4869.
[8] Nandy, Turja, Ronald A. Coutu Jr., and Cristinel Ababei. "Carbon monoxide sensing technologies for next-generation cyber-physical systems." *Sensors* 18, no. 10 (2018): 3443.
[9] Islam, M. Saif, and Logeeswaran Vj. "Nanoscale materials and devices for future communication networks." *IEEE Communications Magazine* 48, no. 6 (2010): 112–120.
[10] Noah, Naumih M. "Design and synthesis of nanostructured materials for sensor applications." *Journal of Nanomaterials* 2020 (2020).
[11] Thirumaran, Sridarshini, Shanmuga Sundar Dhanabalan, and Indira Gandhi Sannasi. "Design and analysis of photonic crystal ring resonator based 6× 6 wavelength router for photonic integrated circuits." *IET Optoelectronics* 15, no. 1 (2021): 40–47.
[12] Notomi, Masaya. "Nanophotonics for optical communications." In *2011 37th European Conference and Exhibition on Optical Communication*, pp. 1–1. IEEE, 2011.
[13] Baig, Nadeem, Irshad Kammakakam, and Wail Falath. "Nanomaterials: A review of synthesis methods, properties, recent progress, and challenges." *Materials Advances* 2, no. 6 (2021): 1821–1871.
[14] Pomerantseva, Ekaterina, Francesco Bonaccorso, Xinliang Feng, Yi Cui, and Yury Gogotsi. "Energy storage: The future enabled by nanomaterials." *Science* 366, no. 6468 (2019): eaan8285.
[15] Riley, Parand R., and Roger J. Narayan. "Recent advances in carbon nanomaterials for biomedical applications: A review." *Current Opinion in Biomedical Engineering* 17 (2021): 100262.
[16] Wu, Ziping, Yonglong Wang, Xianbin Liu, Chao Lv, Yesheng Li, Di Wei, and Zhongfan Liu. "Carbon-nanomaterial-based flexible batteries for wearable electronics." *Advanced Materials* 31, no. 9 (2019): 1800716.
[17] Tan, Yifei, Wenhe Yang, Kohtaroh Yoshida, and Soemon Takakuwa. "Application of IoT-aided simulation to manufacturing systems in cyber-physical system." *Machines* 7, no. 1 (2019): 2.
[18] Jamaludin, Juliza, and Jemmy Mohd Rohani. "Cyber-physical system (cps): State of the art." In *2018 International Conference on Computing, Electronic and Electrical Engineering (ICE Cube)*, pp. 1–5. IEEE, 2018.
[19] Sun, Shengjing, Xiaochen Zheng, Bing Gong, Jorge Garcia Paredes, and Joaquín Ordieres-Meré. "Healthy operator 4.0: A human cyber–physical system architecture for smart workplaces." *Sensors* 20, no. 7 (2020): 2011.
[20] Vaseashta, Ashok. "Cyber-physical systems—nanomaterial sensors based unmanned aerial platforms for real-time monitoring and analysis." In *International Conference on Nanotechnologies and Biomedical Engineering*, pp. 685–689. Springer, 2019.
[21] Mehta, Kapil, Vandana Mohindru Sood, Chamkaur Singh, and Pratham Chabra. "Machine learning based intelligent system for safeguarding specially abled people." In *2022 7th International Conference on Communication and Electronics Systems (ICCES)*, pp. 1199–1206. IEEE, 2022.
[22] Rai, Vinayak, Karan Bagoria, Kapil Mehta, Vandana Mohindru Sood, Kartik Gupta, Lakshya Sharma, and Manav Chauhan. "Cloud computing in healthcare industries: Opportunities and challenges." *Recent Innovations in Computing* (2022): 695–707.
[23] Gatouillat, Arthur, Youakim Badr, Bertrand Massot, and Ervin Sejdić. "Internet of medical things: A review of recent contributions dealing with cyber-physical systems in medicine." *IEEE Internet of Things Journal* 5, no. 5 (2018): 3810–3822.

[24] Shu, Hong, Ping Qi, Yongqing Huang, Fulong Chen, Dong Xie, and Liping Sun. "An efficient certificateless aggregate signature scheme for blockchain-based medical cyber physical systems." *Sensors* 20, no. 5 (2020): 1521.
[25] Mohindru, Vandana, Sunidhi Vashishth, and Deepak Bathija. "Internet of Things (IoT) for healthcare systems: A comprehensive survey." *Recent Innovations in Computing* (2022): 213–229.
[26] Deng, Hsien-Wen, Mizanur Rahman, Mashrur Chowdhury, M. Sabbir Salek, and Mitch Shue. "Commercial cloud computing for connected vehicle applications in transportation cyberphysical systems: A case study." *IEEE Intelligent Transportation Systems Magazine* 13, no. 1 (2020): 6–19.
[27] Mohindru, Vandana, Ravindara Bhatt, and Yashwant Singh. "Reauthentication scheme for mobile wireless sensor networks." *Sustainable Computing: Informatics and Systems* 23 (2019): 158–166.
[28] Tiganasu, Alexandru, Corneliu Lazar, Constantin Florin Caruntu, and Constantin Dosoftei. "Comparative analysis of advanced cooperative adaptive cruise control algorithms for vehicular cyber physical systems." *Journal of Control Engineering and Applied Informatics* 23, no. 1 (2021): 82–92.
[29] Ahmad, Md, Mohd Abdul Ahad, M. Afshar Alam, Farheen Siddiqui, and Gabriella Casalino. "Cyber-physical systems and smart cities in India: Opportunities, issues, and challenges." *Sensors* 21, no. 22 (2021): 7714.
[30] Garg, Kamal Deep, Shashi Shekhar, Ajit Kumar, Vishal Goyal, Bhisham Sharma, Rajeswari Chengoden, and Gautam Srivastava. "Framework for handling rare word problems in neural machine translation system using multi-word expressions." *Applied Sciences* 12, no. 21 (2022): 11038. https://doi.org/10.3390/app122111038.
[31] Khan, Firoz, R. Lakshmana Kumar, Seifedine Kadry, Yunyoung Nam, and Maytham N. Meqdad. "Cyber physical systems: A smart city perspective." *International Journal of Electrical and Computer Engineering* 11, no. 4 (2021): 3609.
[32] Mazumder, Sudip K., Abhijit Kulkarni, Subham Sahoo, Frede Blaabjerg, H. Alan Mantooth, Juan Carlos Balda, Yue Zhao et al. "A review of current research trends in power-electronic innovations in cyber–physical systems." *IEEE Journal of Emerging and Selected Topics in Power Electronics* 9, no. 5 (2021): 5146–5163.
[33] Jha, Amitkumar Vidyakant, Bhargav Appasani, Abu Nasar Ghazali, Prabina Pattanayak, Devendra Singh Gurjar, Ersan Kabalci, and D. K. Mohanta. "Smart grid cyber-physical systems: Communication technologies, standards and challenges." *Wireless Networks* 27, no. 4 (2021): 2595–2613.
[34] Zhang, Qingxia, Chao Tang, Tian Bai, Zihao Meng, Yuhao Zhan, Junyu Niu, and M. Jamal Deen. "A two-layer optimal scheduling framework for energy savings in a data center for cyber–physical–social systems." *Journal of Systems Architecture* 116 (2021): 102050.
[35] Narang, Sushil Kumar, and Neha Kishore. "Issues in credit card transactional data stream: A rational review." In *Proceedings of Third International Conference on Computing, Communications, and Cyber-Security*, pp. 775–789. Springer, 2023.

2 Experimental Investigation on Wear Performance of Different Materials for Stamping of Mild Steel

Santosh Kumar, Harvinder Singh

2.1 INTRODUCTION

Recently, in forming and machining industry, distinct tool steels were developed to enhance production economic efficiency, owing to their increased mechanical characteristics, such as high resistance against wear, strength, toughness, ductility, and hardness [1]. In many industries, metal part production depends on metal machining and forming. For instance, the demand for tool steels has increased as a result of an increase in car manufacturing in the automotive sector. Other industries, like aircraft, transportation, and precision industries, have also seen a growth in demand. The metal-forming industry has significant growth, growing at a pace of 3% through the year 2019 [2]. Six kinds of tool steels have been established: cold work, hot work, shock resistance, mould, high speed, and special purpose tool steels [3]. Cold work tool steels are the most crucial category. The AISI D2, AISI D3, AISI H11, and AISI O1 steels were the subjects of this study's investigation. These metals are employed in several applications, including dies and punching and cutting tools. They are affordable, have great wear resistance, and high hardness [4–6]. The AISI O1 steels steel was found to have great machinability, although AISI D2 had higher wearing resistance in earlier experiments testing the two metals to assess their wearing characteristics and resistance [7–10].

Further, due to the increasing pressure from regulations governing fuel efficiency, the depletion of fossil fuel supplies, and environmental concerns, light weighing of motor vehicles is one of the issues facing the automobile industry. In addition, the light weighting of vehicles needs to be handled in a cost-effective manner without sacrificing passenger safety. Ultra-high-strength steels (UHSS) are one of the options that are thought to be the material of preference for lightweight auto bodies and structures in the near future. But these steels need different processing than ordinary steels. Some of the expertise areas related to UHSS transformation that still

 DOI: 10.1201/9781003367154-2

needs improvement is our understanding of the tribological link between the forming tools and the UHSS components at elevated temperature during the sheet-metal-forming process. Due to rising die cost of maintenance and waste rates, the tool steel wear in sheet-metal-forming continues to be a major problem for the automotive sector. Cold rolling tools are subjected to considerable wear strains because of elevated contact pressures produced by sliding contact between the die and the sheet materials. As a consequence, a substantial quantity of frictional heat is produced, which affects the tool steel's wear characteristics [11]. Despite being subjected to high temperatures, tool steel wear and typical wear mechanisms have been carefully studied [12–13]. Several writers have also considered wear performance in friction operations to explain distinct tool steel grades (uncoated and coated) at high temperatures [13–15], thus reducing the level of tool wear, while hot stamping requires an understanding of the elements that affect the wear operations. The lifespan of the die materials utilized in hot stamping may be extended by using this information to direct tool material selection and die design [16, 17]. Understanding wear failure processes properly is crucial to preventing the surface degradation of hot stamping tools [18–20]. Aspects directly connected to the mechanical characteristics of the materials serve to establish the wear processes responsible for tool degradation. Alterations to the martensitic structure are thought to improve the response to wear [21–30]. In order to determine which tool steel resists wear during press hardening the best, this chapter examines four sample tool steels. In light of the microstructural properties, the results are analysed.

2.2 EXPERIMENTAL WORK

The wear and friction monitor TR 201 was used in the present investigation to examine the wear on pins manufactured of AISI D2, D3, H11, and O1 steel. The main purpose of the TR 201 Series tribometer's creation was to measure fundamental wear and friction. In this device, a test pin is pressed on a rotating disc by means of an identified force. There is a capability intended for monitoring multifaceted wear and frictional force. The chemical configuration of steels and mild steel is depicted in Tables 2.1 and 2.2, respectively.

Further, distinct variables and their levels are given in Table 2.3.

TABLE 2.1
Elemental Configuration of Steel (% Weight)

Grade	C	Mn	P	S	Si	Cr	V	W	Mo
D2	1.52	0.401	0.038	0.03	0.43	12.426	0.029	0.035	0.083
D3	2.135	0.275	0.025	0.014	0.506	11.463	0.047	0.034	0.02
H11	0.35	0.4	0.03	0.03	1.09	5.24	0.45	–	1.54
O1	0.92	1.29	0.03	0.03	0.65	0.79	0.3	0.56	–

TABLE 2.2
Elemental Configuration of Mild Steel (% Weight)

Element	C	Mn	P	S	Si	V	Cr	Mo	Fe
%	0.102	0.456	0.028	0.016	0.192	0.011	0.048	0.053	Balance

TABLE 2.3
Variable and Their Levels

Factor	Level			
Load (N)	50.0	50.0	50.0	50.0
Speed (m/s)	1.16	1.67	2.2	2.56
Time (min.)	10	10	10	10

2.2.1 Weight Loss Measurement

Wear is the gradual loss of material as a result of relative action between a disc and a pin being examined. To within 10^{-4} g, the pins are weighed before and after the test was done to determine the weight decrease. The change in pin masses throughout the test may be used to calculate the volumetric loss of pin material.

$$\text{WEAR VOLUME LOSS IN mm}^3 = \frac{weight\ loss}{Density}$$

2.3 RESULTS AND DISCUSSIONS

For further examination, the entire findings of 16 experiments were performed. Plots of different kinds were used to examine the impact of each parameter. SEM observations were used to show various kinds of wear processes.

2.3.1 Hardness Measurement

The Rockwell hardness values for AISI D2, D3, H11, and O1 are shown in Figure 2.1. Oil hardening is the cause of the high hardness of AISI O1.

2.3.2 Specific Wear Rates

Table 2.4 displays the specific weight loss and wear rate of AISI D2, D3, H11, and O1 at various speeds. High wear performance is shown by lower values of certain wear rates.

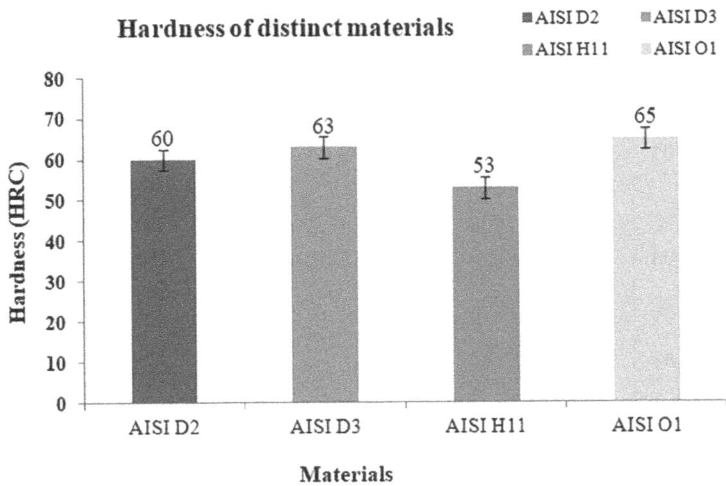

FIGURE 2.1 Measured hardness values of distinct materials.

TABLE 2.4
Experimental Results (at a Load and Time of 50 N and 10 Min, Respectively)

	Material							
	AISI D2		AISI D3		AISI H11		AISI Ol	
Speed (*m/s*)	**Weight loss (*g*)**	**Specific wear (*mm³/Nm*)**	**Weight loss (*g*)**	**Specific wear (*mm³/Nm*)**	**Weight loss (*g*)**	**Specific wear (*mm³/Nm*)**	**Weight loss (*g*)**	**Specific wear (*mm³/Nm*)**
1.15	0.6851	2.60E-03	0.0936	3.62E-04	0.1338	5.20E-04	0.2301	8.62E-04
1.67	0.5825	1.52E-03	0.0508	1.35E-04	0.0635	1.70E-04	0.0279	1.15E-04
2.2	0.7249	1.44E-03	0.1013	2.05E-04	0.1801	3.66E-04	0.0445	1.80E-04
2.56	0.7026	1.20E-03	0.1354	2.35E-04	0.1913	3.34E-04	0.0922	4.69E-05

From Table 2.4, it is clear that O1 performs best among the tested materials at speeds of 1.67 m/s, 2.2 m/s, and 2.56 m/s, while AISI D2 performs worst. The four materials chosen were AISI D2, AISI D3, AISI H11, and AISI O1. At all speeds, AISI D2 performs the least well. AISI H11 is inferior to AISI D3. The findings demonstrate that wear resistance is inversely related to hardness. Oil-hardened steel, or AISI O1, has a harder surface than any of the other materials we considered. Exceptional carbon martensite that has been tempered at a low temperature to produce extremely fine carbide deposition provides high hardness in AISI O1. Oil quenching offers a good depth of hardening because to the high-carbon and

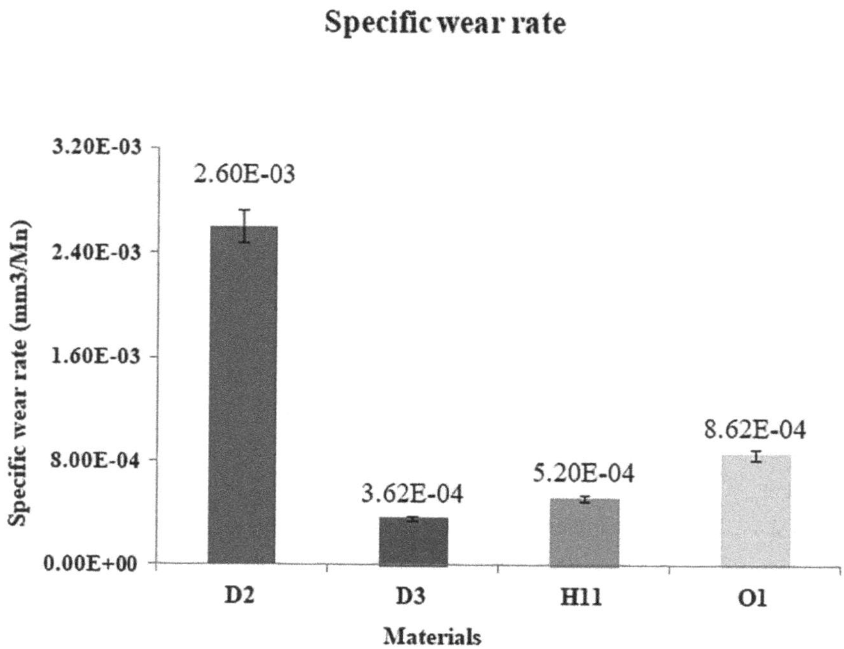

FIGURE 2.2 Plot of specific wear rate at 1.15 m/s.

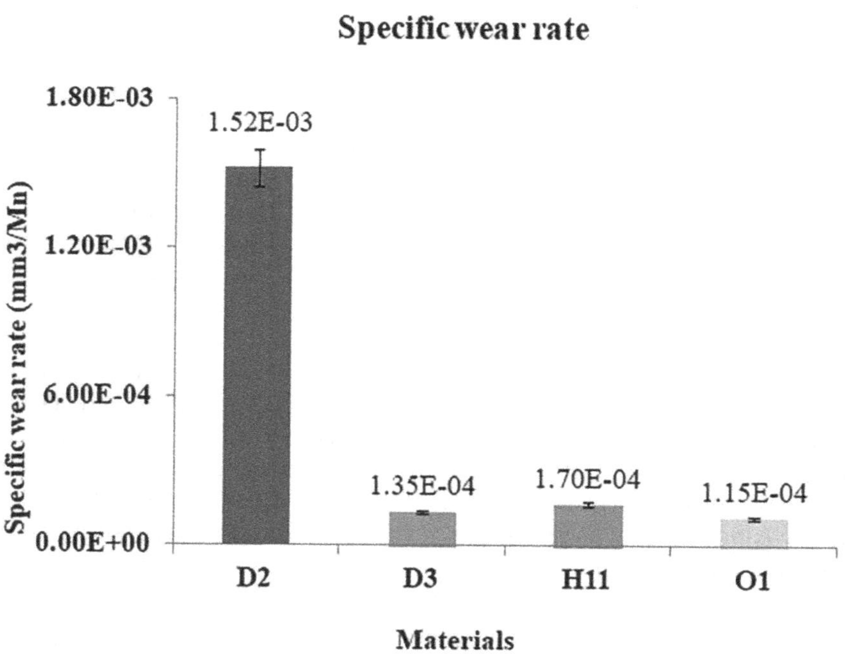

FIGURE 2.3 Specific wear rate at 1.67 m/s.

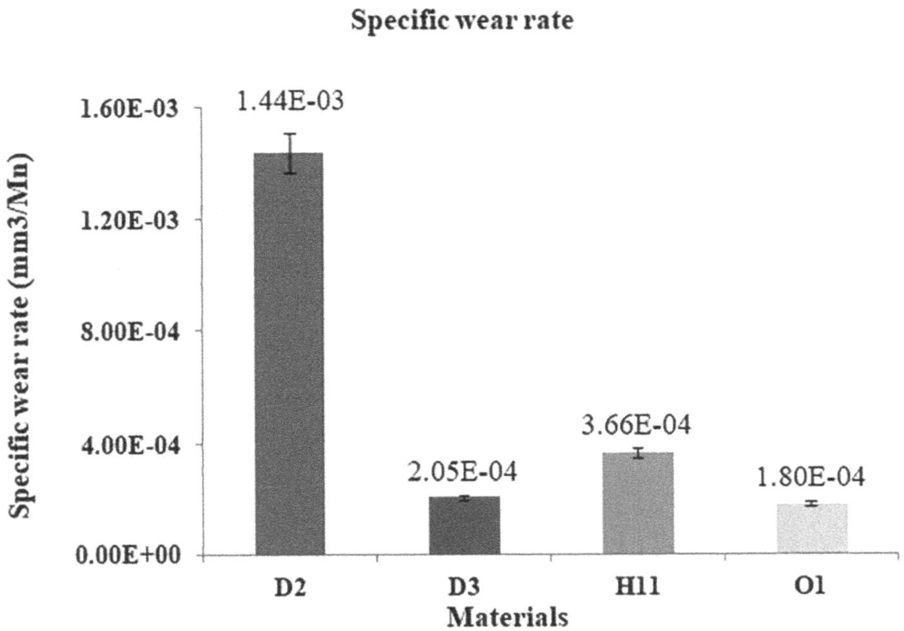

FIGURE 2.4 Specific wear rate at 2.2 m/sec.

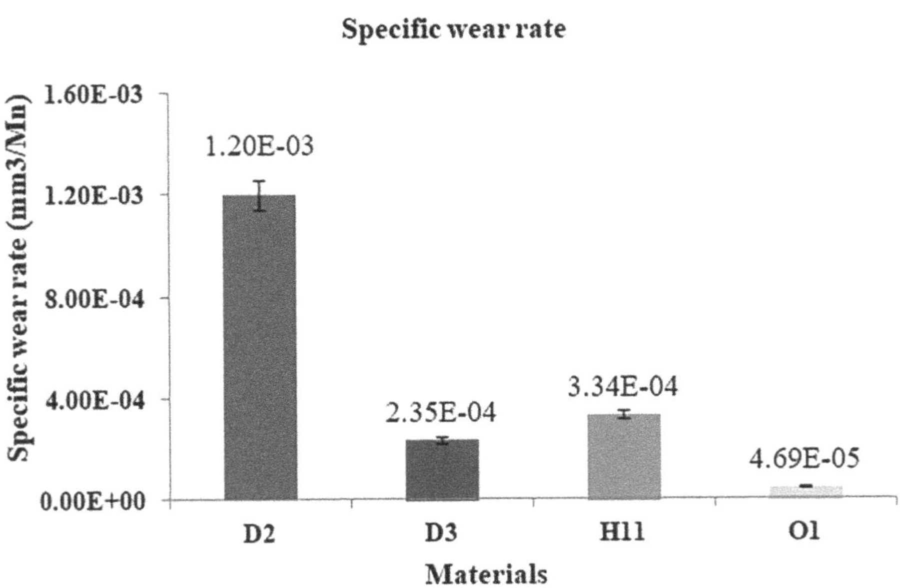

FIGURE 2.5 Specific wear rate at 2.56 m/s.

mild alloy concentrations. According to the aforementioned findings, heat treatment improves a material's qualities even when its composition is lower.

2.3.3 Consequence of Sliding Speed on Weight Loss

Figures 2.6 and 2.7 illustrate how sliding speed affects the weight loss of all materials while subjected to a steady load of 50 N for 10 minutes. Sliding speed has a very erratic effect on weight reduction and often relies on both the materials of the rubbing pairs and the present normal load. The result demonstrates that when sliding speed rises up to 1.67 m/sec, weight loss is reduced. After that, when sliding speed

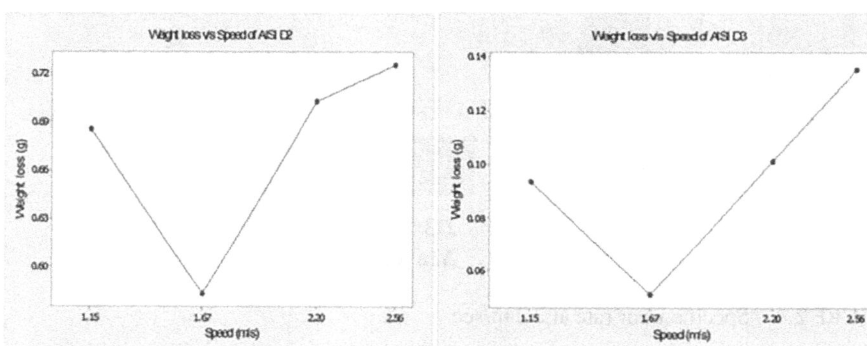

FIGURE 2.6 Plot between sliding speed and weight loss of AISI D2 and D3.

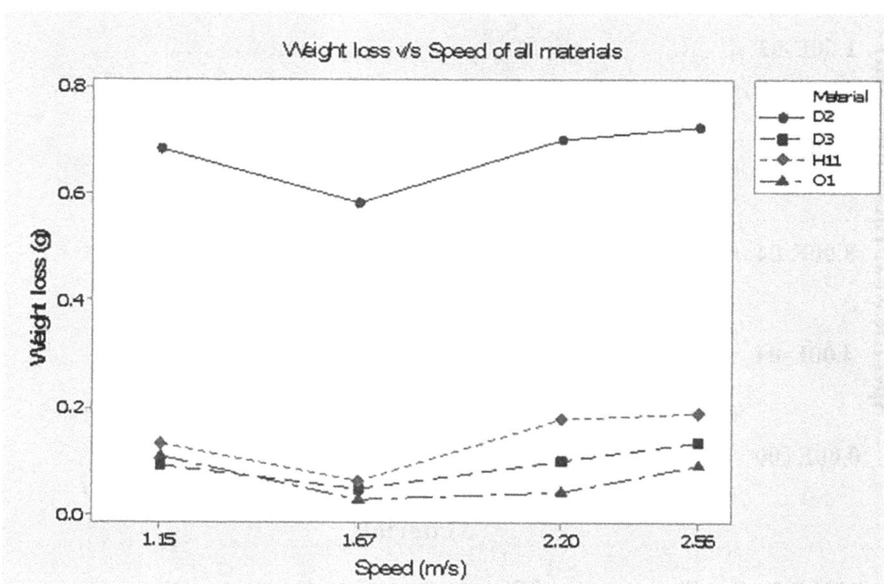

FIGURE 2.7 Plot of sliding speed and weight loss of all tested samples.

rises, weight loss continues to grow. The weight loss reaches its lowest point at 1.67 m/s. The rate of frictional heat production increases as a result of the increase in sliding speed when the normal load is large, which elevates the exterior temperatures. The substrate of the rubbing materials becomes softer as the contact temperature increases, hastening the delamination process.

2.3.4 Surface Roughness Results

All die samples underwent a surface roughness test after the wear test. The measured line surface roughness values were averaged to get an average surface roughness value. Figure 2.8 displays the average Ra value for each die test sample. Ra comparisons revealed that 8 AISI H11 outperformed all other materials; its average surface roughness (Ra) is 0.156 m, while D2's Ra is the lowest at 0.426 m. It's interesting to notice that AISI D3, although suffering substantial mass loss, has a lower average Ra value than D2 and certain other pricey and highly alloyed tool materials.

2.3.5 Wear Behaviour

The wear behaviour of all steels was analysed by SEM. The SEM micrographs and wear behaviour of all tested die samples are explained here.

2.3.5.1 Wear Performance of AISI D2

The SEM of worn surfaces of AISI D2 steel is shown in Figure 2.9(a–d). At 50 N and 1.15 m/s, the SEM findings reveal that the damaged surface of the D2 steel pin looks rougher, as illustrated in Figure 2.9 (a). The impact of load is clearly seen in

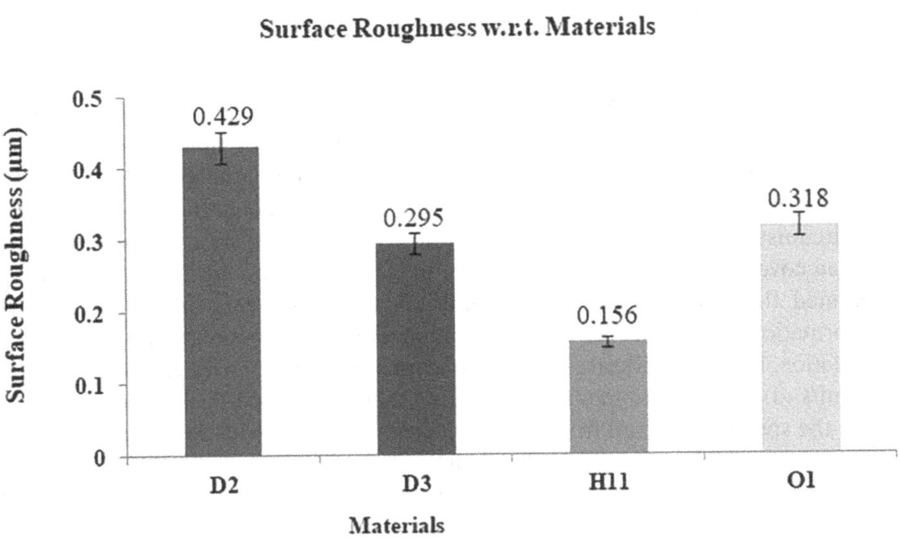

FIGURE 2.8 Average R_a value for tested die materials

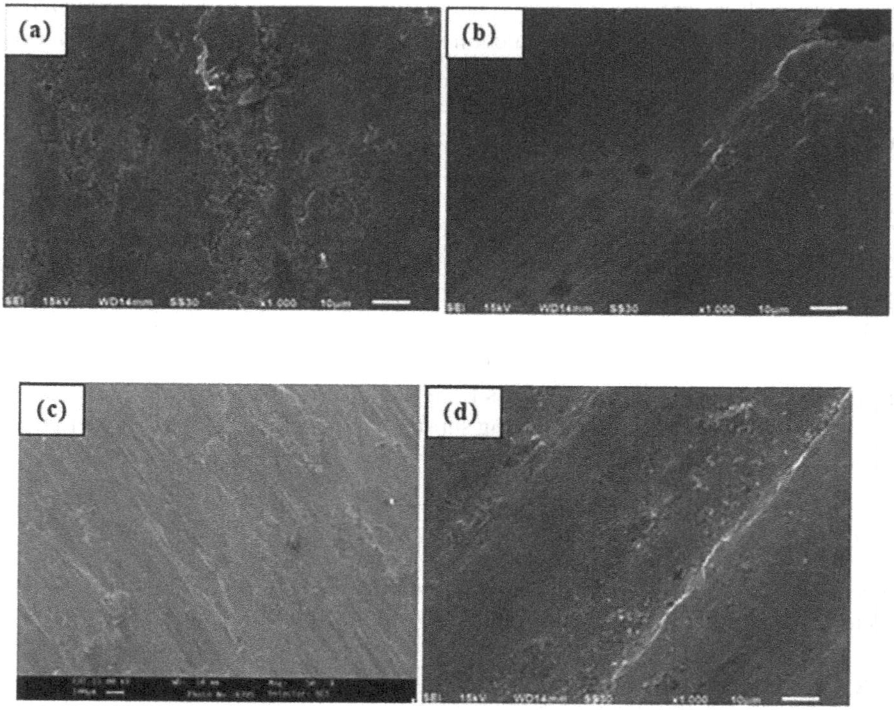

FIGURE 2.9 SEM image of damaged surfaces of AISI D2 at distinct speed in m/s: (a) 1.15, (b) 1.67, (c) 2.2, (d) 2.56.

this micrograph. When the load was high, the wear mechanism changed, and the surface of D2 steel began to spall and plough, which eventually took over as the principal wear mechanism as the traces of abrasion disappeared and the surface became rougher. Figure 2.9 depicts how the worn surface of the D2 steel pin at 1.67 m/s seems to be comparatively smooth (b). The micrograph and measures of weight reduction qualitatively correspond well. This micrograph clearly shows the evidence of abrasion wear combined with small areas of ploughing; however, under the circumstances, abrasion wear is the dominant wear. Layers of worn debris can be seen covering the sliding surface when the speed was 2.2 m/s. Before becoming separated, these layers must achieve a threshold thickness, which ultimately leads to the formation of wear debris. The micrograph and weight loss measures had a good correlation. Figure 2.9 illustrates how the worn scar surface no longer looks clean at 2.56 m/s (d). This micrograph clearly shows that the surface visibly looks rougher when the speed was raised from the previous level. Along with adhesion wear indicators, there are abrasion wear telltales. It is often seen that material is moved from the softer to the harder surface in adhesive wear of different metal couples. The D2 pins should be properly protected from wear by the extensive material transfer. This discovery may help to explain why the D2 steel pins saw less weight loss since the transfer happened faster.

2.3.5.2 Wear Behaviour of AISI D3

Figure 2.10(a–d) depicts the scanning electron micrograph of AISI D3 surfaces with worn-out surfaces. As shown in figures, the compacted wear debris transfer layers can be seen across the surface when the speed is 1.15 m/s. Before being severed from the surface, these layers build up to a certain thickness, which causes more weight loss. On the surface, there are indications of both adhesive and abrasive wear. With an increase in speed, the worn surface becomes smoother and shows signs of abrasive wear, but no layer has developed as a result of material transfer. The transfer layers now develop when the speed is raised to a greater level, owing to high heat production, and the wear behaviour is sticky. Higher weight loss results from the layers becoming dislodged when they approach a critical thickness.

2.3.5.3 Wear Behaviour of AISI H11

The SEM images in Figure 2.11(a–d) depicts the worn surfaces of AISI H11 steel. The compacted wear debris transfer layers and wear track may be seen across the sliding surface, according to SEM observations. Before being separated, these layers build up to a certain thickness, which finally causes the formation of heat and wear debris. Due to higher frictional heating and thus improved compaction, the amount of cover offered by these transfer layers rises with increasing load and sliding speed. The qualitative correlation between the micrograph and the weight loss

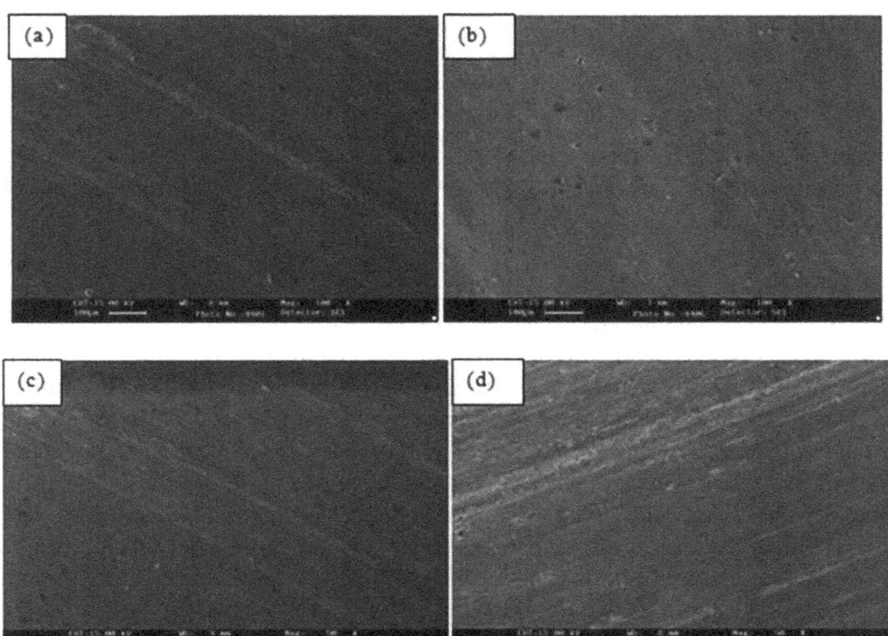

FIGURE 2.10 SEM micrographs of an AISI D3 steel pin's worn surface at distinct speed in m/s: (a) 1.15, (b) 1.67, (c) 2.2, (d) 2.56.

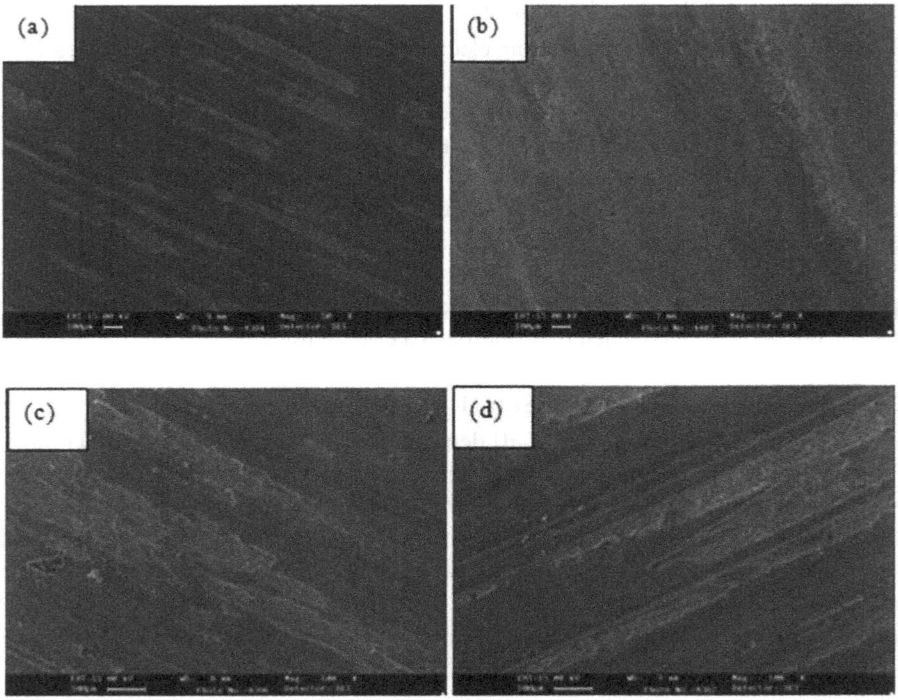

FIGURE 2.11 SEM image of worn-out surfaces of AISI H11 at distinct speed in m/s: (a) 1.15, (b) 1.67, (c) 2.2, (d) 2.56.

measures is good. This micrograph clearly shows indications of abrasion wear combined with patches from ploughing; however, under these circumstances, abrasion wear predominates.

2.3.5.4 Wear Performance of AISI O1

The SEM image of damaged surfaces of AISI O1 steel is shown in Figure 2.12(a–d). The findings of the SEM micrographs of AISI O1 steel at 1.15 m/s are almost identical to those of AISI H11. The transferred metal layers in this fashion follow the wear track and separate once they reach a certain thickness. Thus, the weight reduction is greater at this rate. In this case, the wear rate has been impacted more than sliding speed. On the surface, you can also see wear debris. On the surface, there are indications of both adhesive and abrasive wear. The smooth surface and abrasive wear behaviour are demonstrated to occur at 1.67 m/sec. As the speed is raised, considerable heat is produced from the high temperature, which causes oxide scales to develop. The development of an oxide layer on the surface stops wear. The four samples with the least wear were those where oxidative wear had formed.

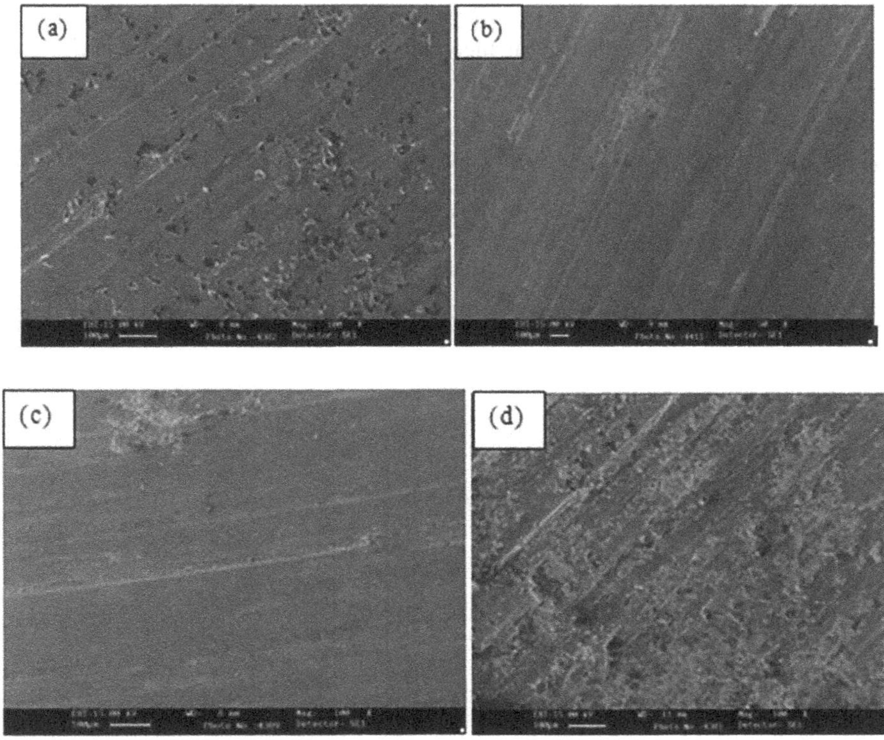

FIGURE 2.12 SEM image of worn-out surface of AISI O1 at distinct speed in m/s: (a) 1.15, (b) 1.67, (c) 2.2, (d) 2.56.

2.4 CONCLUSIONS

Based upon experimental following conclusion were made:

1. The AISI D2 samples experienced the greatest weight loss, whereas the AISI O1 samples experienced the least weight loss. The high hardness rating of AISI O1 steel is the cause of this. The relationship between wear resistance and hardness is direct. Even at low composition, the oil hardening has enhanced the hardness of AISI O1 steel.
2. According to the test findings, the least amount of weight decrease occurs at a velocity of 1.67 m/s. The rate of weight loss increases as speed does. This is due to the fact that at high temperatures, large frictional heats develop. The substrate of the rubbing materials becomes softer as the surface temperature rises, hastening the delamination process.
3. AISI H11 steel has the best surface finish ever seen. The surface finish value of AISI D2 is lower.

4. Depending on the present speed values, several wear mechanisms were seen. The most common wear mechanisms seen in the research via SEM analyses include abrasion, adhesion, galling, and surface ploughing.

2.5 DATA AVAILABILITY STATEMENT

The authors confirm that the data supporting the findings of this study are available within the book/chapter.

2.6 CONFLICTS OF INTEREST

The authors declare no conflict of interest.

2.7 FUNDING

There is no funding available for this research.

REFERENCES

[1] Algarni, M. Mechanical properties and microstructure characterization of AISI "D2" and "O1" cold work tool steels. *Metals*, 2019, Vol. 9 (11), p. 1169. https://doi.org/10.3390/met9111169
[2] Toboła, D.; Brostow, W.; Czechowski, K.; Rusek, P. Improvement of wear resistance of some cold working tool steels. *Wear*, 2017, Vol. 382, pp. 29–39.
[3] Budinski, K.G.; Budinski, M.K. *Engineering Materials: Properties and Selection.* Prentice Hall: Upper Saddle River, NJ, 2010, p. 756.
[4] Bourithis, L.; Papadimitriou, G.D.; Sideris, J. Comparison of wear properties of tool steels AISI D2 and O1 with the same hardness. *Tribology International*, 2006, Vol. 39, pp. 479–489.
[5] Glaeser, W.A. *Characterization of Tribological Materials.* Momentum Press: New York, 2012.
[6] Kheirandish, S.; Saghafian, H.; Hedjazi, J.; Momeni, M. Effect of heat treatment on microstructure of modified cast AISI D3 cold work tool steel. *Journal of Iron and Steel Research International*, 2010, Vol. 17, pp. 40–45.
[7] Kumar, S.; Singh, H.; Kumar, R.; Chohan, J.S. Parametric optimization and wear analysis of AISI D2 steel components. *Materials Today: Proceedings*, 2023. https://doi.org/10.1016/j.matpr.2023.01.247
[8] Singh, H.; Kumar, S.; Kumar, R.; Chauhan, J.S. Impact of operating parameters on electric discharge machining of cobalt based alloys. *Materials Today: Proceedings*, 2023. https://doi.org/10.1016/j.matpr.2023.01.234
[9] Lansdown, A.R.; Price, A.L. *Materials to Resist Wear.* Pergamon Press: Elmsford, NY, 1986.
[10] Ma, X.; Liu, R.; Li, D.Y. Abrasive wear behavior of D2 tool steel with respect to load and sliding speed under dry sand/rubber wheel abrasion condition. *Wear*, 2000, Vol. 241, pp. 79–85.
[11] Cora, O.N.; Namiki, K.; Koc, M. Wear performance assessment of alternative stamping die materials utilizing a novel test system. *Wear*, 2009, Vol. 267 (5–8), pp. 1123–1129.

[12] Hardell, J.; Prakash, B. High-temperature friction and wear behaviour of different tool steels during sliding against Al–Si-coated high-strength steel. *Tribology International*, 2008, Vol. 41 (7), pp. 663–671.
[13] Kumar, R.; Kumar, M.; Chohan, J.S.; Kumar, S. Effect of process parameters on surface roughness of 316L stainless steel coated 3D printed PLA parts. *Material Today Proceeding*, 2022, Vol. 68 (4), pp. 734–741. https://doi.org/10.1016/j.matpr.2022.06.004
[14] Boher, C.; Le Roux, S.; Penazzi, L.; Dessain, C. Experimental investigation of the tribological behavior and wear mechanisms of tool steel grades in hot stamping of a high-strength boron steel. *Wear*, 2012, pp. 294–295.
[15] Sultan, U.; Kumar, J.; Kumar, S. Experimental investigations on the tribological behaviour of advanced aluminium metal matrix composites using grey relational analysis. *Material Today Proceeding*, 2022. https://doi.org/10.1016/j.matpr.2022.12.171
[16] Deng, L.; Mozgovoy, S.; Hardell, J.; Prakash, B.; Oldenburg, M. Press-hardening thermo-mechanical conditions in the contact between blank and tool. In *Proceedings of 4th International Conference on Hot Sheet Metal Forming of High-Performance Steel (CHS2)*, Lule°a, Sweden, 2013, pp. 293–300.
[17] Ghiotti, A.; Bruschi, S.; Borsetto, F. Tribological characteristics of high strength steel sheets under hot stamping conditions. *Journal of Materials Processing Technology*, 2011, pp. 1694–1700.
[18] Fontalvo, G.A.; Mitterer, C. Effect of oxide-forming alloying elements on the high temperature wear of a hot work steel. *Wear*, 2005, Vol. 258 (10), pp. 1491–1499.
[19] Pelcastre, L.; Hardell, J.; Prakash, B. Galling mechanisms during interaction of tool steel and Al–Si coated ultra-high strength steel at elevated temperature. *Tribology International*, 2013, Vol. 67, pp. 263–271.
[20] Dohda, K.; Boher, C.; Rezai-Aria, F.; Mahayotsanun, N. Tribology in metal forming at elevated temperatures. *Friction*, 2015, Vol. 3 (1), pp. 1–27.
[21] Fontalvo, G.A.; Humer, R.; Mitterer, C.; Sammt, K.; Schemmel, I. Microstructural aspects determining the adhesive wear of tool steels. *Wear*, 2006, Vol. 260 (9–10), pp. 1028–1034.
[22] Hussainova, I.; Hamed, E.; Jasiuk, I. Nanoindentation testing and modeling of chromium-carbide-based composites. *Mechanics of Composite Materials*, 2011, Vol. 46 (6), pp. 667–678.
[23] Liu, Y.Z.; Jiang, Y.H.; Feng, J.; Zhou, R. Elasticity, electronic properties and hardness of MoC investigated by first principles calculations. *Physica B: Condensed Matter*, 2013, Vol. 419, pp. 45–50.
[24] Wu, L.; Yao, T.; Wang, Y.; Zhang, J.; Xiao, F.; Liao, B. Understanding the mechanical properties of vanadium carbides: Nano-indentation measurement and first-principles calculations. *Journal of Alloys and Compounds*, 2013, Vol. 548, pp. 60–64.
[25] Mahajan, A.; Singh, H.; Kumar, S.; Kumar, S. Mechanical properties assessment of TIG welded SS 304 joints. *Material Today Proceedings*, 2022, Vol. 56 (5), pp. 3073–3077. https://doi.org/10.1016/j.matpr.2021.12.133
[26] Kumar, A.; Kumar, R.; Kumar, S.; Verma, P. A review on machining performance of AISI 304 steel. *Material Today Proceedings*, 2022, Vol. 56 (5), pp. 2945–2951. https://doi.org/10.1016/j.matpr.2021.11.003
[27] Kumar, S. Influence of processing conditions on the mechanical, tribological and fatigue performance of cold spray coating: A review. *Surface Engineering*, 2022, Vol. 38 (4), pp. 324–365. https://doi.org/10.1080/02670844.2022.2073424
[28] Kumar, S.; Kumar, M.; Handa, A. Combating hot corrosion of boiler tubes—A study. *Journal of Engineering Failure Analysis*, 2018, Vol. 94, pp. 379–395. https://doi.org/10.1016/j.engfailanal.2018.08.004

[29] Kumar, P.; Kumar, R.; Kumar, S. An experimental investigation on tribological performance of graphite grease. *A Journal of Composition Theory*, 2019, Vol. 12 (7), pp. 853–859.

[30] Kumar, S.; Kumar, R. Influence of processing conditions on the properties of thermal sprayed coating: A review. *Surface Engineering*, 2021, Vol. 37 (11), pp. 1339–1372. http://doi.org/10.1080/02670844.2021.1967024

3 Paving Materials and Cooling Strategies for Outdoor Thermal Comfort in Open Public Spaces

Mehardeep Kaur, Priyanka Agarwal

3.1 INTRODUCTION

Open public spaces (OPSs) like roads, streets, squares, plazas, and parks are an integral part of urban life, considering the opportunities, activities, and experiences that they have to offer. They are sites of civic, cultural, and social activity and reflect the interaction between physical, social, political, and economic realities (Jelena Djekic, 2017). With human-centric cities bagging the limelight in the recent decades, making outdoor spaces attractive for people and ultimately usable has increasingly been recognized as an urban planning and design goal (Liang Chen, 2012), whereas a large part of these open spaces belongs to the so-called pedestrian movement (Jelena Djekic, 2017).

The history of architecture in the cities and more specifically of the OPS has always been tied to the idea of durability and performance of the materials, especially the paving materials which are used for the construction of the surfaces of the OPS. The paving materials of OPS is a feature that numerous researchers have found to be extremely important for determining the surface and air temperatures in urban canyons as well as the thermal comfort levels of pedestrians (Faragallah & Ragheb, 2022).

The inherent properties of the paving materials have certain impact on the surface temperature and human thermal comfort, which further defines their influence on the environment at the macro level and the user experience in these places. Thus, the choices of paving materials play a significant role in defining the quality of the public realm, given their considerable effect on the micro and macro climate.

Though mechanical operations such as air conditioning makes it simple to control the temperature of an inside space, there are limited ways to create a comfortable outdoor space.

However, the importance of the latter is being increasingly recognized with changing climate and heat stress in cities (Honjo, 2009). Recent studies have stated

DOI: 10.1201/9781003367154-3

that the amount and intensity of the activities conducted in OPSs depend on the level of happiness or discontent under the current climatic circumstances, making the thermal comfort conditions in outdoor spaces crucial to people's well-being and the triumph of pedestrian public spaces (Cortesão et al., 2016). Existing studies also indicate that people may become distressed and avoid using these open urban areas due to poor thermal comfort factors (Marialena Nikolopoulou, 2001). Moreover, given that paved surfaces have greater effects on urban thermal balancing and that a high percentage of urban surfaces are covered by pavement, human thermal comfort conditions will be negatively impacted as rapid urbanization occurs around the world (Irmak et al., 2017).

The study thus aims to understand the relationship between the pavement materials and human thermal comfort with respect to the formers' properties, in order to gain a better comprehension of their interlinkages and establish the choice and use of the materials in different types of OPS in a more effective way. Moreover, since recent studies indicate that minimizing urban heat islands (UHIs) in OPSs is an important factor in achieving balanced thermal comfort in outdoor urban environments (Riham Nady Faragallah, 2022); the research also presents an overview of appropriate cooling strategies to enhance outdoor thermal comfort by mitigating the UHI effect.

The paper shall present the research structured under five sections: Section 1 is the introduction, which is followed by Section 2, which discusses the existing literature. Further, Section 3 presents the adopted research methodology along with the selected paving materials for the current research; this section also illustrates the data collection of the materials, followed by their analysis and interpretation. Section 4 brings out the findings, which include discussions about the effective application of paving materials in different OPS along with suggestive strategies to mitigate the UHI effect. Finally, the conclusion and the implications of the study have been discussed in Section 5 and Section 6, respectively.

3.2 LITERATURE REVIEW

3.2.1 Role of Paving Materials in Thermal Comfort of Outdoor Spaces

The most major metric that affects outdoor human comfort is thermal, which is crucial and may be used to assess the value of an outdoor place (Yee Yong Lee, 2017). Thermal comfort is a condition in which people choose the optimal temperature, which is neither warmer nor cooler (Nasir et al., 2013). Some other sources describe thermal comfort as the condition of mind that expresses satisfaction with the thermal environment (Chun et al., 2004). Existing research validates that the thermal behaviour of different urban surfaces plays an important role as it is directly linked to the UHI phenomenon and environmental issues such as heat stress and air pollution (Yee Yong Lee, 2017). Therefore, there has been a growing interest in using appropriate materials to reduce UHIs and enhance the thermal properties of the urban environment (Doulos et al., 2004).

Previous studies have demonstrated the significance of the microclimate of OPSs for the intensity of use since people's behaviour and use of outdoor spaces are influenced by thermal conditions (Aleksandra Djukic, 2014). On the other hand, an important spatial characteristic that influences the urban microclimate is the surface materials (Jelena Djekic, 2017).

3.2.2 Properties of Paving Materials for Outdoor Thermal Comfort

Previously published research indicates that the physical characteristics of paving materials, such as heat absorption capacity, emissivity, and albedo, influence the surface temperature, which in turn affects the temperature in outdoor urban areas (Jelena Djekic, 2017). Nevertheless, existing research also shows that the most prominent parameters affecting the surface temperatures of the materials are albedo and emissivity (Ferguson, 2008). Albedo (or solar reflectivity) is a dimensionless fraction, measured on a scale of 0 to 1, defined as the ratio of reflected solar radiation to solar radiation incident on the surface (Li et al., 2012). In simple terms, Albedo is a solar reflection coefficient that defines that amount of solar energy that gets reflected off of the Earth and lands back in space.

The positive effect of higher albedo on the UHI effect reduction has been demonstrated through previous studies (Morini et al., 2016), thus indicating its significant impact on level of heat on Earth, and further onto the outdoor thermal comfort. Higher the albedo value, the less heat it absorbs and reflects back more.

Some existing studies have confirmed the relationship between albedo and the colour of a material; while the albedo value of a white surface equals to 1, the albedo value of a black surface which inevitably absorbs all the received energy is 0 (Muñoz, 2012). Thus, albedo also depends on colour of the material—light-coloured materials are known to have higher albedo value.

Prior research studies have also demonstrated that surface roughness has a significant influence in raising surface temperatures (Chatzidimitriou et al., 2006). In a study conducted by Jelena Djekic, black rough granite turned out to be the hottest material compared to its smoother variations, demonstrating the significant impact of colour and surface texture on a material's thermal properties (Jelena Djekic, 2017).

Additionally, existing studies have also proven that designing OPSs for thermal comfort solely based on physical model is insufficient with reference to the urban setting. While the physical environment has a significant role in outdoor thermal comfort, psychological adaptation also plays an important role (Marialena Nikolopoulou, 2003). All individuals perceive the environment differently, and the way they react to physical stimuli depends more on their level of "knowledge" about the circumstance than it does on the strength of the stimulus (Nikolopoulou, 2004). Therefore, psychological factors such as naturalness, expectations, experience, time of exposure, perceived control, and environmental stimulation influence the thermal perception of a space and the changes occurring in it.

Figure 3.1 shows that the parameter of naturalness influences other parameters without getting influenced by them and is part of the character of a place; naturalness thus emerges as the most prominent psychological factor affecting outdoor thermal comfort.

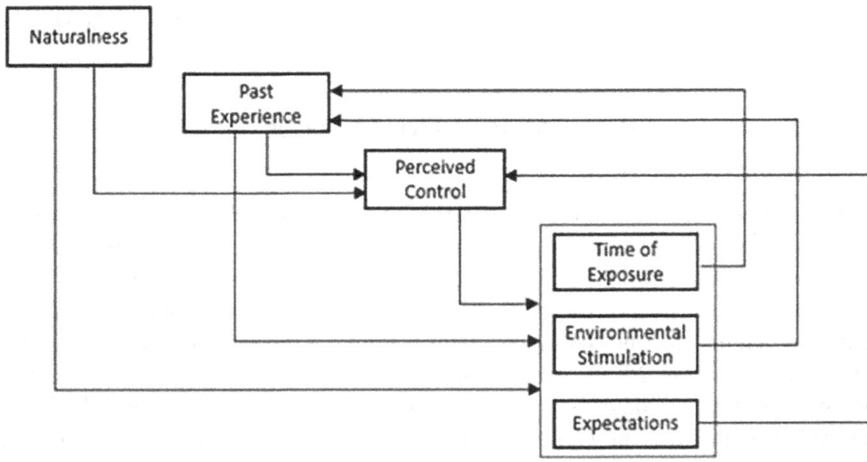

FIGURE 3.1 Network demonstrating interrelationships between the different parameters of psychological adaptation (Liang Chen, 2012).

Naturalness refers to a setting free of artificiality, where it appears that humans can withstand significant changes in the physical environment as long as they are produced naturally, as evidenced by expanding research (Marialena Nikolopoulou, 2003).

The hypotheses is thus framed:

The physical parameters of paving materials such as albedo, colour, and texture influence outdoor thermal comfort.

The naturalness of a paving material impacts the psychological aspect of outdoor thermal comfort.

The UHI effect influences outdoor thermal comfort.

3.3 METHODS AND MATERIALS

3.3.1 Research Approach

The current research investigates the relationship between the pavement materials and outdoor thermal comfort in OPSs in order to establish their applicability in most effective and suitable ways. To carry out this study, the four main properties of paving materials with respect to outdoor thermal comfort were derived from the existing literature/studies, which were albedo, colour, texture (physical parameters), and naturalness (psychological parameter). Based on these identified parameters, this relationship was established between paving materials and outdoor thermal comfort, which was used as the framework to evaluate the identified paving materials and carry out further studies.

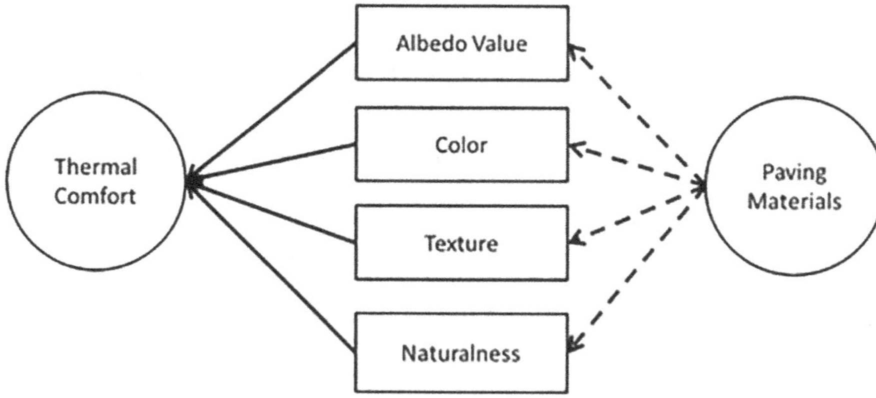

FIGURE 3.2 Relationship between paving materials and outdoor thermal comfort.

3.3.2 Selection of Paving Materials for Study

In order to carry out the current examination, the commonly used paving materials in public places as identified from literature resources and first-hand observations are as follows:

1. Asphalt
2. Concrete
3. Red brick
4. White marble
5. Granite
6. Pebble/gravel
7. Grass

3.3.3 Data Collection

Based on the identified parameters and materials for the current research study, data was compiled and tabulated (Table 3.2) to draw the results and findings. Table 3.2 was formulated, which represents each material's information documented from valid resources. The albedo values of the different materials were derived from relevant existing studies (Mohammad Taleghania, 2018; Hendel et al., 2018; Fisch et al., 1994). The information of the colour and textures of the paving materials were also derived from previous studies (Doulos et al., 2004) or on the basis of common sense. To evaluate the naturalness of a material, we developed a naturalness scale; the same was assessed based on visual perception and common sense with respect to the materials' formation and fabrication and indexed to a scale ranging from 0 (for artificial or man-made material) to 5 (for most naturally formed material). For instance, white marble and granite have been given an intermediate rating of 3, because though they are naturally formed versions of stone, they need to undergo certain processes before they can be used as paving materials in OPS.

TABLE 3.1
Data Compilation of Paving Materials with Respect to Their Properties

Paving Materials	Physical Properties			Psychological Property
	Albedo	Colour	Texture	Naturalness
Asphalt	0.05–0.2 (Mohammad Taleghania, 2018)	Dark grey/ black	Rough	1
Concrete	0.1–0.35 (Mohammad Taleghania, 2018)	Grey	Smooth/rough	1
Red brick	0.3 (Mohammad Taleghania, 2018)	Red	Rough	4
White marble	0.55 (Mohammad Taleghania, 2018)	White	Smooth/rough	3
Granite	0.26 (Hendel et al., 2018)	Grey	Smooth/rough	3
Pebble	0.72 (Mohammad Taleghania, 2018)	Red/grey	Rough	5
Grass	0.2 (Fisch et al., 1994)	Green	Smooth/rough	5

3.3.4 Analysing the Paving Materials

Asphalt—Asphalt is one of the most commonly used paving materials, especially for the roads, considering its strength and load-bearing capacity that allows it to withstand the high volumes of traffic without collapsing. However, as is evident from Table 3.1, the low albedo value of asphalt (0.05–0.2) represents that the material absorbs most of the incoming radiations and very little is reflected back. This eventually increases the temperature during nighttime when the surface tends to attain equilibrium state. Moreover, its physical properties such as dark grey colour and rough texture further increases its amount of heat absorption. In terms of naturalness, asphalt has been rated 1, given the material's manual fabrication, thus implying that it does not support human thermal comfort in terms of psychology. However, the positive attributes of asphalt such as strength and durability cannot be neglected. Therefore, the use of same is relevant for OPS such as roads and parking lots, however, needs rethinking while using for pedestrian areas like plazas in front of malls, shopping complexes, temple porch, walkways in gardens.

Concrete—This versatile material is readily available, affordable, and easy to install and maintain. In terms of thermal characteristics, concrete has low albedo value (0.1–0.35) and contributes to heat retention in the material that results in reduced outdoor thermal comfort. Though the absorption rate can be slightly reduced by using the smoother textures over the rough variations of the material, it still remains higher compared to other materials. Due to its man-made nature, concrete has been rated 1 on the parameter of naturalness. Therefore, the use of concrete is suitable for OPS, such as roads with heavy traffic like airports, highways, and loading unloading decks, but not where human traffic supersedes the vehicular traffic.

Red brick—The brick pavers should be more commonly used for pedestrian pathways, plazas, and garden walkways. Despite of brick's low albedo value (0.3), the material has less negative impact on outdoor temperature because of its high permeability and porosity, which also accounts for its rough texture. Moreover, the red colour is symbolic of Earth, which tends to appear more comfortable to the human eye. Brick is a material made directly from clay and sand (natural materials) and therefore has a higher rating of naturalness (4) due to its constituent materials. It can be laid in varied compositions giving an interesting and engaging experience throughout. However, it should not be used for OPS with heavy traffic due to its tendency of cracking or chipping.

White marble—This material is most commonly used for paving outside temple complexes, malls, and shopping plazas. Its white colour helps in achieving higher albedo (0.55) and therefore does not tend to increase the outdoor temperature. Its smooth texture gives a richer look, thus positively adding to the aesthetics of the spaces. It also is a good material in terms of naturalness (3) as found in nature and therefore positively adds on to the overall user experience. However, its usability is restricted for the areas with walkways and light vehicles. The material has some limitations at the same time as it requires proper care and maintenance and easily prone to acidic damage. Also, it is not an ideal material for a colder climate.

Granite—The natural stone has an albedo value of 0.26 and is available in many colours and texture per need of the user and space. Its strength and density are high and also easy to maintain. But the stone is costlier and therefore not a good option to be used as a paving material. However, the material can be used for street furniture as it does not get too heated up and people can easily use the space. Its naturalness (3) again adds up to the aesthetic character of outdoor elements when used wisely.

Pebble—The material has a very high albedo value (0.72), which does not make the surface warm and easy to walk. The pebbles are available in many colour ranges and are maintenance free. The material is naturally available (5) and offers porosity when loosely packed. This material adds warmth and luxurious texture to patios, pool decks, and walkways. Also, no maintenance or cleaning is required, which makes it favourable material to be used in public places.

Grass—It has low albedo value (0.2); however, its dark colour helps in retaining or absorbing more temperature, and it is imperative to consider that the heat retained by grass is ultimately utilized throughout the day during the photosynthesis process; hence, it does not increase the outdoor temperature. The grass material can thus be used in large scale OPS, such as grounds and fields, to bring down the temperature naturally. It is known to have positive and therapeutic effects and thus has a good impact on subjective parameter affecting outdoor thermal comfort (psychology) due to the material's strong association with naturalness. However, when it comes to using it as a paving material, it can easily get eroded; therefore, it should not be used in OPS with heavy vehicular traffic.

3.4 FINDINGS AND DISCUSSIONS

Per the presented data analysis, the findings in this section majorly focus upon reducing the thermal discomfort and the UHI effect in outdoor spaces. Therefore, the discussions and findings have further been divided in two sections. The first section discusses the applications of the studied paving materials in different OPS per their suitability with respect to the space's functions and thermal comfort. This strategy will help us choose wisely which material can be replaced with the other so that the unnecessary increase in temperature can be controlled. Further, the second section presents an overview of cooling strategies for reducing the UHI effect that can ultimately enhance the user's experience.

3.4.1 Application of Paving Materials in Different OPS

As presented in Section 3, the varying properties of individual paving material tend to define the suitability of its application in different types of OPS.

From literature review, albedo emerges as one of the key elements that influences how well a pavement absorbs or reflects thermal energy from the sun's rays during the day. Although literature studies suggest that materials with high albedo values should be preferred over the ones with lower albedo value for paving of surfaces, this may not hold true in case of some materials for their applicability in OPSs. For case, despite of higher side albedo value of white marble (0.55), its other disadvantageous parameters like dazzling effect and disability to withstand heavy urban traffic limits its use for extensive paving of OPSs.

Though higher albedo unquestionably lowers the surface temperature of pavements, there have been limited studies conducted about the extra reflective pavements that have a negative environmental impact and can affect glare, thermal comfort, and citizen health (Abbas Mohajerani, 2017). The amount of reflected radiation is due to higher albedo because using light-coloured material reaches the human level, which raises concerns about their suitability for usage in OPSs, particularly with relation to pedestrian comfort (Chatzidimitriou et al., 2006). The same has been refocused by Taleghani in his study, who also concluded that putting high-albedo materials on the ground surface reflects solar radiation back onto pedestrians, making them uncomfortable in the heat (in spite of reducing air temperature) (Taleghani, 2018). Therefore, even though high albedo makes ground surfaces colder, it also has another consequence that cannot be ignored—it reflects back a greater portion of solar radiation.

Another interesting thing to note here is that despite low albedo value (0.2) of grass (majority of rays are absorbed through photosynthesis), it is actually one of the most desirable materials to be used due to its naturalness and positive impact on outdoor temperature, which invalidates the otherwise established relation between the albedo and UHI effect that states that material with less albedo value have negative impact on the UHI. However, since grass surfaces cannot withstand heavy vehicular traffic, the material can be combined with some other more durable paving material for added strength, thereby increasing its scope of applications in different OPS.

Furthermore, it can be observed that the colour and roughness of the materials have a significant impact on the albedo value. In a previously conducted research study, Hao Wu experimented albedo values of the specimens made with cement concrete and asphalt wherein he concluded that the relatively darker colour of the asphalt was the reason for its relatively strong radiation-absorbing properties. (Wu et al., 2018). This validates that the colour of a material is a strong determinant for its respective albedo value.

TABLE 3.2
Research Findings: Choice of Materials per Their Use in Different Outdoor Public Spaces

Material	Most Suitable OPS for Application	Least Suitable OPS for Application	State-of-the-Art Alternative Solutions for Substitution in Least-Suitable Areas for Enhanced Effect on OTC
Asphalt	Transportation corridors, arterial roads, ring roads, airport runways	Pathways, plazas, parking lots	Porous asphalt pavement
Concrete			Pervious concrete, granitecrete
Red brick	Plazas outside temples, malls, garden walkways, street furniture	Low-volume Roads (LVRs), local streets	A-class bricks for elevated durability
White marble	Outdoor furniture	Sidewalks, plazas outside temples, malls	Natural stone tiles, such as sandstone and granite
Granite	Outdoor furniture	Sidewalks, plazas outside temples, malls	Granitecrete
Pebble	Patios, pool decks, garden walkways	Pathways, plazas, parking lots	Crushed stone
Grass	Parks, lawns, gardens, fencing for water elements, pathways	Parking lots	Grasscrete, turfstone

These inferences thus establish the need of holistic approach towards selection of paving materials in different types of OPSs. The selection of the material for paving or surface conditioning depends on the purpose of these open spaces as well as the frequency of use.

Hence, based on the studies parameters of the identified commonly used paving materials, the following table has been formulated to present their applications—that is, their choice and use in different types of OPS in a more effective way.

3.4.2 Review of Cooling Strategies for Mitigating UHI Effect

Considering the strongly negative impact of UHI effect on outdoor thermal comfort, the current research study also presents an overview of UHI mitigation strategies in this section. To lessen the net radiant heat, a number of heat mitigation strategies have been executed in various cities. Theoretically, the lower the resulting radiation, the lower is the city's heat. Based on this principle, the employment of vegetation, water in the surroundings, and reflective materials are some of widely used passive heat mitigation methods in urban areas (Taleghani, 2018). Existing studies also suggest that cooling strategies must be taken into consideration while talking about the pedestrian circulation and paving materials. Some of the easily adoptable UHI mitigation measures as identified by previous studies are as follows.

3.4.2.1 Urban Greening through Vegetation

- Green wall/green roofs—vegetation can be used on roofs (Lino Sanchez, 2019) or on facades, as green walls (Taleghani, 2018). The walls could either be of wooden panels or wire mesh jaalis to hang vegetative pots. This reduces the air temperature and creates a cooling effect, which further adds on to user's experience.
- Vegetated pavements involve using blocks made of clay, plastic, or concrete that have been filled with soil and covered in grass or other plants as paving material (Sharifi, 2017).
- Previous studies suggest that the addition of trees have a large impact on both microclimate as well as comfort of pedestrians (Chatzidimitriou & Yannas, 2016).

3.4.2.2 Cool Pavements and Reflective Materials

- Among the existing methods for reducing UHIs, the expansion of green spaces and highly reflecting "cool" surfaces is considered the most effective method of reducing (1) overheating of building surfaces, (2) energy consumption needed to cool the building, and (3) CO_2 emissions to the atmosphere (Federica Rosso, 2017).
- Using combination of hardscape and softscape—rather than using independent grass, it can be clubbed with hollow paver blocks to increase the porosity and permeability of the substructure and also help in reducing the surface temperature of the material.
- Using light-coloured materials per suitability for better albedo effect.
- The recent experiment in Doha, the capital city of Qatar, revolved around coating of the roads with blue paint to cool the tarmac surface and reduce the temperature of surrounding areas. Moreover, the material of paint also resistant to the wear and tear from the vehicular movement.

3.4.2.3 Other Elements

- Canopy—the shaded canopies created by wooden pergolas has a significant impact; it improves the thermal conditions in OPS (Chatzidimitriou & Yannas, 2016). His studies further focus on how these partial open spaces covered with trees results in improved air temperature, which is just concentrated below the foliage's of the tree.
- Water—replacing a hard surface with a water surface can be advantageous because the microclimatic effects of water and the evaporation process help in lowering the existing ambient temperature, thus improving the thermal conditions of the environment.
- Using porous versions of materials like asphalt and concrete, which are new engineered materials available in today's time, can also be beneficial in reducing UHI effect, thereby improving outdoor thermal comfort conditions of the OPS.

3.5 CONCLUSION

The paving materials that are employed for the surfaces of the OPSs influence the thermal comfort at micro level, which inevitably defines the user experience and dictates the frequency and intensity of usage of these spaces, as well as at the macro level by impacting the UHI. The intent of this research was thus two-fold: firstly, to establish the relationship between the pavement materials and outdoor thermal comfort with respect to various parameters that were identified through literature review, such as albedo value, colour, texture and naturalness; secondly, to tabulate and further analyse the data corresponding to these parameters.

Research results show that the choice of pavement materials in public places should be context-specific and may vary in accordance to its inherent properties and the purpose of the urban space; some state-of-the-art alternative sustainable versions of the selected paving materials were also suggested, such as porous asphalt, pervious concrete, A-class bricks, granitecrete, grasscrete, and turfstone in order to expand their respective scope of application in otherwise relatively less suitable OPSs. The results also indicated that although literature studies suggest that materials with high albedo values should be preferred over the ones with lower albedo value for paving of surfaces, this may not hold true in case of some materials. For instance, despite of white marble's high albedo value, its usage for extensive paving of OPSs should be constrained due to the material's dazzling impact and inability to handle heavy vehicular traffic, whereas grass, despite of its low albedo value, is a highly favourable paving material due to its naturalness, therapeutic effects, and air-cooling properties.

The second purpose of this research was to present an overview of cooling strategies for mitigating the UHI effect, given its strong association with outdoor thermal comfort. The findings indicate that the adoption of vegetation, reflective materials, canopies, and water on urban surfaces can prove beneficial in reducing the UHI effect, thereby optimizing the levels of outdoor thermal comfort.

3.6 RESEARCH DIRECTIONS

This study will be useful for the professionals involved in construction of OPSs, such as urban designers, architects, and civil engineers, in evaluating the respective effects of the different paving materials and use them accordingly to create outdoor environments that would not only help in reducing negative environmental impacts but also enhance outdoor thermal comfort of the users. The allied future studies can be approached by considering the other physical and psychological parameters of the paving materials, apart from the ones reviewed in the current study so as to form a holistic comprehension of this topic.

REFERENCES

Abbas Mohajerani, J. B.-B. (2017). The urban heat island effect, its causes, and mitigation, with reference. *Journal of Environmental Management*, 522–538.

Aleksandra Djukic, N. N. (2014). Comfort of open public spaces: Case study New Belgrade. In *Places and Technologies 2014 [Elektronski izvor]: Keeping up with Technologies to Improve Places* (pp. 145–150). Belgrade: Faculty of Architecture.

Chatzidimitriou, A., Chrissomallidou, N., & Yannas, S. (2006, September). Ground surface materials and microclimates in urban open spaces. *PLEA2006–The 23rd Conference on Passive and Low Energy Architecture*, Geneva (p. 6).

Chatzidimitriou, A., & Yannas, S. (2016). Microclimate design for open spaces: Ranking urban design effects on pedestrian thermal comfort in summer. *Sustainable Cities and Society*, 26, 27–47.

Chun, C., Kwok, A., & Tamura, A. (2004). Thermal comfort in transitional spaces—Basic concepts: Literature review and trial measurement. *Building and Environment*, 39(10), 1187–1192.

Cortesão, J., Alves, F. B., Corvacho, H., & Rocha, C. (2016). Retrofitting public spaces for thermal comfort and sustainability. *Indoor and Built Environment*, 25(7), 1085–1095.

Doulos, L., et al. (2004). Passive cooling of outdoor urban spaces. The role of materials. *Solar Energy*, 231–249.

Faragallah, R. N., & Ragheb, R. A. (2022). Evaluation of thermal comfort and urban heat island through cool paving materials using ENVI-Met. *Ain Shams Engineering Journal*, 13(3), 101609.

Federica Rosso, I. G. (2017). On the impact of innovative materials on outdoor thermal comfort of pedestrians in historical urban canyons. *Renewable Energy*, 1–33.

Ferguson, B. K. (2008). *Reducing Urban Heat Islands: Compendium of Strategies—Cool Pavements*. The National Academies of Sciences, Engineering, and Medicine, Environmental Protection Agency, Washington, DC.

Fisch, G., Wright, I. R., & Bastable, H. G. (1994). Albedo of tropical grass: A case study of pre-and post-burning. *International Journal of Climatology*, 14(1), 103–107.

Hendel, M., Parison, S., Grados, A., & Royon, L. (2018). Which pavement structures are best suited to limiting the UHI effect? A laboratory-scale study of Parisian pavement structures. *Building and Environment*, 144, 216–229.

Honjo, T. (2009). Thermal comfort in outdoor environment. *Global Environmental Research*, 43–47.

Irmak, M. A., Yilmaz, S., & Dursun, D. (2017). Effect of different pavements on human thermal comfort conditions. *Atmósfera*, 30(4), 355–366.

Jelena Djekic, A. D. (2017). Thermal comfort of pedestrain spaces and the influence of pavement materials on warming up during summer. *Energy and Buildings*, 1–32.

Lee, Y., Kelundapyan, R., Hanipah, H., & Abdullah, A. H. (2017). A review on outdoor thermal comfort evaluation for building arrangement Parameters. In *Sustainable Construction and Building Technology: Total Quality Management (TQM) in Industrialized Building System (IBS)* (pp. 1–14).

Li, H., et al. (2012). The use of reflective and permeable pavements as a potential practice for heat island mitigation and stormwater management. *Environmental Research Letters*, 8(1).

Liang Chen, E. N. (2012). Outdoor thermal comfort and outdoor activities: A review of research. *Cities*, 118–125.

Lino Sanchez, T. G. (2019). Cooling detroit: A socio-spatial analysis of equity in green roofs as an urban heat island mitigation strategy. *Urban Forestry & Urban Greening*, 44.

Marialena Nikolopoulou, N. B. (2001). Thermal comfort in outdoor urban spaces: understanding the human parameter. *Solar Energy*, 227–235.

Marialena Nikolopoulou, K. S. (2003). Thermal comfort and psychological adaptation as a guide for designing urban spaces. *Energy and Buildings*, 95–101.

Mohammad Taleghania, U. B. (2018). The effect of pavement characteristics on pedestrians' thermal comfort in Toronto. *Urban Climate*, 449–459.

Morini, E., Touchaei, A. G., Castellani, B., Rossi, F., & Cotana, F. (2016). The impact of albedo increase to mitigate the urban heat island in Terni (Italy) using the WRF model. *Sustainability*, 8(10), 999.

Muñoz, A. Z. (2012). Albedo effect and energy efficiency of cities—manufacturing and environment. In Ghenai, C. (Ed.), *Sustainable Development—Energy, Engineering and Technologies.* London: IntechOpen Limited.

Nasir, R. A., et al. (2013). Physical activity and human comfort correlation in an urban park in hot and humid conditions. *The Procedia—Social and Behavioral Sciences*, 105, 598–609.

Nikolopoulou, M. (2004). Outdoor comfort. *Environmental Diversity and Architecture*, 101–119.

Sharifi, P. O. (2017). Cool paving technologies. In P. O. Sharifi (Ed.), *Guide to Urban Cooling Strategies* (p. 18). Australia: Low Carbon Living CRC.

Taleghani, M. (2018). Outdoor thermal comfort by different heat mitigation strategies-A review. *Renewable and Sustainable Energy Reviews*, 81, 2011–2018.

Wu, H., Sun, B., Li, Z., & Yu, J. (2018). Characterizing thermal behaviors of various pavement materials and their thermal impacts on ambient environment. *Journal of Cleaner Production*, 172, 1358–1367.

4 A Study of Microwave Hybrid Heating Techniques Used in the Processing of Various Materials for Engineering Applications

Shaman Gupta, Ravish Arora, Neeraj Sharma

4.1 INTRODUCTION

In every production/manufacturing processes heating of material are required and it is a major concern for almost every industry, such as aerospace, automobile, and ship-building. Both conventional and non-conventional methods are used to serve this purpose, which have some pros and cons. Some of the limitations which conventional methods possess are higher processing time, higher energy consumption, and poor mechanical and metallurgical properties; apart from those, CO_2 emissions in the environment are also one of the biggest concerns. The microwaves are the waves whose frequency lies between 1 GHz to 300 GHz [1]; the applications of microwaves are shown in Figure 4.1. The working range of frequency as reserved by the FCC is 915 MHz and 2.45 GHz [2]. Some of the researchers have also identified that the working range for processing of materials by microwave hybrid heating (MHH) lies from 915 MHz to 18 GHz [3–6]. There are lots of limitations of conventional processes as compared with microwave heating [7–9]. The credit for this goes to the higher warming rates and higher scattering rates which have allowed modifications in physical and mechanical properties of the microwave-dealt-materials or parts. Further, volumetric heating provides uniform warming, resulting in reduced temperature concentration zones and a lower tendency to material distortion due to heating.

4.2 SCOPE OF MICROWAVE IN PROCESSING OF MATERIAL

It has reported that microwave energy, as a very novel method by the materials, can also be processed and has lots of advantages over other methods, such as reduction

DOI: 10.1201/9781003367154-4

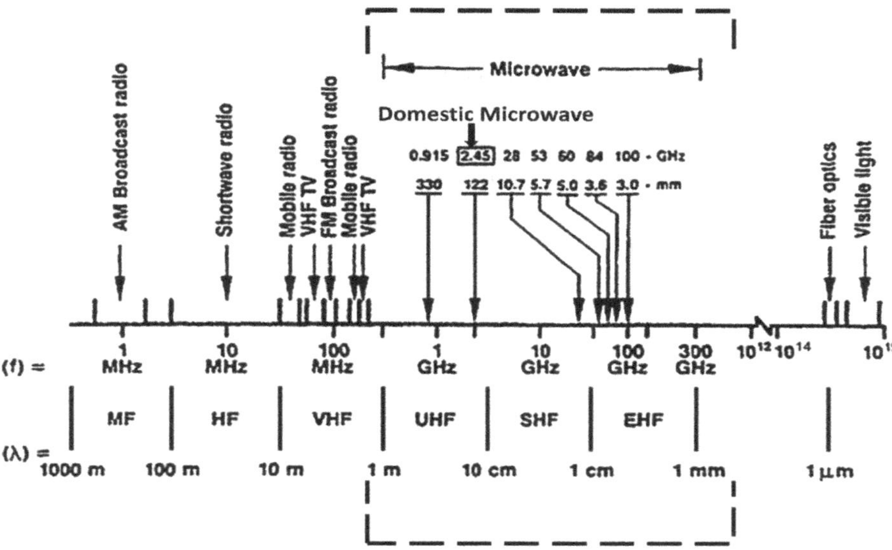

FIGURE 4.1 Spectrum of electromagnetic radiation.

in consumption of energy and cost. In earlier times, applications of microwaves were limited to communication and food processing, but now this field has expanded its horizons into processing various materials [10]. The application of microwaves on processing of Al_2O_3 ceramic has been reported which is very commonly used in the manufacturing industry. It has also been reported that by processing Al_2O_3 with microwave hybrid heating has resulted to attain the same densification with significant decrease in the temperature of about 200 degrees Celsius as compared to conventional methods of sintering [11–12]. Figure 4.2 shows the relationship between the two entities, from the findings.

Appleton et al. reviewed the scope of heating of material with the newer technology, MHH; and also investigated the interaction of various materials when exposed to microwaves. Modelling of microwaves heating based on laws of physics has also been reviewed at a deeper level and also throws light on the applications of variable-frequency microwaves for different materials, like ceramics and composites [13]

In the course of processing materials, Clark and Sutton [14] examined the microwaves. The substance treated in microwaves interacts with frigid microwaves rather than bright warmth, in contrast to conventional furnaces. Since the substance generated heat on its own, the warming is more intense and copious. These highlights bring about better creation consistency, less floor space, quicker creation all through, and decrease in inefficient heating.

4.2.1 Joining of Metals

Badiger et al. investigated certain alloys which have higher melting temperature of the range 1,300–1,400 degrees Celsius and also find affinity with microwave interaction.

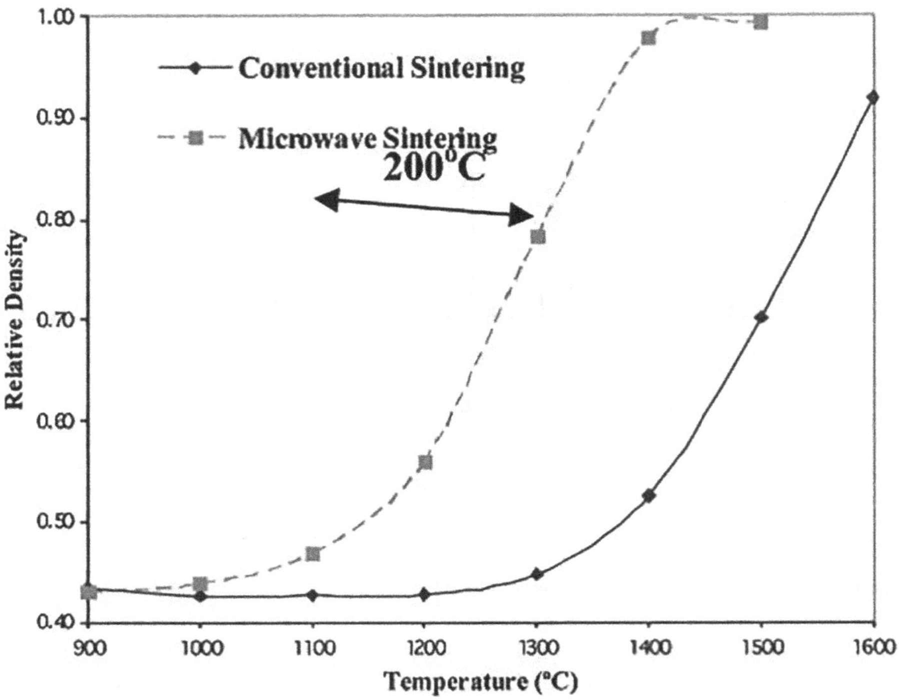

FIGURE 4.2 Relative density versus heating temperature graph for MHH and conventional sintering process.

In this investigation, the joining of super alloy Inconel 625 (IN 625) by MHH process was considered; it was found in the study that the super alloy IN 625 was successfully joined with MHH. Characterizations of the joint was also done, and it was found that the joint produced by MHH have superior mechanical and metallurgical properties [15]

Bansal et al. examined the joining of stainless-steel plates of grade 316; the interfacial material used in the joint was stainless steel-316 powder [16]. The study was carried out using a conventional oven at a frequency of 2.45 GHz and 900 W power. The joints were successfully joined using microwave energy and also showed better strength.

In this examination, copper metals were joined successfully utilizing COMSOL simulation. The total time required for complete joining of copper was five minutes. Homogeneous heating in between the joint was responsible for the significant strength of the joints. Porosity in the joint is also seen in the joint region with decreased significance [17].

Srinath et al. [18] studied the scope of joining of bulk metals through MHH. Domestic microwave applicator was used in this experimentation. Copper plates were successfully joined by the said technique. The joint was furnished in 15 minutes. Copper powder was used as interfacial material between the joints. The susceptor material used in the study was chosen as charcoal powder. After the processing of copper through microwave heating, a change in the atomic structure of copper powder was observed; for example, Cu (311) transformed into Cu (111). Also one-fourth of the copper powder material transformed into oxides of copper, and good coupling of microwaves with copper has been seen due to the presence of oxides. The joints also showed significant tensile strength, processed through microwaves. Bansal et al. [19] investigated the joining of IN 718 plates with a butt joint configuration of the 30 mm × 10 mm × 4 mm dimensions. The filler material used in this experimentation is IN 718 powder. A post-weld heat treatment was provided at 981 degree Celsius treated and aged (981STA) and 1,080 degrees Celsius treated and aged (1080STA). The effect of using these heat treatments showed that there is an increase in tensile strength and improved microhardness. The joints made were also free from any cracks, the bonding between the joint is also very good. The microstructural features are also investigated by FESEM. The findings are shown in Figure 4.3. Figure 4.3(a) shows the findings at 981STA condition, and Figure 4.3(b) shows at 1080STA conditions. The average grain size of second condition is larger than the first and hence due to this coarsening of grains is seen in the second condition. Thus, 1080STA's condition results in grain coarsening of the material, which further degrades the mechanical properties of the base material.

Arora et al. [20] also used microwave energy in the form of MHH to join hard-to-join material, such as superalloys like Inconel 600, which has excellent mechanical properties at elevated temperatures and also possess excellent mechanical properties in the corrosive environmental application viz. ships and turbines etc. The problem in welding of this superalloy with the other conventional processes like TIG welding is that it deteriorates the joint strength due to excessive heat applications which compromises the microstructural characteristics of the joints; this problem has been drastically improved by the microwave hybrid welding of Inconel 600.

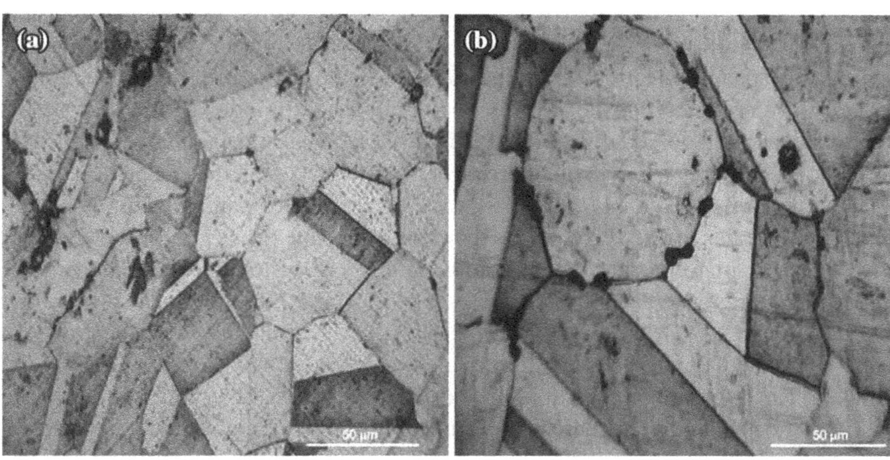

FIGURE 4.3 (a) 981STA condition FESEM; (b) 1080STA condition FESEM.

4.2.2 Cladding of Metals through Microwave

Corrosion and wear of metals are very big concerns in the various industries as they can decrease material strength. Due to this kind of failure in metals, there is a lot of pressure on industries to reduce this because it has created a lot of financial burden on industries. To handle these problems, various methods and techniques have been adopted, like carburizing, cladding, and cyaniding. This segment gives a portion of the significant and late audits, which are completed by analysts in the field of surface coatings and treatment utilizing microwaves.

MHH undertakes both conventional and microwave (MW) heating approaches for mitigating thermal runaway issues. Furthermore, MW cladding shares other properties with MW heating, such as uniform heating and considerably lower thermal distortion. However, MW cladding is a new concept, and it has been present in current academia for approximately ten years. Before 2010, there were some initiatives which enhanced technological progression towards MW cladding. Chiu et al. [21] indicated a study involving steel cladding with nickel-titanium involving MW-assisted brazing. Thus, the present study indicates that MW-assisted technology proved to be useful for cladding since it provided better economic output. MHH includes volumetric heating of substances, which helps in keeping a uniform temperature while conducting its operations. According to Kaushal et al. [22], MW cladding indicates volumetric heating as well due to its similarity with MHH. Furthermore, MW cladding indicates other features as well, such as significantly low thermal distortions in substance, low processing time, low energy consumption, and more. Due to its similarity with MHH process, integration of MW cladding and MHH becomes a plausible option. As suggested by Prasad et al. [23], the cladding of Tribaloy T400 through MHH process is prime focus of this discussion. Nevertheless, MW cladding is better than any other conventional cladding methods. Cladding is a process of partial dilution of substrate, which forms a bonding between deposits and substrates. MW cladding is a new concept. Fitzpatrick et al. [24] indicates that phosphor-doped polymer is better than cladding. However, modern concepts of cladding are proving to be extensively useful in present scenarios. As mentioned by Gupta and Sharma [25], MW cladding is one of most effective and popular surface modification technologies. MW cladding has different applications, such as developing composite clads, metallic alloys, and more. Unlike MW sintering, this process is inspired by MHH and undertakes basic principles of the MHH process. By adopting a domestic microwave oven with a 2.45 GHz frequency, Gupta and Sharma developed a unique cycle for the fabrication of metallic compounds on metallic substrates in 2010 [26]. It was seen that it is viable to use microwave technology to process materials because it is a financially savvy strategy and requires very small processing time and also requires very less energy too. It was also observed that clad prepare by MHH has shown lesser imperfections and lower defects like pores and cracks due to solidification, which may lead to significant change in the strength. The domestic kitchen appliance (microwave oven) is largely used for heating foodstuffs. However, researchers extended the employment of the kitchen appliance to material processing also. Lin et al. [27] developed a completely unique method for surface coatings known as cladding through microwaves. Coating of

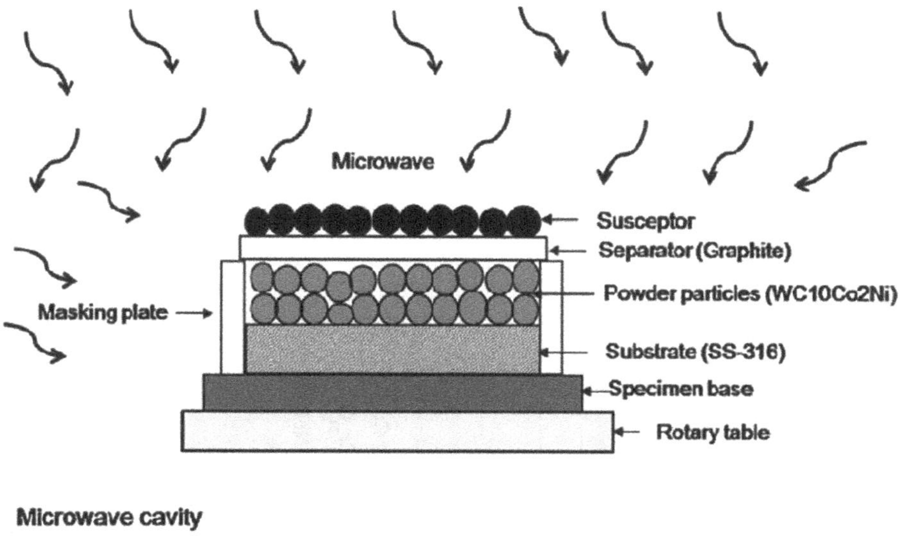

FIGURE 4.4 Arrangement of cladding through microwave hybrid heating.

W-based powder was successfully added on the surface of steel. The exposure of microwaves was for two minutes only. The preheated powder of ($WC_{10}Co_2Ni$ powder) was placed over steel, and microwaves was imposed over it. The arrangement for the same is shown in Figure 4.4. The X-ray diffractogram (XRD) of the clad is also shown in Figure 4.5.

Zhou et al. [28] introduced a novel MHH innovation for quickly applying a uniform, very thin, and crack-less cladding on titanium inserts. The technique utilized for applying clad is made up of conventional biomimetic cladding together with microwaves. Microwave lights sped up the Ca-P core arrangements and covered the entire surface of implants with cladding material. The water insoluble coatings developed through MHH have showed more noteworthy potential in biomedical applications.

Mazur et al. [29] described the investigation into In-Sn oxide coatings. Magnifying lens slides and silicon plates were used as storage surfaces for these coatings, and the interaction of microwave-aided receptive magnetron faltering was employed. Nuclear power microscopy analysis of the stored coatings revealed homogenous and smooth covering surfaces, whereas the XRD data revealed that the coatings were poorly defined. During steel wool scratch tests, the mechanical characteristics of the applied claddings revealed great hardness and fewer scratches.

Using MHH, Gupta et al. [30] produced an EWAC 20% $Cr_{23}C_6$ cladding on an austenitic steel SS316 substrate. The theory behind the microwave cladding technique is illustrated in Figure 4.6. By using XRD, FESEM, and EDS, the generated cladding on steel substrate was further evaluated. Its microhardness was also noted. The results of an XRD of the clad revealed the existence of nickel silicide, nickel iron, and chromium carbide phases, which improved the clad's microhardness. The clad

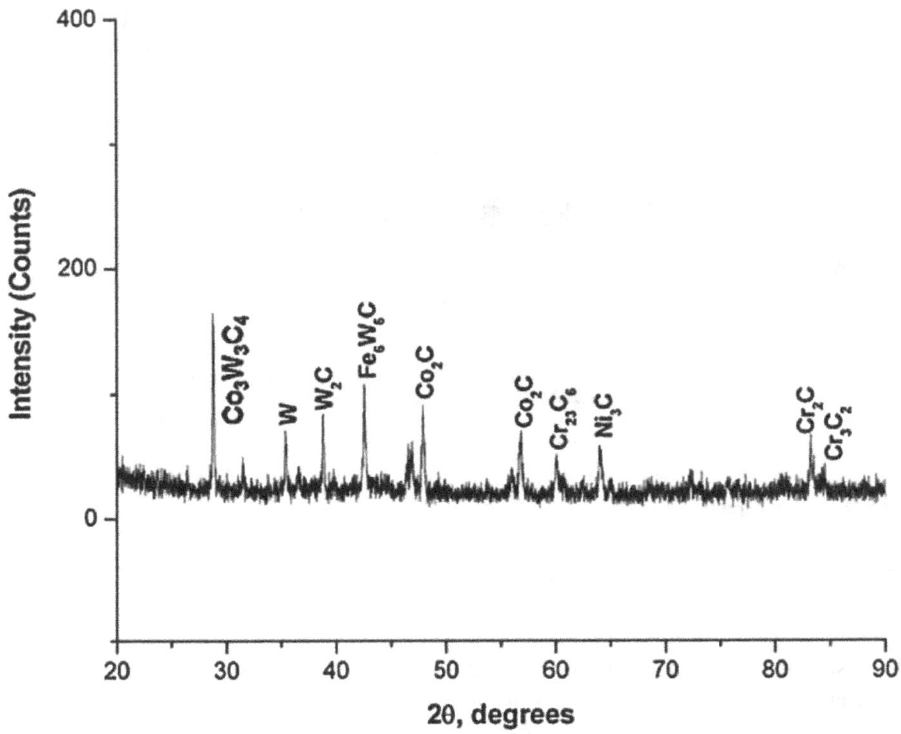

FIGURE 4.5 The XRD of the clad formed through MHH.

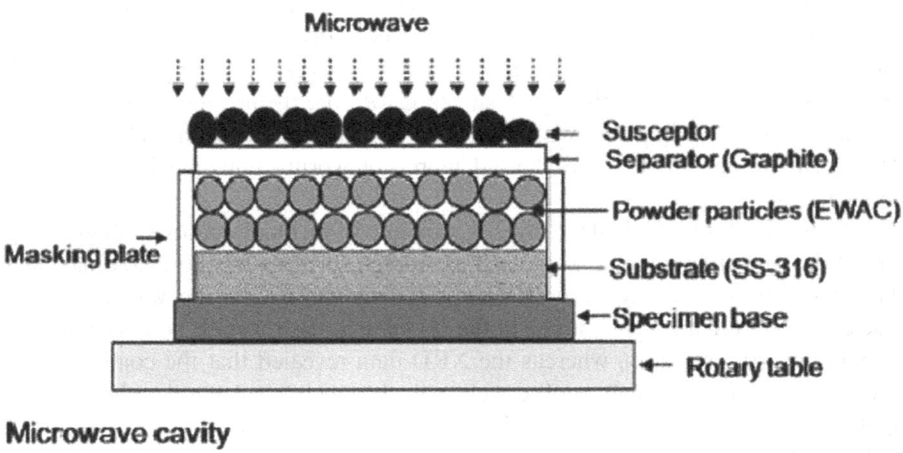

FIGURE 4.6 Cladding setup of EWAC 20% $Cr_{23}C_6$ on the substrate of austenitic steel SS316.

was manufactured with a 1 mm thickness and no obvious interfacial fractures. The clad has a microstructure that resembles a cell. The developed clad is more microhard than the underlying material. The implication was that these claddings would apparently work well in wear opposition applications, it was assumed.

4.2.3 Sintering of Metals through Microwave

Heating and compressing a dense mass of material without liquefying it is the process of sintering. Sintering frequently takes place in material stockpiles or as a metal assembly process. The creation of pottery, plastics, and other materials. Researchers from many fields have also contributed to the body of knowledge regarding the microwave-assisted sintering of tungsten-copper composites. Specialty ceramics materials have also been produced by microwave technology. By the use of this particular technology, a great amount of time has been saved to sinter the material, and also there is a drastic improvement in the bending strength and hardness values of the products processed through MHH. In the event of an irregularity in the structure of the material, microwave coupling creates highly-rapid-response- and new-response-shaped materials at considerably lower temperatures than is typically achieved by ordinary heating processes. [31].

Several scientists have successfully sintered metallic powders using microwave energy. Due to their widespread use in industrial applications, electromagnetic waves are still required to heat or sinter metals even though ceramics and other non-metallic materials may now be heated or sintered using microwave radiation. They got the chance to sinter a variety of typical powdered metals from commercial sources using MHH and superior mechanical properties were obtained as compared to the conventional procedures. Many pure metal powders have been sintered using microwave radiation [32].

Two-directional sintering was applied by some of the investigators to produce an enhancement in the mechanical behaviour [33]. Primarily Al, Mg, and Pb free solder was used as a material. Gupta and Wong were successful to sinter the said material. The apparatus chosen for this was a 2.45 GHz microwave oven with the power capacity of 900 W. As there is requirement of the susceptor material to start heating, in this study the susceptor material used was SiC for the coupling of microwaves with the material.

There was a study about the application of microwave radiation to sinter Fe, Co, Ni, Cu, and hardened steel powders. The effects of conventional sintering and microwave-aided sintering were compared. A recurrence of 2.45 GHz was used for the microwave used for the sintering. Microwave radiation accelerated the sintering of metal powders but had no effect on the sintering's activation energy. When compared to conventional ones, samples of microwave-sintered materials had excellent mechanical characteristics [34].

Apart from sintering various researchers have also used the technology of microwave processing for brazing of materials and for joining of different materials too. In microwave, a multitude of conventional steel pieces, pure metals, and refractory materials have been sintered to nearly full potential with better mechanical qualities. Various industrial metals, such as Fe, Cu, Al, and Ni, were used in this experiment, and their amalgams were sintered in a microwave. In addition, microwave energy was used to braze and join these components. Various metallic materials were produced in the oven. Microwave-sintered powders have been identified to make a common commodity. Microwaves have also been used to sinter materials such as Al and Cu, as well as their compounds; the microwave-processed materials were produced around fully thick bodies [35].

Some studies have been done which created Mg/Cu nano composites with the help of microwave sintering. Mg composites comprising varied measures of nano-sized Cu particles were successfully integrated throughout the dip-coating approach in this investigation. Cu particles increase hardness, yield strength, elastic modulus, and UTS by 0.2%. Microwave energy was used to enhance mechanical characteristics. [36]

4.3 CONCLUSIONS

Due to environmental concerns and energy inadequacy, microwave material processing has become more crucial. This technology saves energy since it is utilized for a variety of applications that provide bulk heating and targeted heating based on microwave-material coupling instead of traditional heating methods, such as resistive heating. There are a few points that pertain to the process's future scope.

1. Microwave cladding opens up the opportunity of using additional metallic materials to improve surface properties.
2. Microwave cladding technique with additional developed appropriate metallic powders with acceptable characteristics can be employed.
3. This approach can greatly enhance wear and corrosion properties.
4. It is possible to investigate the use of composite cladding in different fields.
5. Other machining operations, such as drilling and casting, can also be done with microwave processing.

REFERENCES

[1]. Sutton, W.H. *Theory and Application in Materials Processing II.* American Ceramic Society: Westerville, OH, 1993; 3 pp. ISBN-13: 978-0944904664

[2]. Lauf, R.J.; Bible, D.W.; Johnson, A.C.; Everliegh, C.A. 2–18GHz broadband microwave heating systems. *Microwave Journal* 1993, 36, 24–27.

[3]. Thostenson, E.T.; Chou, T. Microwave processing: Fundamentals and applications. *Composites: Part A* 1999, 30, 1055–1071.

[4]. Appleton, T.J.; Colder, R.I.; Kingman, S.W; Lowndes, I.S.; Read, A.G. Microwave technology for energy-efficient processing of waste. *Applied Energy* 2005, 81, 85–113.

[5]. Das, S.; Mukhopadhyay, A.K.; Datta, S.; Basu, D. Prospects of microwave processing: An overview. *Bulletin of Material Science* 2009, 32 (1), 1–13.

[6]. Ku, H.S.; Siores, E.; Ball, J.A.R. Review—microwave processing of materials: Part I. *The Honking Institution of Engineers Transactions* 2001, 8, 31–37.

[7]. Leonelli, C.; Veronesi, P.; Denti, L.; Gatto, A.; Iuliano, L. Microwave assisted sintering of green metal parts. *Journal of Materials Processing Technology* 2008, 205, 489–496.

[8]. Rao, R.B.; Patnaik, N. Microwave energy in mineral processing—a review. *IE(I) Mineral Journal* 2004, 84, 56–61.

[9]. Omer, Y. The effect of heat treatment on colemanite processing: A ceramics application. *Powder Technology Journal* 2004, 142, 7–12.

[10]. Agrawal, D. Latest global developments in microwave materials processing. *Materials Research Innovations* 2010, 14 (1), 3–8

[11]. Brosnan, K. H.; Messing, G. L.; Agrawal, D. K. Microwave sintering of alumina at 2.45 GHz. *Journal of the American Ceramic Society* 2003, 86 (8), 1307–1312.

[12]. Fang, Y.; Agrawal, D.; Roy, R. *Materials Letters* 2004, 58, 498–501.

[13]. Chandrasekaran, S.; Ramanathan, S.; Basak, T. Microwave material processing—a review. *AIChE Journal* 2012, 58, 330–363.

[14]. Clark, D.E.; Sutton, W. Microwave processing of materials. *Material Science* 1996, 26, 299–331.

[15]. Badiger, R.I.; Narendranath, S.; Srinath, M.S. Joining of Inconel-625 alloy through microwave hybrid heating and its characterization. *Journal of Manufacturing Processes* 2015, 18, 117–123.

[16]. Bansal, A.; Sharma, A.K.; Kumar, P.; Das, S. Characterization of bulk stainless steel joints developed through microwave hybrid heating. *Materials Characterization* 2014, 91, 34–41.

[17]. Srinath, M.S.; Murthy, P.S.; Sharma, A.K.; Kumar, P.; Kartikeyan, M.V. Simulation and analysis of microwave heating while joining bulk copper. *International Journal of Engineering, Science and Technology* 2012, 4 (2), 152–158.

[18]. Srinath, M.S.; Sharma, A.K.; Kumar, P. A new approach to joining of bulk copper using microwave energy. *Materials & Design* 2011, 32 (5), 2685–2694.

[19]. Bansal, A.; Sharma, A.K.; Kumar, P.; Das, S. Structure–property correlations in microwave joining of Inconel 718. *Jom* 2015, 67 (9), 2087–2098.

[20]. Arora, R.; Kapoor, J.; Sharma, R.C. Development of microwave hybrid heating welded joints of inconel-600 superalloys using grey relational analysis. *International Journal of Vehicle Structures & Systems (IJVSS)* 2021, 13 (1).

[21]. Chiu, K.Y.; Cheng, F.T.; Man, H.C. A preliminary study of cladding steel with NiTi by microwave-assisted brazing. *Materials Science and Engineering: A* 2005, 407 (1–2), 273–281.

[22]. Kaushal, S.; Gupta, D.; Bhowmick, H. Investigation of dry sliding wear behavior of Ni–SiC microwave cladding. *Journal of Tribology* 2017, 139 (4), 41603–41612.

[23]. Prasad, C.D.; Joladarashi, S.; Ramesh, M.R.; Srinath, M.S.; Channabasappa, B.H. Development and sliding wear behavior of Co-Mo-Cr-Si cladding through microwave heating. *Silicon* 2019, 11 (6), 2975–2986.

[24]. Fitzpatrick, C.; Lewis, E.; Al-Shamma'a, A.; Pandithas, I.; Cullen, J.; Lucas, J. An optical fiber sensor based on cladding photoluminescence for high power microwave plasma ultraviolet lamps used in water treatment. *Optical Review* 2001, 8 (6), 459–462.

[25]. Gupta, D.; Sharma, A.K. A method of cladding/coating of metallic and non-metallic powders on metallic substrate by microwave irradiation. *Indian Patent* 527/Del/2010, 2010.

material used in the study was chosen as charcoal powder. After the processing of copper through microwave heating, a change in the atomic structure of copper powder was observed; for example, Cu (311) transformed into Cu (111). Also one-fourth of the copper powder material transformed into oxides of copper, and good coupling of microwaves with copper has been seen due to the presence of oxides. The joints also showed significant tensile strength, processed through microwaves. Bansal et al. [19] investigated the joining of IN 718 plates with a butt joint configuration of the 30 mm × 10 mm × 4 mm dimensions. The filler material used in this experimentation is IN 718 powder. A post-weld heat treatment was provided at 981 degree Celsius treated and aged (981STA) and 1,080 degrees Celsius treated and aged (1080STA). The effect of using these heat treatments showed that there is an increase in tensile strength and improved microhardness. The joints made were also free from any cracks, the bonding between the joint is also very good. The microstructural features are also investigated by FESEM. The findings are shown in Figure 4.3. Figure 4.3(a) shows the findings at 981STA condition, and Figure 4.3(b) shows at 1080STA conditions. The average grain size of second condition is larger than the first and hence due to this coarsening of grains is seen in the second condition. Thus, 1080STA's condition results in grain coarsening of the material, which further degrades the mechanical properties of the base material.

Arora et al. [20] also used microwave energy in the form of MHH to join hard-to-join material, such as superalloys like Inconel 600, which has excellent mechanical properties at elevated temperatures and also possess excellent mechanical properties in the corrosive environmental application viz. ships and turbines etc. The problem in welding of this superalloy with the other conventional processes like TIG welding is that it deteriorates the joint strength due to excessive heat applications which compromises the microstructural characteristics of the joints; this problem has been drastically improved by the microwave hybrid welding of Inconel 600.

FIGURE 4.3 (a) 981STA condition FESEM; (b) 1080STA condition FESEM.

[32]. Roy, R.; Agarwal, D.; Cheng, J.; Gedevanishvili, S. Full sintering of powdered-metal bodies in a microwave field. *Nature* 1999, 399, 668–670.
[33]. Gupta, M.; Wong, W.L.E. Enhancing overall mechanical performance of metallic materials using two-directional microwave assisted rapid sintering. *Scripta Materialia* 2005, 52, 479–483.
[34]. Saitou, K. Microwave sintering of iron, cobalt, nickel, copper and stainless steel powders. *Scripta Materialia* 2006, 54, 875–879.
[35]. Agarwal, D. Microwave sintering, brazing and melting of metallic materials. *Advanced Processing of Metals and Materials* 2006, 4.
[36]. Wong, W.L.E.; Gupta, M. Development of Mg/Cu nanocomposites using microwave assisted rapid sintering. *Composites Science and Technology* 2007, 67, 1541–1552.

current academia for approximately ten years. Before 2010, there were some initiatives which enhanced technological progression towards MW cladding. Chiu et al. [21] indicated a study involving steel cladding with nickel-titanium involving MW-assisted brazing. Thus, the present study indicates that MW-assisted technology proved to be useful for cladding since it provided better economic output. MHH includes volumetric heating of substances, which helps in keeping a uniform temperature while conducting its operations. According to Kaushal et al. [22], MW cladding indicates volumetric heating as well due to its similarity with MHH. Furthermore, MW cladding indicates other features as well, such as significantly low thermal distortions in substance, low processing time, low energy consumption, and more. Due to its similarity with MHH process, integration of MW cladding and MHH becomes a plausible option. As suggested by Prasad et al. [23], the cladding of Tribaloy T400 through MHH process is prime focus of this discussion. Nevertheless, MW cladding is better than any other conventional cladding methods. Cladding is a process of partial dilution of substrate, which forms a bonding between deposits and substrates. MW cladding is a new concept. Fitzpatrick et al. [24] indicates that phosphor-doped polymer is better than cladding. However, modern concepts of cladding are proving to be extensively useful in present scenarios. As mentioned by Gupta and Sharma [25], MW cladding is one of most effective and popular surface modification technologies. MW cladding has different applications, such as developing composite clads, metallic alloys, and more. Unlike MW sintering, this process is inspired by MHH and undertakes basic principles of the MHH process. By adopting a domestic microwave oven with a 2.45 GHz frequency, Gupta and Sharma developed a unique cycle for the fabrication of metallic compounds on metallic substrates in 2010 [26]. It was seen that it is viable to use microwave technology to process materials because it is a financially savvy strategy and requires very small processing time and also requires very less energy too. It was also observed that clad prepare by MHH has shown lesser imperfections and lower defects like pores and cracks due to solidification, which may lead to significant change in the strength. The domestic kitchen appliance (microwave oven) is largely used for heating foodstuffs. However, researchers extended the employment of the kitchen appliance to material processing also. Lin et al. [27] developed a completely unique method for surface coatings known as cladding through microwaves. Coating of

5 Impact Response of Fibre-Reinforced Polymer Composite Materials

A Review

T. Jambhulkar, R. K. Sahu

5.1 INTRODUCTION

A composite material is an assemblage of at least two materials that possess qualities that are superior to the properties of the component materials. In other words, a composite material mainly contains a matrix and a reinforcement, as shown in Figure 5.1, when combined produces properties that are superior to those of the components alone. The properties of the resultant composite can be altered depending on how the reinforcement fibres are arranged. A type of resin called matrix keeps the reinforcement in the required orientation. The primary function of the matrix is to safeguard the reinforcement from a chemical and hazardous environment and also to bind the reinforcement so that the applied loads may be transferred effectively [1]. Examples of composite materials include wood, concrete, and human bone. Composites can be categorized according to the type of reinforcement and matrix. Matrices can be categorized into metals, polymers, and ceramics, and reinforcements can be categorized into fibres, particulates, flakes, whiskers, and fillers. Carbon, glass, and Kevlar fibres are often used in polymer-matrix composite materials [2].

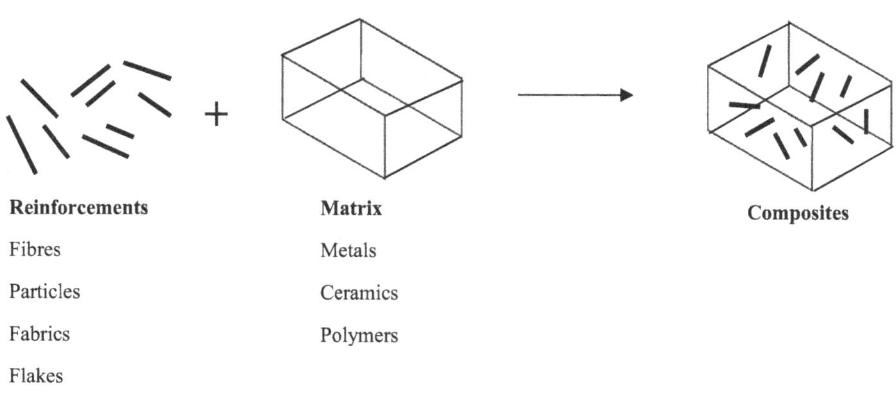

FIGURE 5.1 Composite material [8].

DOI: 10.1201/9781003367154-5

In recent times, the utilization of composite materials has steadily increased. Several examples demonstrate how composite materials have become alternative to traditional materials in the industry. Although the aerospace and defence industries were early adopters of composite materials, their attractive performance features eventually led to their use in the fabrication of sports equipment, boats and reinforcement parts in construction and energy applications [3]. The transportation industry is leading the way in aircraft design, with modern aircraft comprising up to 60% of their weight from composite materials. For a few years, the automobile industry has used composite in luxury and racing vehicles, and it is currently employing them in the mass manufacture of passenger automobiles [4]. The main benefit of using composite materials over standard materials is their superior qualities, such their lighter weight and better strength compared to steel or aluminium [5]. Composites have far better toughness compared to commonly used materials, and they do not corrode like steel since they have good corrosion-resistant properties. Another advantage of this material is that it enables the creation of features that would be impossible to achieve with metals, ceramics, or polymers alone [6]. The performance of composite materials can be improved by several parameters, including fibre-matrix interface, matrix factors, and fibre property characteristics. Properties of fibre include the length of the fibre, the orientation of fibre, the shape of the fibre, and the material of the fibre. Matrix factors include transverse modulus and strength, compressive strength, and fatigue strength [7]. Fibre-matrix interfaces include chemical bonding, mechanical bonding, and reaction bonding. Composite materials mainly undergo three types of failure: fibre-level damage mechanism, matrix-level damage mechanism, and coupled fibre-matrix level damage mechanism [7].

Composites which are made with polymer matrix are more common and are widely used in aircraft industries. In the mid-1960s and early 1970s, this type of composite started its use in aircraft industries. Polymer-matrix composites (PMCs) are mainly used in high-performance structural applications [9]. PMC consists of excellent mechanical properties, high specific stiffness, high fracture toughness, high specific strength, increased fatigue life, and high corrosion resistance. PMCs find their application in many fields, some of them in bulletproof vests, aerospace structures, and automobiles [10]. Recently, there has been a great deal of study done on the impact sensitivity of composite materials. The impact reaction might be dynamic or quasi-static, depending on the properties of the structure and impactor [11]. There can be several occasions of impact loading, such as in lightweight fibre-reinforced PMC, which is applied in aerospace applications because of its good strength, corrosion resistance, and fatigue resistance.

Therefore, this review discusses the impact response of PMC, which is majorly observed in the industry by taking various parameters into account. In addition, this provides a pathway for researchers by stating the importance of PMC and the class of fibres which can be used as reinforcement. This chapter intends to indirectly state the various types of velocity impact, along with different parameters affecting the impact response. This review analysis will also look at some of the advanced engineering applications subjected to impact load along with its conclusion and future scope. It will provide the readers with a brief overview for carrying out their work in the field of impact tests against a fibre-reinforced PMC.

5.2 POLYMER-MATRIX COMPOSITES

A polymer-matrix composite (PMC) is a material made up of a polymer (resin) matrix and a fibre-reinforcing dispersion phase [12]. The polymer, which is regarded as one of the greatest materials due to its distinct properties, has gained relevance in a variety of applications. Polymers are typically reinforced with additional materials to produce composites to effectively possess the requisite qualities for high performance in a variety of industries. As a result, the production of PMCs has remained among the most effective means of influencing polymer characteristics [13]. PMCs are made up of various organic polymers with continuous or short fibres and various reinforcing agents. This improves the qualities of composite materials, such as improved strength, fracture toughness, and rigidity [14].

PMCs consist of two types of fibres: natural fibre and synthetic fibre. Some examples of natural fibres are jute, kenaf, wood, corn, and many more. Synthetic fibres include carbon fibre, glass fibre, and Kevlar fibre. By adding fibres to the matrix, fibre-reinforced PMCs can be strengthened. Some common synthetic fibres included in the polymer matrix are glass, carbon, and Kevlar, as shown in Figure 5.2, [15]. Apart from its higher tensile strength and tolerance to high temperatures, it also has exceptional electrical insulation. In contrast, glass fibre is brittle. Natural rubber and cotton are two frequent examples of polymer-based materials that we encounter in our daily lives [16]. There are several production processes available for making PMCs; a few of them are resin

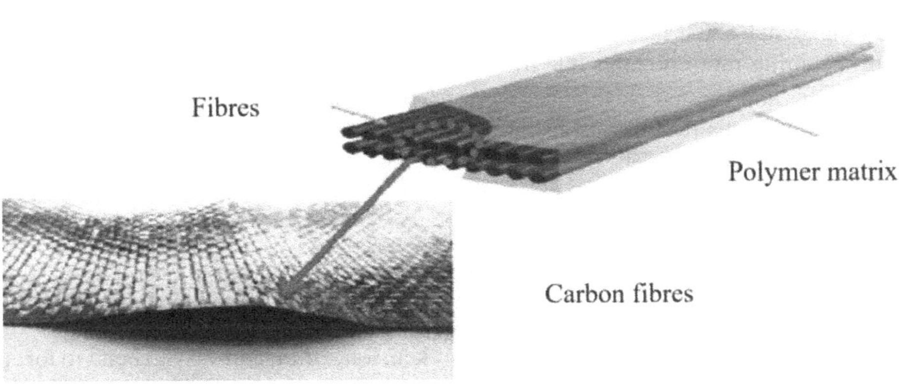

FIGURE 5.2 Polymer-matrix composite [18].

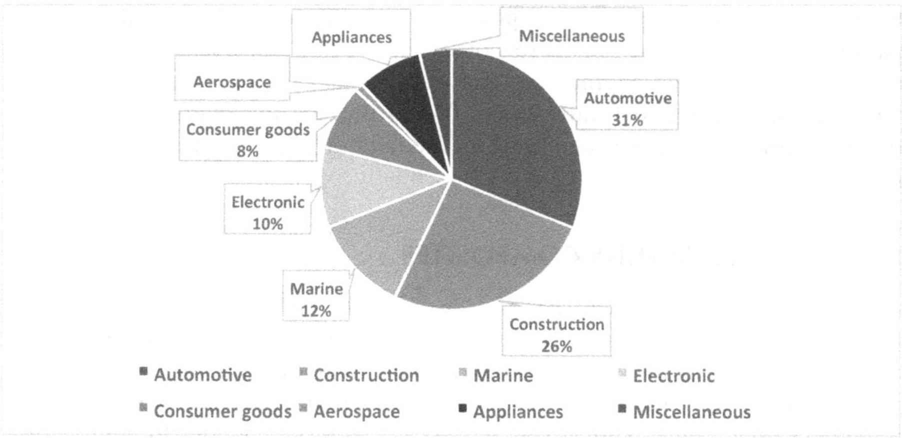

FIGURE 5.3 Market share distribution of fibre-reinforced polymer composite by application [21].

transfer moulding, injection moulding, extrusion, pultrusion, compression moulding, filament winding, and prepreg tape layup [17]. PMCs offer several advantages, including low density, cheap cost, and lower abrasiveness. The properties of the final composite material can be altered depending on the type of constituent employed.

By varying the constituent, we can increase the strength, decrease the weight, and create cost-effective composite materials which can be applied in applications and components, such as boat bodies, bulletproof vests and other armour parts, space shuttle and satellite systems, graphite/epoxy for many structural parts, and drive shafts, doors, racing car bodywork, leaf springs, bumpers, and other automobile components, as shown in Figure 5.3, [19]. Within biomedical field, PMCs find their use in medical implants and X-ray tables. In the electrical field, PMCs are used in making switchgears, insulators, panels, and connectors. Bridges made up of PMCs are becoming more popular due to their low weight, corrosion resistance, fatigue resistance, and limited earthquake damage. In the sports industry, PMCs are used for making tennis rackets and fishing rods. Apart from these, PMCs find their use in making pressure vessels, pump bodies, valves, and chemical storage tanks [20].

5.2.1 Fibres

Fibres are long, fine forms of matter having a diameter in the range of 1 micron to 10 microns and length ranging from a few millimetres to practically continuous. In other words, fibre is a substance that is unbroken, long, thin, and easy to bend to form elongated tissue [23]. For example, hair has a diameter of roughly 80 mm, making it somewhat coarse in comparison to many fibres. However, only a few types of fibres are significantly thicker, most notably the boron and silicon carbide fibres created as the first examples of very-high-performance fibres [24]. Reinforcements do not have to be lengthy fibres. They come in the shape of particles, flakes, whiskers,

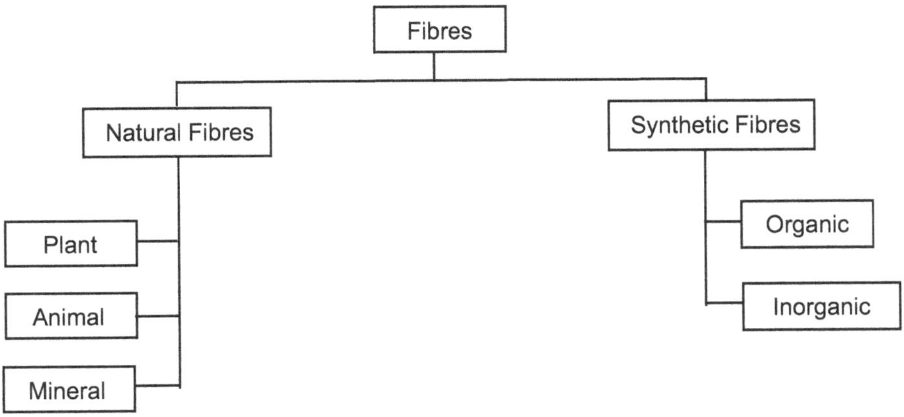

FIGURE 5.4 Classification of fibres [22].

discontinuous fibres, continuous fibres, and sheets. It turns out that the large bulk of materials is stronger and stiffer in their fibrous state than in any other form, hence the popularity of fibrous reinforcements [25]. Fibres are further classified into two types, natural fibres and synthetic fibres, which are explained in Figure 5.4 [26].

5.2.1.1 Natural Fibres

Since non-renewable resources are growing limited, more renewable plant resources have been identified and utilized in recent years. This fibre is strong, light, inexpensive, and renewable. When high elastic modulus is not needed, these low-cost natural fibres can be a viable alternative to expensive and non-renewable synthetic fibres (e.g. e-glass [alumina-borosilicate glass] and carbon fibres) [28]. Natural fibre is more plentiful and less expensive than synthetic fibre, with low density and energy requirement, renewability, no skin irritation, higher strength-to-weight ratio, higher aspect ratio length to diameter (L/D) of around 100, and higher strength and elastic modulus, indicating great potential as glass, carbon, or other synthetic fibre replacements [28]. Furthermore, the benefits of natural fibre have led to its usage for both human needs and industrial raw materials, such as textiles, pulp and paper, accessories, biocomposites, and crafts. Fibre selection, which includes the kind of fibre utilized, harvest time, extraction technique, aspect ratio, treatment, and fibre content, has the greatest influence on the efficiency of natural fibre composites. Matrix selection, fibre interfacial strength, fibre dispersion, fibre orientation, composite production process, and porosity are all important considerations [29].

There are many natural fibres present in an environment which can be used by many industries per property requirements. A typical classification of natural fibres is shown in Figure 5.5. Natural fibres can be obtained from animals or plants. Animal-based fibres include wool and silk. There are many different parts of plants from which fibres can be obtained; for example, bast fibres can be obtained from flax, hemp, jute, ramie, and kenaf; leaf fibres, from sisal, banana, and abaca. Seed fibres

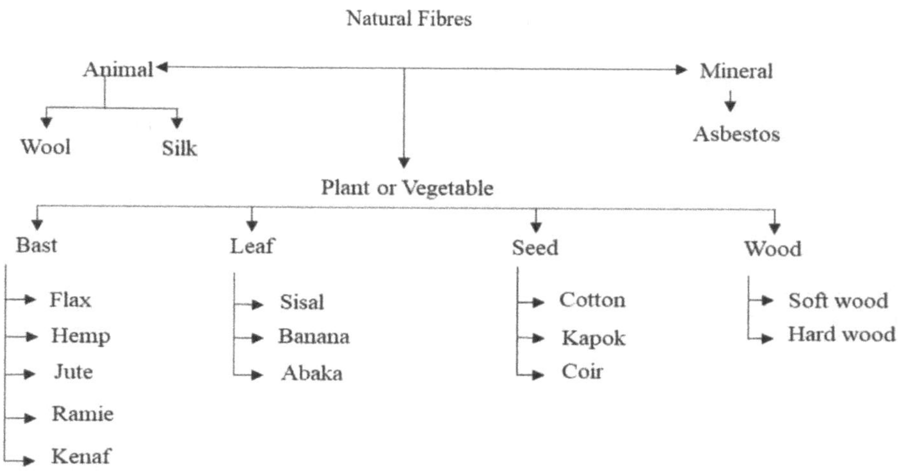

FIGURE 5.5 Types of natural fibres [26].

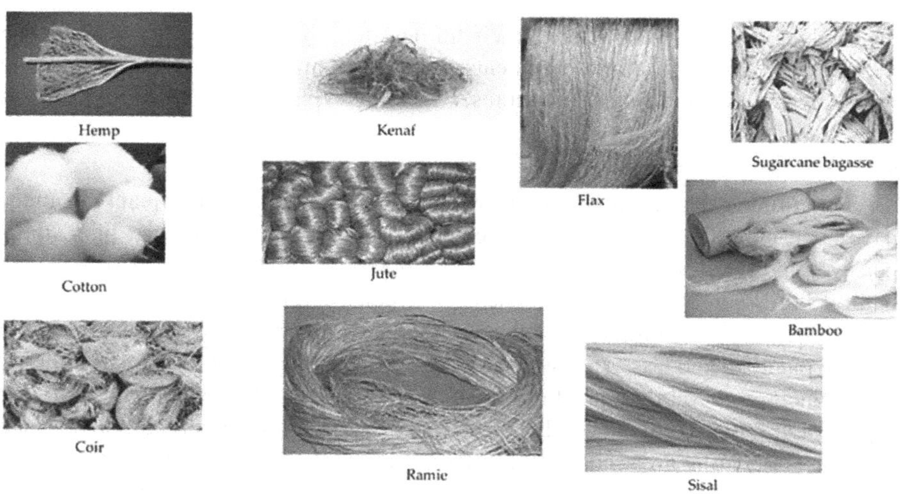

FIGURE 5.6 Natural fibres from plants [27].

include cotton, kapok, and coir. Figure 5.6 show images of natural fibre obtained from various plants. From the wood, we can obtain softwood and hardwood; this all-natural fibre can be used as reinforcement with a matrix for improving properties, such as strength, toughness, and so on [30].

Natural fibre composites outclass synthetic-fibre-based composites in terms of performance. Natural fibres have several benefits over synthetic fibres, including low density, good modulus-to-weight proportion, great acoustic damping, lower industrial fuel costs, lower carbon footprint, and the capacity to decompose [31]. Another group of researchers provides even more compelling evidence for their key benefits, such as the fact that they are significantly less expensive than other products during reinforced composites and require significantly less energy than conventional strengthening fibres, such as glass and carbon.

Along with several advantages, they have some drawbacks, including moisture absorption, manufacturing variance, lower temperature resistance, and low polymer fibre compatibility. Natural fibres have a strong moisture absorption capacity because they contain hydroxyl and polar groups. The dimensional stability and mechanical characteristics of biocomposites decline as moisture content increases. Moisture absorption characteristics are determined by the fibre chosen for reinforcing. Coir and pineapple leaf fibres were reinforced individually into epoxy resin in an experimental investigation for moisture behaviour [32]. Low moisture absorption was observed for coir fibre composites compared to pineapple leaf fibres. As a result of the experiment, it is obvious that moisture absorption in composites reduces mechanical properties to a greater extent [33]. The drawback of using natural fibres as reinforcement is that they do not adhere well to polymers. Chemical treatments on the fibre surface, on the other hand, can minimize this. Alkalinization, acetylation, silanization, acrylation, coupling agents, and other chemical treatments have been tested by researchers. Fibre surface treatment procedure significantly increases the adhesion between the polymer and fibre, resulting in improved material properties [34].

5.2.1.2 Synthetic Fibres

Synthetic fibres (SFs) are becoming increasingly important as a material for fibre-reinforced composite constructions across the world. Because of their exceptional properties, SFs are in high demand as a result of the rising desire for lighter and novel composite materials [36]. Synthetic fibres and natural fibres (NFs) are strong competitors. Ceramic textiles, carbon fibres, glass fibres, basalt fibres, and polymeric fibres have attracted significant attention over the last 20 years. SFs are frequently employed for advanced composite materials manufacturing in aircraft and

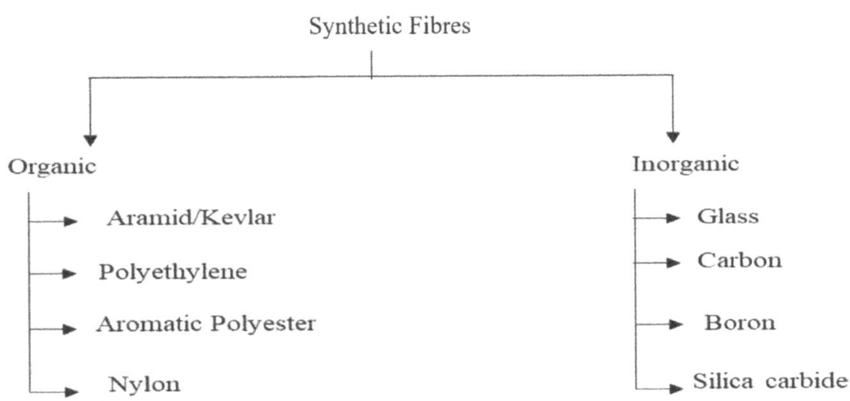

FIGURE 5.7 Classification of synthetic fibres [35].

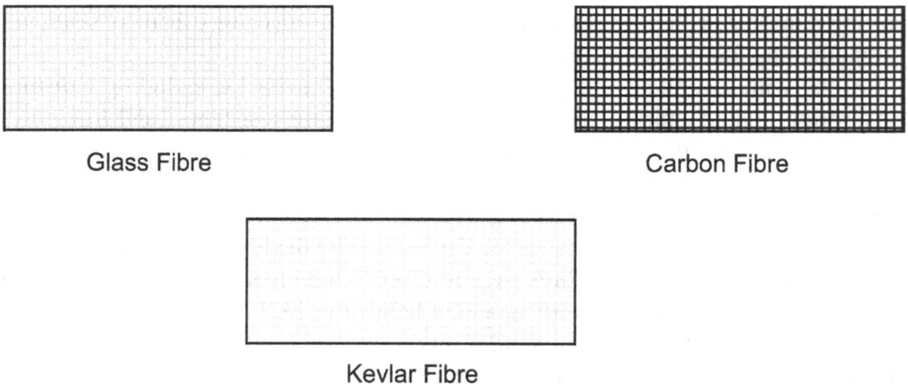

FIGURE 5.8 Synthetic fibres.

automotive industries because of their strong rigidity and strength; they have frequently been employed in the development of innovative and powerful composite materials [37]. Figures 5.7 and 5.8 show the classification and texture of popular synthetic fibres, respectively.

Synthetic fibres can last for long period than natural fibres. The characteristics of synthetic fibres and textiles may also be easily modified to meet the requirements by adjusting their chemical properties and the conditions under which they are made. They can be elastic, flexible, hard, and strong, and they can absorb less or more water depending on their qualities.

Synthetic fibres are often more resistant to water, stains, heat, and chemical damage than natural fibres [39].

Natural fibres, in general, are more reactive to chemical degradation than synthetic fibres as these are biodegradable and can be attacked by a variety of bacteria and fungi. Most synthetic fibres, on the other hand, are not biodegradable and break and wear down with time, whereas natural fibres decompose rapidly. Synthetic fibres can be classified into organic and inorganic fibres, which are further subclassified.

5.2.1.2.1 Organic Fibres

Organic fibres are produced from a product which grows inside the soil, is taken from animal skin, or is produced by an insect. Some of the biological fibres are abaca, jute, camel hair, coir, hemp, silk, and aramid. The softness of items created from these fibres varies greatly.

5.2.1.2.1.1 Aramid/Kevlar Fibres Kevlar fibre (KF) has reinforcing capability among synthetic fibres due to its high specific strength and modulus, low density, and chemical inertness, as well as its good dimensional and thermal stability. KF is frequently employed in aerospace, military, and mechanical applications [40]. KF, however, is incompatible with thermoplastic matrices. As a result, KF must be treated before applying to any application. Many types of treatment, such as the ionomer matrix, plasma treatment, chemical treatments, and the use of coupling agents, have been used [41].

5.2.1.2.1.2 Polyethylene Fibres Polyethylene is a multipurpose and lightweight synthetic resin derived from the polymerization of ethylene. Polyethylene is the most used plastic in society, with uses ranging from shopping bags to containers, oil spill remediation, and automotive fuel tanks [38]. Polyethylene fibre aggregates can be of minimum density, higher density, linear, or ultrahigh molecular weight (UHMWPE) [42].

5.2.1.2.1.3 Aromatic Polyester Fibres Aromatic polyester fibres offer a large range of industrial applications due to their high crystallinity and difficulty in processing. These polyester fibres have extremely high softening points, great mechanical properties, and good heat stability. They are created by poly-condensing aromatic dicarboxylic acids, and various chemicals are mixed to enhance the characteristics of aromatic polyesters [43].

5.2.1.2.1.4 Nylon Fibres Nylon is one of the most efficient synthetic materials, having applications varying from daily life to industry. It is a type of plastic which can be pulled into threads or moulded into everyday goods to create amenities. Nylon fibre is amongst the most often used synthetic fibres in everyday life. Nylon fibre is resistant to wear, has high tensile strength, and absorbs a lot of water [44]. With the rapid expansion of society, demand for synthetic fibre is increasing, and the manufacture of a substantial quantity of fibre unavoidably generates energy consumption and environmental degradation. Every year, industrial production and daily life create a substantial number of waste fibre textiles all over the world, including waste nylon fibre fabrics that must not be ignored, such as socks, ropes, and garments [45].

5.2.1.2.2 Inorganic Fibres

Inorganic fibres are becoming increasingly popular these days. Glass, boron, carbon, silica carbide, potassium titanate, alumina, and ceramics are the most frequent inorganic fibres made.

5.2.1.2.2.1 Glass Fibre Glass fibre (GF) has proven to be a particularly effective reinforcement in the constantly developing fibre-reinforced polymer composites industry. GF products now account for more than 95% of fibre reinforcements used in the composites industry, owing to their highly attractive performance-to-price ratio. GFs are created by melting silica sand, boric acid, limestone, and other ingredients at an extremely high temperature of around 1,200°C [46]. GF is a highly fine, lightweight, high-strength, and long-lasting material. GFs have lesser strength than carbon fibre, but they are less costly [46].

5.2.1.2.2.2 Carbon Fibre Carbon fibres are fibres composed primarily of carbon atoms with diameters ranging from 5 to 10 micrometres. Carbon fibres, a unique high-strength material, are generally used as reinforcement in composite materials such as carbon-fibre-reinforced plastics, carbon-carbon composites, carbon-fibre-reinforced polymers, and carbon-fibre-reinforced cement [47]. Among reinforcing fibres, carbon fibres have the greatest strength and specific modulus. Carbon fibre applications are classified into two types: high technology (aerospace and nuclear engineering) and general engineering and transportation (engineering components, such as bearings, gears, cams, fan blades, and so on, as well as car bodywork). The demands of the two industries, however, are fundamentally different. The widespread usage of carbon fibre in aeroplanes and aerospace is motivated by the need for optimal performance and fuel economy [48].

5.2.1.2.2.3 Boron Fibre Boron fibre (BF) is an elemental boron product that is amorphous and is frequently utilized in aircraft because of its excellent strength and lightweight properties. It was created by performing chemical vapour deposition (CVD) of boron on a substrate at a temperature of 1,000°C. The main application of BF is in bicycle frames, fishing rods, golf club shafts, and so on [49].

5.2.1.2.2.4 Silica Carbide Fibre Silicon carbide (SiC) is a high-temperature and radiation-resistant ceramic material with good chemical, mechanical, thermal, and electrical properties. SiC fibres come in both alpha- and beta-SiC configurations. They are very fine, flexible, and continuous fibres created from numerous woven textiles at the same time [50]. SiC fibres exhibit distinct properties as compared to organic and certain ceramic fibres, such as strong tensile strength, modulus, stiffness, chemical resistance, minimum thermal expansion, low-weight, and higher-temperature tolerance. SiC fibres are much more durable than other fibres, such as glass, carbon, alumina, and alumina silicate [51].

5.3 IMPACT VELOCITY

Impact velocity is the velocity with which the impactor makes a collision with the target plate. For example, the bullet fired from the gun travels at 100 m/s and strikes the target. So the velocity with which the bullet got impacted the target is known as impact velocity. There are mainly four types of impact velocities: low-velocity impact, high-velocity impact, ballistic velocity impact, and hypervelocity impact; all these are explained in detail in the following sections as shown in Figure 5.9.

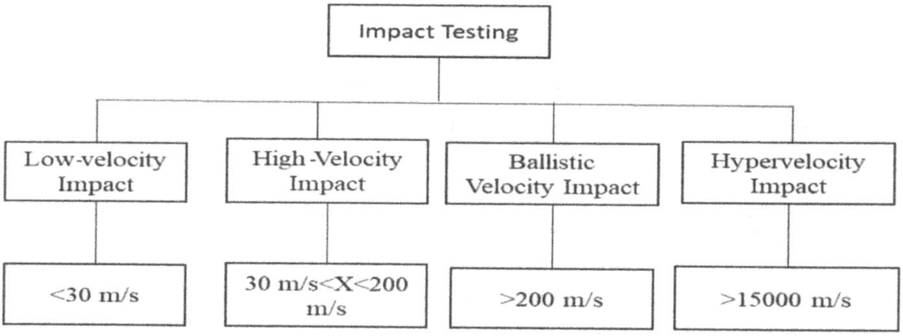

FIGURE 5.9 Types of impact velocities [52].

5.3.1 Types of Impact Velocity

5.3.1.1 Low-Velocity Impact

Low-velocity impact generally takes place when a projectile is impacting with a velocity <30 m/s on any specimen. Low-velocity impacts could be observed in services or during the testing process, and this is amongst the most destructive loads encountered on composite laminates [53]. There are various methods to test composite using low-velocity impact testing. Some of the low-velocity impact testers are the Izod Charpy impact test and the drop weight test. The main function of the Izod Charpy impact test is to identify the impact toughness of the material, and the function of the drop weight test is to determine the behaviour of composite plates.

5.3.1.2 High-Velocity Impact

The most common damage is due to high-velocity impact, which leads to immediate failure of the material, and it generally occurs within a range of 30 to 200 m/s. A high-velocity impact can occur from a variety of sources, including runway debris striking the fuselage during take-off or landing, ice from propeller impacts, and bird attacks [54]. Such materials must be much more resistant to high-velocity projectiles. It is essential to get familiar with the term "high-velocity impact testing." In high-velocity impact testing, impact refers to a collision between two or more bodies, and a projectile is any object that is launched. Any fixed or moving object that can be hit by a projectile is referred to as a target.

5.3.1.3 Ballistic Impact

The impact taking place in a short period as compared to low-velocity impact time is known to be a ballistic impact. An example of a ballistic impact situation could be shooting a bullet from a gun at a high-speed surpassing 200 m/s. Ballistic impact applications can generally be seen in military operations. To put it another way, a ballistic impact usually involves a low-mass, high-velocity impact that is brought on by a propelling force.

Understanding the ballistic behaviour under high-velocity impact is necessary for the efficient usage of composite materials in structural applications [55].

5.3.1.4 Hypervelocity Impact

A hypervelocity collision event occurs when orbital debris flows at speeds of up to 15,000 m/s with extremely high stress in outer space. Hypervelocity impact can be observed in the aerospace industry, such as in hypersonic flight or military applications, such as in antiballistic missile technology. In general, this impact could be observed when spaceship structures get hit by space debris at high speeds.

5.4 INFLUENCE OF PARAMETERS ON IMPACT RESPONSE OF FIBRE-REINFORCED POLYMER COMPOSITES

5.4.1 Influence of Impactor Nose Shape on Glass/Epoxy Composite

In the study of ballistic impact analysis on fibre-reinforced polymer composite material, the most important thing is the shape of the impactor hitting the target plate or any structure. The shape of the impactor is an important factor which can affect parameters such as energy absorption capacity, ballistic limit velocity, residual velocity, the contact force between the impactor and target, and many more. In this chapter, we have highlighted one such research detailing the effect of shape change of impactor on glass/epoxy composite plate. Few researchers have performed ballistic impact simulation against a glass/epoxy composite target plate against a velocity of 100 m/s in ABAQUS software. Different types of impactor shapes, such as conical, ogival, flat, hemispherical, and spherical, have been used against a glass/epoxy composite plate which consists of different orientations of fibres, such as [0°/90°/0°/90°-rep-], [45°/45°/45°/–45°-rep-], [0°/90°/0°/90°-sym-], [45°/–45°/45°/–45°-sym-]. The conclusion drawn from this work is that in the case of conical impactor, maximum energy absorption is with [0°/90°/0°/90°-sym-] layup configuration. In the case of a flat impactor [0°/90°/0°/90°-rep-] was found to be best followed by [0°/90°/0°/90°-sym-] layup configuration. A hemispherical shape impactor [0°/90°/0°/90°-rep-] layup configuration proved to be best in terms of energy absorption capacity, whereas spherical impactor [0°/90°/0°/90°-rep-] layup configuration has the maximum capacity to absorb energy. Among all these parameters, finally, a conclusion drawn is that the flat-shaped impactor with [0°/90°/0°/90°-rep-] layup configuration proved to be best in terms of its ability to absorb maximum energy. So with this, we can see how the nose shape of the impactor becomes an important parameter influencing composite material reaction subjected to impact loading [56]. A brief description of influence of parameters on impact response of Fibre-Reinforced Polymer Composites is listed in Table 5.1.

5.4.2 Influence of Fibre Orientation and Fibre Stacking Sequence on Fibre-Reinforced Polymer Composite

Stacking the sequence of fibre and orientation of fibre can also affect the impact response of fibre-reinforced polymer composite material in many ways, such as in terms of ballistic limit velocity, residual velocity, and energy absorption capacity of the target plate. There is much research done in this field one such research work has been discussed in this chapter which will prove the importance of this parameter. The effect of fibre orientation on the high-velocity impact behaviour of fibre-reinforced polymer composites was investigated by some researchers using finite element analysis. In this work, firstly they performed impact analysis with the help of a 9 mm hemispherical-ended projectile impacting with 373 m/s against both glass/epoxy and carbon/epoxy composite material with fibre orientation of [0°/90°], [0°/90°/30°/–60°], [0°/90°/45°/–45°], [30°/–60°/60°/–30°] in LS-DYNA software. They observed that laminates with a fibre orientation of [0°/90°], often known as cross-ply orientation, offer the highest energy absorption as compared to other fibre orientations for both carbon/epoxy and glass/epoxy composite materials. Based on this orientation, they obtain the best stacking sequence by considering non-hybrid and hybrid fibre-reinforced polymer composite laminates, such as G/G/G, C/C/C, C/G/C, G/C/G, G/G/C, and C/G/C, were used with [0°/90°] orientation where G represents glass fibre and C represents carbon fibre. The conclusion drawn from this work is that [0°/90°] layup was the best orientation as compared to other orientations. Among stacking sequences, the G/C/G composite exhibited

TABLE 5.1
Influence of Parameters on Impact Response of Fibre-Reinforced Polymer Composites

Sr. No.	Parameters	Details
1.	Influence of impactor nose shape on glass/epoxy composite	Few researchers have carried out their work to study the influence of impactor nose shape on glass/epoxy composite with different fibre orientations, and results revealed that few parameters such as residual velocity, energy absorption, and ballistic limit velocity get changed when impacted with different shapes of the impactor. So the impactor nose shape plays a major role in varying certain parameters of any composites [56].
2.	Influence of fibre orientation and fibre stacking sequence on fibre-reinforced polymer composite	Orientation of fibres and fibre-stacking sequence can also affect the fibre-reinforced polymer composite impact response in many ways, such as in terms of ballistic limit velocity, residual velocity of the impactor, and energy absorption capacity of the target plate. Some studies revealed that with varying orientations of fibres and fibre-stacking sequences, the optimum energy absorption capacity of the target plate can be easily achieved [57].
3.	Influence of plate thickness and impactor incidence angle on fibre-reinforced polymer composite	Few studies noted that varying plate thickness and impactor incidence angle have affected the residual velocity of the impactor, ballistic limit velocity, and energy absorption capacity of the target plate. It was observed that the energy absorption capacity and ballistic limit velocity increase with an increase in target thickness and impactor incidence angle [58].

the highest energy absorption. So with this, we came across how the fibre orientation and stacking sequence can play a major role in impact applications [57].

5.4.3 Influence of Plate Thickness and Impactor Incidence Angle on Fibre-Reinforced Polymer Composite

Another parameter which may affect the fibre-reinforced polymer composite impact response is plate thickness and the angle of incidence of the impactor. This has been proven by many researchers of which few have done a numerical study on the influence of target thickness and projectile or impactor incidence angle on ballistic resistance of glass-fibre-reinforced polymer composite. In this, the response of glass-fibre-reinforced PMC has been investigated against a 2.88 g hemispherical-nosed projectile. The target thickness and impact angle were varied, and the effects on fibre failure, delamination of plies, matrix failure, energy absorption, residual velocities, and ballistic limit have been analysed. The conclusion drawn is that the performance of the glass-fibre-reinforced polymer target increased when increasing the target thickness. The impact angle has contributed to the damage and performance of the plate. The energy absorbing capacity and ballistic limit velocity of laminate increased with target thickness and impact angle. This study provides enough information on the effect of target thickness and impactor incidence angle on the impact response of fibre-reinforced polymer composite [58].

5.5 APPLICATIONS OF IMPACT LOAD IN DIFFERENT SECTORS

5.5.1 Automobile Sector

As we can see, the utilization of fibre-reinforced polymer composite in many applications, also undergoes different types of impacts. Similarly, in the automobile field, different parts of the vehicles, such as the bonnet and the windshield, can undergo impact loading in many cases, such as during vehicle accidents, the crash test of vehicles, and so on [59]. So the study of impact response has become crucial for improving resistance against impact load.

5.5.2 Aerospace Sector

Impact damage is a major threat, especially with the new generation of aircraft coming to serve as major portions of their structures are made of composites and are subjected to external sources of impact. There are many possibilities where different parts of aircraft can get damaged due to impact with any object [60]. The main events include plane crashes, accidents during landing, and so on. The impact damage can also be observed in space when two satellites crash; they can shatter apart into thousands of new pieces, creating lots of new debris.

5.5.3 Bulletproof Vest

A bulletproof vest act as a barrier and at the same time absorbs the impact energy of a projectile shot from a gun or firearm. To minimize the impact due to projectile shot many fibre-reinforced composites have been used to prepare these vests for improving their impact resistance [61]. There are many applications in which this composite undergoes impact damage; some of them are military helmets, ballistic shields, and other defence equipment [62].

5.5.4 Wind Turbine

A wind turbine, which is used to generate electrical energy by converting kinetic energy into electric energy, is often installed in an open environment [63]. Due to this, many flying birds or any foreign object in an environment may strike wind turbine blades, which may cause injury to wildlife. It may get damaged due to lightning strikes, and also any object present in the environment may cause impact damage to the wind turbine blade [64].

5.6 CONCLUSION AND FUTURE SCOPE

In this chapter, a detailed discussion of fibre-reinforced polymers for high-impact applications is presented. The important parameters responsible for impact damage of polymer composite are discussed. Some of the key findings are as follows.

- Among various shapes of impactors impacting against fibre-reinforced polymer composite, it was found that the best-performing composite in terms of highest energy absorption capacity is having a [°0/90°] (i.e. cross-ply fibre) orientation when impacting against flat-nose-shaped impactor.
- Another factor which affects the impact response of fibre-reinforced polymer composite is the variation in fibre orientation and fibre-stacking sequence. It was found that a [0°/90°] (i.e. cross-ply fibre) orientation and G/C/G stacking sequence proved to improve the overall impact performance of fibre-reinforced composite and also make the composite cost-effective without affecting its performance.
- Variations in the thickness of the targeted specimen and changing the impactor angle can also improve impact resistance. From the review, it was found that increasing the thickness of the target specimen or increasing the impactor angle was found effective in terms of providing better impact resistance and improved energy absorption capacity.

The current review will provide good exposure to readers in carrying out their work in the field of impact tests against fibre-reinforced composite materials as it will help them in understanding different parameters which have to be kept in mind before performing impact analysis against fibre-reinforced polymer composite. Also, information related to different types of impact velocities with their range has been provided along with the impact damage applications.

In future, further work can be carried forward by performing ballistic impact tests with changing impactor nose shapes against the non-hybrid and hybrid composite laminate. This can be done for investigating the capability of designed composite laminate to resist impacts from various shapes of the impactor and to find suitability in particular ballistic applications.

REFERENCES

[1] D. Kumar, D. D. Pagar, R. Kumar, and C. I. Pruncu, "Recent progress of reinforcement materials: A comprehensive overview of composite materials," *Integr. Med. Res.*, vol. 8, no. 6, pp. 6354–6374, 2019, http://doi.org/10.1016/j.jmrt.2019.09.068.

[2] Y. Yang, R. Boom, B. Irion, D. Van Heerden, P. Kuiper, and H. De Wit, "Chemical engineering and processing: Process intensification recycling of composite materials," *Chem. Eng. Process. Process Intensif.*, vol. 51, pp. 53–68, 2012, http://doi.org/10.1016/j.cep.2011.09.007.

[3] L. Zhu, N. Li, and P. R. N. Childs, "Light-weighting in aerospace component and system design," *Propuls. Power Res.*, vol. 7, no. 2, pp. 103–119, 2018, http://doi.org/10.1016/j.jppr.2018.04.001.

[4] A. Treviso, B. Van Genechten, D. Mundo, and M. Tournour, "Damping in composite materials: Properties and models," *Compos. Part B*, vol. 78, pp. 144–152, 2015, http://doi.org/10.1016/j.compositesb.2015.03.081.

[5] A. K. Sharma, R. Bhandari, A. Aherwar, R. Rimašauskiene, and C. Pinca-Bretotean, "A study of advancement in application opportunities of aluminum metal matrix composites," *Mater. Today Proc.*, vol. 26, pp. 2419–2424, 2020, http://doi.org/10.1016/j.matpr.2020.02.516.

[6] C. R. Dandekar and Y. C. Shin, "International journal of machine tools & manufacture modeling of machining of composite materials: A review," *Int. J. Mach. Tools Manuf.*, vol. 57, pp. 102–121, 2012, http://doi.org/10.1016/j.ijmachtools.2012.01.006.

[7] S. Huang, Q. Fu, L. Yan, and B. Kasal, "Characterization of interfacial properties between fibre and polymer matrix in composite materials—A critical review," *J. Mater. Res. Technol.*, vol. 13, pp. 1441–1484, 2021, http://doi.org/10.1016/j.jmrt.2021.05.076.

[8] A. B. Nair and R. Joseph, "Eco-friendly bio-composites using natural rubber (NR) matrices and natural fiber reinforcements," *Chem. Manufact. Appl. Nat. Rubber,* 249–283 2014.

[9] R. Yadav, M. Tirumali, X. Wang, M. Naebe, and B. Kandasubramanian, "Polymer composite for antistatic application in aerospace," *Def. Technol.*, vol. 16, no. 1, pp. 107–118, 2020, http://doi.org/10.1016/j.dt.2019.04.008.

[10] A. Sayam, A. N. M. M. Rahman, M. S. Rahman, et al., A review on carbon fiber-reinforced hierarchical composites: mechanical performance, manufacturing process, structural applications and allied challenges. *Carbon Lett.* 32, 1173–1205 (2022). https://doi.org/10.1007/s42823-022-00358-2

[11] W. Harizi, S. Kaidi, A. Monnin, N. El Hajj, Z. Aboura, and M. Benzeggagh, "Study of the dynamic response of polymer-matrix composites using an innovative hydraulic crash machine," *J. Dyn. Behav. Mater.*, vol. 1, no. 4, pp. 359–369, 2015, http://doi.org/10.1007/s40870-015-0032-4.

[12] S. Chauhan and R. Kumar Bhushan, "Study of polymer matrix composite with natural particulate/fiber in PMC: A review," *Int. J. Adv. Res.*, vol. 3, pp. 1168–1179, 2017 [Online]. Available: www.IJARIIT.com.

[13] I. O. Oladele, T. F. Omotosho, and A. A. Adediran, "Review article polymer-based composites: An indispensable material for present and future applications," vol. 2020, Article ID 8834518, Page 1–12. 2020.

[14] S. Sajan and D. Philip Selvaraj, "A review on polymer matrix composite materials and their applications," *Mater. Today Proc.*, vol. 47, pp. 5493–5498, 2021, http://doi.org/10.1016/j.matpr.2021.08.034.

[15] T. P. Sathishkumar, J. Naveen, and S. Satheeshkumar, "Hybrid fiber reinforced polymer composites—A review," *J. Reinf. Plast. Compos.*, vol. 33, no. 5, pp. 454–471, 2014, http://doi.org/10.1177/0731684413516393.

[16] A. Kumar Sharma, R. Bhandari, C. Sharma, S. Krishna Dhakad, and C. Pinca-Bretotean, "Polymer matrix composites: A state of art review," *Mater. Today Proc.*, vol. 57, pp. 2330–2333, 2022, http://doi.org/10.1016/j.matpr.2021.12.592.

[17] F. F. Zhang, Z. H. Yuan, and D. D. Shi, "Research and development on polymer composites in BIAM," *Adv. Perform. Mater.*, vol. 2, no. 3, pp. 321–334, 1995, http://doi.org/10.1007/BF00705455.

[18] N. I. Khan and S. Halder, "Self-healing fiber-reinforced polymer composites for their potential structural applications," *Self-Healing Polym. Syst.*, no. December, pp. 455–472, 2020, http://doi.org/10.1016/b978-0-12-818450-9.00015-5.

[19] I. O. Oladele, T. F. Omotosho, and A. A. Adediran, "Polymer-based composites: An indispensable material for present and future applications," *Int. J. Polym. Sci.*, vol. 2020, 2020, http://doi.org/10.1155/2020/8834518.

[20] S. Kangishwar, N. Radhika, A. A. Sheik, A. Chavali, and S. Hariharan, *Material selection, fabrication, and application*, no. 0123456789. Springer, 2022.

[21] A. Zyjewski, J. Chróścielewski, and Ł. Pyrzowski, "The use of fibre-reinforced polymers (FRP) in bridges as a favourable solution for the environment," *E3S Web Conf.*, vol. 17, no. June, 2017, http://doi.org/10.1051/e3sconf/20171700102.

[22] T. Raja et al., "Thermal and flame retardant behavior of neem and banyan fibers when reinforced with a bran particulate epoxy hybrid composite," *Polymers (Basel).*, vol. 13, no. 22, 2021, http://doi.org/10.3390/polym13223859

[23] T. Sabir, *Fibers used for high-performance apparel.* Elsevier Ltd., 2017.

[24] A. R. Bunsell, *Introduction to the science of fibers.* Elsevier Ltd, 2018.

[25] H. Estrada and L. S. Lee, "FRP composite constituent materials," *Int. Handb. FRP Compos. Civ. Eng.*, pp. 31–49, 2013, http://doi.org/10.1201/b15806-5.

[26] M. Saxena, A. Pappu, A. Sharma, R. Haque, and S. Wankhede, "Composite Materials from Natural Resources: Recent Trends and Future Potentials," *Adv. Compos. Mater.—Anal. Nat. Man-Made Mater.*, no. September, 2011, http://doi.org/10.5772/18264.

[27] K. M. F. Hasan, P. G. Horváth, and T. Alpár, "Potential natural fiber polymeric nanobiocomposites: A review," *Polymers (Basel).*, vol. 12, no. 5, 2020, http://doi.org/10.3390/POLYM12051072.

[28] A. Karimah et al., "A comprehensive review on natural fibers: Technological and socio-economical aspects," *Polymers (Basel).*, vol. 13, no. 24, 2021, http://doi.org/10.3390/polym13244280.

[29] K. L. Pickering, M. G. A. Efendy, and T. M. Le, "Composites: Part A a review of recent developments in natural fibre composites and their mechanical performance," *Compos.—A: Appl. Sci. Manuf.*, vol. 83, pp. 98–112, 2016, http://doi.org/10.1016/j.compositesa.2015.08.038.

[30] S. R. Tridico, *Natural animal textile fibres: Structure, characteristics and identification.* Woodhead Publishing Limited, 2009.

[31] J. Ahmad and Z. Zhou, "Mechanical properties of natural as well as synthetic fiber reinforced concrete: A review," *Constr. Build. Mater.*, vol. 333, no. April, p. 127353, 2022, http://doi.org/10.1016/j.conbuildmat.2022.127353.

[32] A. Ali et al., "Hydrophobic treatment of natural fibers and their composites—A review," *J. Ind. Text.*, vol. 47, no. 8, pp. 2153–2183, 2018, http://doi.org/10.1177/1528083716654468.

[33] C. S. Jawalkar and S. Kant, "Critical review on chemical treatment of natural fibers to enhance mechanical properties of bio composites," *Silicon*, vol. 14, pp. 5103–5124, 2022.

[34] M. S. Rabbi, T. Islam, and G. M. S. Islam, "Injection-molded natural fiber-reinforced polymer composites—a review," *Int. J. Mech. Mater. Eng.*, pp. 1–21, 2021.

[35] D. K. Rajak, P. H. Wagh, and E. Linul, "A review on synthetic fibers for polymer matrix composites: Performance, failure modes and applications," *Materials*, vol. 15, no. 14, 2022, http://doi.org/10.3390/ma15144790.

[36] J. Andrzejewski, B. Gapiński, A. Islam, and M. Szostak, "The influence of the hybridization process on the mechanical and thermal properties of polyoxymethylene (POM) composites with the use of a novel sustainable reinforcing system based on biocarbon and basalt fiber (BC/BF)," *Materials (Basel).*, vol. 13, no. 16, pp. 1–24, 2020, http://doi.org/10.3390/MA13163496.

[37] D. K. Rajak, P. H. Wagh, and E. Linul, "A review on synthetic fibers for polymer matrix composites: Performance, failure modes and applications," *Materials (Basel)*, vol. 15, no. 14, 2022, http://doi.org/10.3390/ma15144790.

[38] E. Rand, "Product packaging," *Ellis Isl. Snow Globe*, pp. 207–238, 2021, http://doi.org/10.2307/j.ctv11cw45p.12.

[39] F. I. Mahir, K. N. Keya, B. Sarker, K. M. Nahiun, and R. A. Khan, "A brief review on natural fiber used as a replacement of synthetic fiber in polymer composites," *Mater. Eng. Res.*, vol. 1, no. 2, pp. 88–99, 2019, http://doi.org/10.25082/mer.2019.02.007.
[40] M. Akibul Islam, M. A. Chowdhury, M. Arefin Kowser, M. Osman Ali, K. Azad, and M. Ramjan Ali, "Enhancement of thermal properties of Kevlar 29 coated by SiC and TiO2 nanoparticles and their binding energy analysis," *Arab. J. Chem.*, vol. 15, no. 8, p. 103959, 2022, http://doi.org/10.1016/j.arabjc.2022.103959.
[41] Y. F. Ou and R. X. I. E. Y. Wang, "Reinforcing effects of modified Kevlar ® fiber on the mechanical properties of wood-flour/polypropylene composites," *J. For. Res.*, vol. 24, pp. 149–153, 2013, http://doi.org/10.1007/s11676-013-0335-z.
[42] L. A. Pruitt, *Structural biomedical polymers (Nondegradable).* Elsevier Ltd., 2011.
[43] J. Pascault and A. Rousseau, "Bio-based alternatives in the synthesis of aliphatic-aromatic polyesters dedicated to biodegradable film applications Bio-based alternatives in the synthesis of aliphatic e aromatic polyesters dedicated to biodegradable film applications," *Polymer*, vol. 59, no. December, 2014, http://doi.org/10.1016/j.polymer.2014.12.021.
[44] S. Soltani, G. Naderi, and S. Mohseniyan, "Mechanical, morphological and rheological properties of short nylon fiber reinforced acrylonitrile-butadiene rubber composites," *Fibers Polym.*, vol. 15, no. 11, pp. 2360–2369, 2014, http://doi.org/10.1007/s12221-014-2360-8.
[45] Y. Qin et al., "Effects of nylon fiber and nylon fiber fabric on the permeability of cracked concrete," *Constr. Build. Mater.*, vol. 274, p. 121786, 2021, http://doi.org/10.1016/j.conbuildmat.2020.121786.
[46] J. L. Thomason, "Glass fibre sizing: A review. *Compos.—A: Appl. Sci. Manuf.*, vol. 127, no. August, 2019, http://doi.org/10.1016/j.compositesa.2019.105619.
[47] E. P. Koumoulos et al., "Research and development in carbon fibers and advanced high-performance composites supply chain in Europe: A roadmap for challenges and the industrial uptake," *J. Compos. Sci.*, vol. 3, no. 3, 2019, http://doi.org/10.3390/jcs3030086.
[48] C. Science and K. A. Publishers, "Carbon fibers for composites," vol. 5, pp. 1303–1313, 2000.
[49] D. E. Görgün, S. Kumartaşlı, O. Avinc, and N. Atar, "Boron fibers and their applications," *Text. Sci. Econ.*, no. January 2022, p. 56, 2021.
[50] G. Agarwal, A. Patnaik, and R. K. Sharma, "Thermo-mechanical properties of silicon carbide filled chopped glass fiber reinforced epoxy composites," *Int. J. Adv. Struct. Eng.*, vol. 5, no. 1, pp. 1–8, 2013, http://doi.org/10.1186/2008-6695-5-21.
[51] U. Santoro, E. Novitskaya, K. Karandikar, H. E. Khalifa, and O. A. Graeve, "Phase stability of SiC/SiC fiber reinforced composites: The effect of processing on the formation of a and b phases," *Mater. Lett.*, vol. 241, pp. 123–127, 2019, http://doi.org/10.1016/j.matlet.2019.01.055.
[52] T. M. Loganathan, M. T. H. Sultan, M. K. Gobalakrishnan, and G. Muthaiyah, *Ballistic impact response of laminated hybrid composite materials.* Elsevier Ltd, 2018.
[53] M. O. W. Richardson and M. J. Wisheart, "Review of low-velocity impact properties of composite materials," *Compos. Part A Appl. Sci. Manuf.*, vol. 27, no. 12 PART A, pp. 1123–1131, 1996, http://doi.org/10.1016/1359-835X(96)00074-7.
[54] S. Gholizadeh, "A review of impact behaviour in composite materials," *Int. J. Mech. Prod. Eng.*, no. 7, pp. 2321–2071, 2019 [Online]. Available: http://iraj.in.
[55] L. Liu, S. Yin, G. Luo, Z. Zhao, and W. Chen, "The influences of projectile material and environmental temperature on the high velocity impact behavior of triaxial braided composites," *Appl. Sci.*, vol. 11, no. 8, 2021, http://doi.org/10.3390/app11083466.

[56] M. Kumar, "Numerical simulation of ballistic impact response on composite materials for different shape of projectiles," *Sādhanā*, vol. 0123456789, 2022, http://doi.org/10.1007/s12046-022-01889-0.
[57] C. Stephen, S. R. Behara, B. Shivamurthy, R. Selvam, S. Kannan, and M. Abbadi, "Finite element study on the influence of fiber orientation on the high velocity impact behavior of fiber reinforced polymer composites," *Int. J. Interact. Des. Manuf.*, vol. 16, no. 2, pp. 459–468, 2022, http://doi.org/10.1007/s12008-021-00808-7.
[58] P. Karthick and K. Ramajeyathilagam, "Numerical study of influence of target thickness and projectile incidence angle on ballistic resistance of the GFRP composites," *Mater. Today Proc.*, vol. 47, pp. 992–999, 2021, http://doi.org/10.1016/j.matpr.2021.05.459.
[59] K. Friedrich and A. A. Almajid, "Manufacturing aspects of advanced polymer composites for automotive applications," *Appl. Compos. Mater.*, vol. 20, no. 2, pp. 107–128, 2013, http://doi.org/10.1007/s10443-012-9258-7.
[60] M. Sreejith and R. S. Rajeev, *Fiber reinforced composites for aerospace and sports applications*. Elsevier Ltd., 2021.
[61] Z. Benzait and L. Trabzon, "A review of recent research on materials used in polymer–matrix composites for body armor application," *J. Compos. Mater.*, vol. 52, no. 23, pp. 3241–3263, 2018, http://doi.org/10.1177/0021998318764002.
[62] J. Naveen, K. Jayakrishna, M. T. Bin Hameed Sultan, and S. M. M. Amir, "Ballistic performance of natural fiber based soft and hard body armour—A mini review," *Front. Mater.*, vol. 7, no. December, pp. 1–6, 2020, http://doi.org/10.3389/fmats.2020.608139.
[63] M. Hafner and G. Luciani, *The Palgrave handbook of international energy economics*, Springer Nature, Cham, 2022.
[64] W. Finnegan, R. Allen, C. Glennon, J. Maguire, M. Flanagan, and T. Flanagan, "Manufacture of high-performance tidal turbine blades using advanced composite manufacturing technologies," *Appl. Compos. Mater.*, vol. 28, no. 6, pp. 2061–2086, 2021, http://doi.org/10.1007/s10443-021-09967-y.

6 Design, Fabrication, and Performance Evaluation of Metamaterial-Inspired Dual-Band Microwave Absorber

Atipriya Sharma, Aarti Bansal

6.1 INTRODUCTION

Metamaterial (MM) structures are quasi-periodic or periodic in nature and have distinct shape, size, or periodicity in structure that defines its electromagnetic properties [1]. MMs have been widely known over half a century due to their unique electromagnetic properties and pragmatic applications like antenna, radome and stealth, cloaking, perfect lens, etc. [2]–[3]. Recently, MM-based absorbers have been developed for different frequency ranges [4]–[6], providing different characteristics like multiband, dual-band, single-band with wide angle, polarization, insensitivity, and so on [7]–[9].

The structure of MM absorber incorporates three layers: the first is the topmost copper layer, after that is the substrate layer, and the last one is the ground copper layer. The principle of MM absorber is based on the perfect matching layer in order to enlarge the absorbance. The reflection would be close to zero, which implies maximum absorbance of the incoming wave will occur in case its free space impedance gets closer to the surface impedance. Frequency-selective surfaces (FSSs) shape and size will decide either it will block or pass electromagnetic waves [10]. FSSs with hexagonal [8] and octagonal [11]–[13] geometrical configurations have become more in fashion due to polarization-independent characteristics [11]. Costa et al. proposed high-absorption bandwidth-ranging broadband absorber typically based on MM configuration, obtained from 6.86 GHz to 15.16 GHz [10] and a value of 8.30 GHz. Li et al. have obtained good absorbance with the help of lumped resistance and octagonal-ring-shaped MM [12], while an octagonal split-ring resonator with the multi-band performance has been proposed by Kollatou et al. [13]. An MM circular SRR based ultra-wideband and ultrathin absorber proposed in [14] exhibited absorption bandwidth (7.85 GHz—12.25 GHz). Bakir et al. have presented an ultrathin MM absorber integrated with high-impedance surface (HIS). The proposed absorber exhibited both wideband and narrowband behaviours, which majorly depend upon

 DOI: 10.1201/9781003367154-6

selected shape and substrate material of the FSS structure [15]. Researchers have faced challenges in the development of an absorber having a good MM absorber with wide bandwidth. The proper fabrication and performance evaluation of MM absorber is also a challenging task. Therefore, MM-inspired absorber with dual-resonance is designed, proposed, and optimized for the range of 2 to 18 GHz. Further, the designed dual band absorber was also fabricated to evaluate its measurement performance.

6.2 DESIGN AND OPTIMIZATION

The side view in Figure 6.1(a) displays the proposed absorber's unit cell structure. The geometry composed of a ground layer and top layer with periodic FSS structure etched on a substrate layer (FR4 having permittivity (ε_r) = 4.3 and tan δ = 0.025). The 0.035 mm copper metallic layer is etched on top and bottom whereas, the substrate is

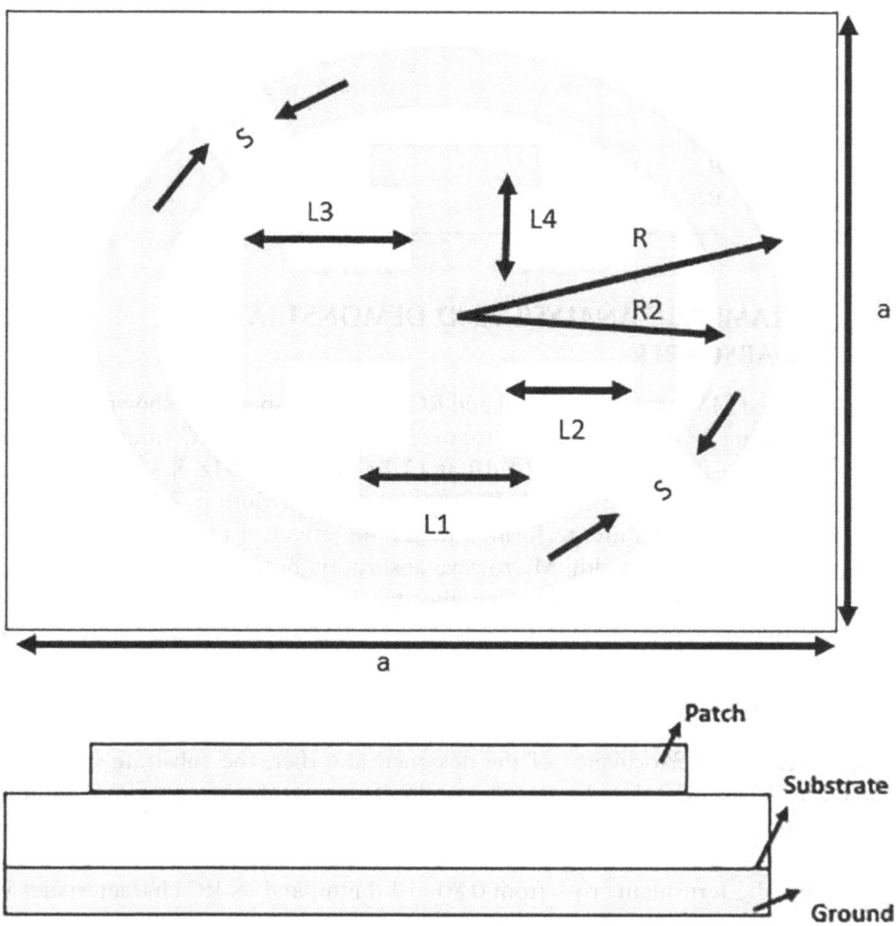

FIGURE 6.1 Schematic of the (a) top view and (b) side view of the absorber unit cell.

2.0 mm thick. The Figure 6.1(b) illustrates the structure's top view. Here, the yellow colour represents the FSS layer, which is made up of copper, whereas the remaining part shows the dielectric substrate. The CST Microwave Studio is used for its simulation followed by its optimization through critical parametric analysis.

To excite the incident wave, the periodic boundary conditions are assigned in the *x* and *y* directions, whereas in the *z* direction, the Floquet port is assigned. Table 6.1 summarizes the dimensions of design variables in described unit cell structure and their symbols:

TABLE 6.1
Geometry Variables of Presented FSS Structure

Symbols	Dimension in mm
A	8.00
S	0.25
L2	1.40
L4	1.40
L1	1.20
L3	1.60
R	3.14
R2	2.24

6.3 PARAMETRIC ANALYSIS AND DEMONSTRATION OF ABSORBER

The proposed MM absorber's simulated RC frequency spectra is shown in Figure 6.2 (a) covering 2 GHz to 18 GHz. Proposed structure's peak RC values are −28.6 dB, −20.6 dB, −19.9 dB, and −17.4 dB at 12.2 GHz, 15.4 GHz, 8.8 GHz, and 9.3 GHz, respectively. The structure's RC absorption bandwidth is 7.9 GHz. Further parametric simulations are performed to see the effect of change in structure's dimensions on its bandwidth. Microwave absorption property is mainly affected by substrate thickness [16]. Therefore, the structure's RC properties corresponding to different substrate layer thickness is analysed through a plot, as can be depicted from Figure 6.2 (b). The design utilizes hybrid structure composed of circle and cross structure, which adds to little computational complexity while simulating.

To check the performance of the designed absorber, the substrate's width is varied. It is observed that the structure exhibits best performance with a 2.0 mm thick substrate. Thus, optimized thickness of 2.0 mm is chosen for the further studies. In the next step, the gap between the circle and Jerusalem cross is varied by scaling the Jerusalem cross from 0.80 to 1.0 mm, and its RC characteristics is studied.

Figure 6.3 (a) illustrates its frequency-dependent RC characteristics with respect to gap between the circle and the Jerusalem cross. Mutual coupling gets affected

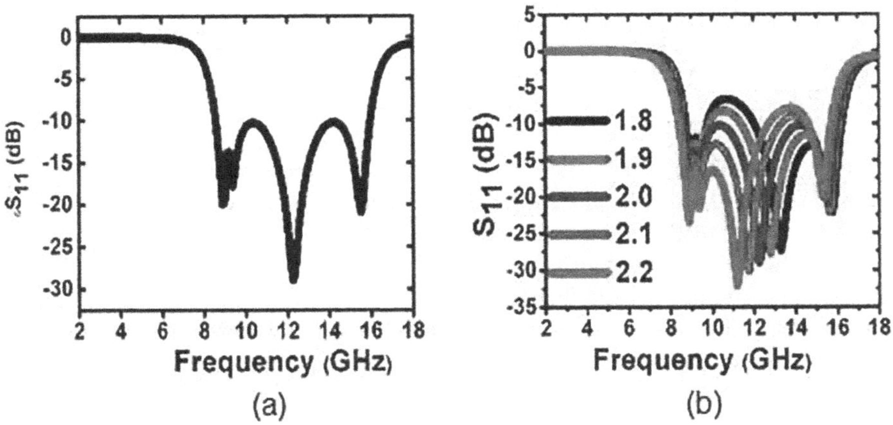

FIGURE 6.2 (a) Reflection constant characteristics of proposed MAS and for (b) varying substrate thickness with respect to frequency (2–18 GHz).

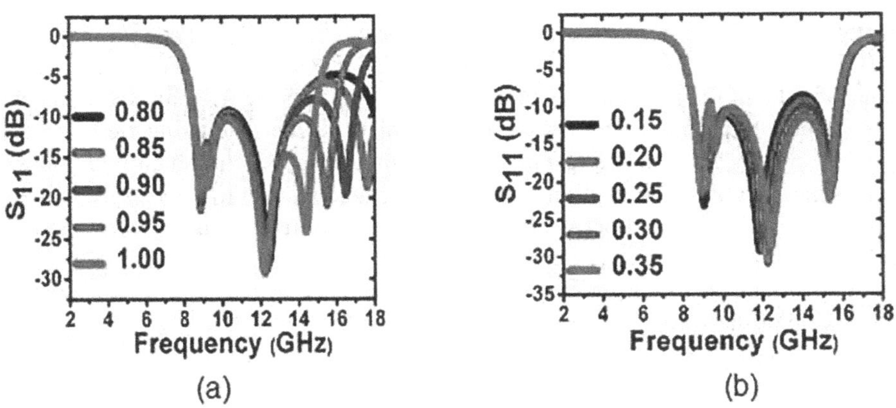

FIGURE 6.3 Study of varying the geometry parameters from the 2–18 GHz (a) gap between the circle and the Jerusalem cross. (b) Gap between the split ends.

due to the gap between both of them [14]. Meanwhile, the study on the space between the split ends over RC characteristics of the absorber is also analysed. Figure 6.3 (b) demonstrates the study on the gap over RC characteristics. The gap between the split ends varies from 0.15 to 0.35 mm in steps of 0.05 mm. It is depicted that the best results are obtained at 0.25 mm. The RC characteristics are found to be influenced by the gap, which may be because of the corresponding variation in the capacitance.

Figure 6.3 (a) exhibits the dependency of RC characteristics of the absorber corresponding to change in the gap between the circle and the Jerusalem cross. Mutual coupling gets affected due to the gap between both of them [15–16].

Figure 6.3 (b) demonstrates the performance of structure's RC characteristics as the gap varies. The gap between the split ends varies from 0.15 to 0.35 mm. It is analysed that the best results are obtained at 0.25 mm. The RC characteristics are found to be influenced by the gap, which may be because of the corresponding variation in the capacitance.

Further, the angle of incidence (θ) is varied to see the polarizations (TM, TE) of the structure. The *y* direction signifies the magnetic field, whereas the *x* direction is assigned electric field TE and TM modes. In the TE polarization scenario, variation is seen in the H field's direction, whereas uniformity is seen in the E field's direction. Similar to this, with TM polarization, variation is shown in the electric field's direction, whereas uniformity is seen in the magnetic field's direction. Figures 6.4 (a) and 6.4 (b) represents RC values for TM and TE polarization under oblique incidences. The shift in resonance frequency is observed towards up with an increase of oblique incident.

MMs are mainly categorized with respect to the complex magnetic permeability (μ_r) and dielectric permittivity (ε_r) and complex negative values [17–18]. The S-parameters are utilized to extract the characteristics, such as the structure's impedance, frequency-dependent complex magnetic permeability, and complex dielectric permittivity. The detail of the adopted parameter retrieval method is explained in [19]. Figure 6.5 (a) shows the normalized impedance's imaginary part, whereas the Figure 6.5 (b) depicts the normalized impedance's real part. It is observed that the impedance's real part is close to 1 at the resonant frequencies. This implies that maximum absorption take place at the resonating frequencies. Figure 6.6 shows the structure's normalized impedance (Z) graphically. It shows the mechanism of the absorption in the absorber with respect to impedance matching. The permittivity and permeability with respect to frequency are summarized in Table 6.2. The permittivity and permeability's native values clearly indicating the MM nature of the proposed MMA, which also controls the electromagnetic absorption of the wave.

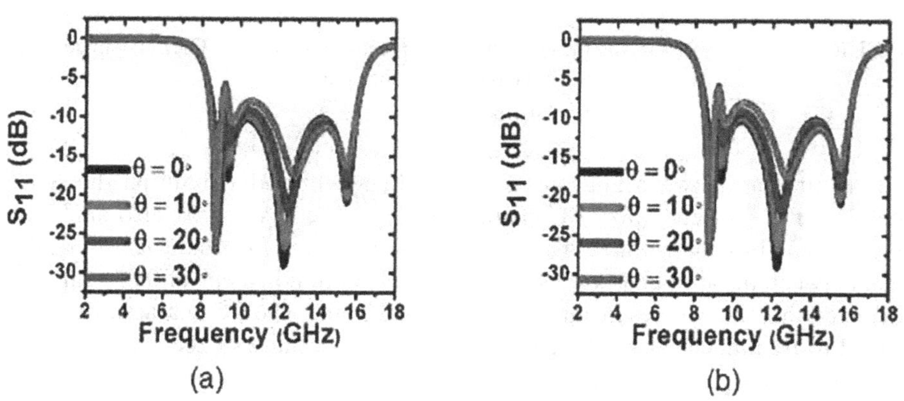

FIGURE 6.4 (a) TE polarization. (b) TM polarization simulations at various incidence angles.

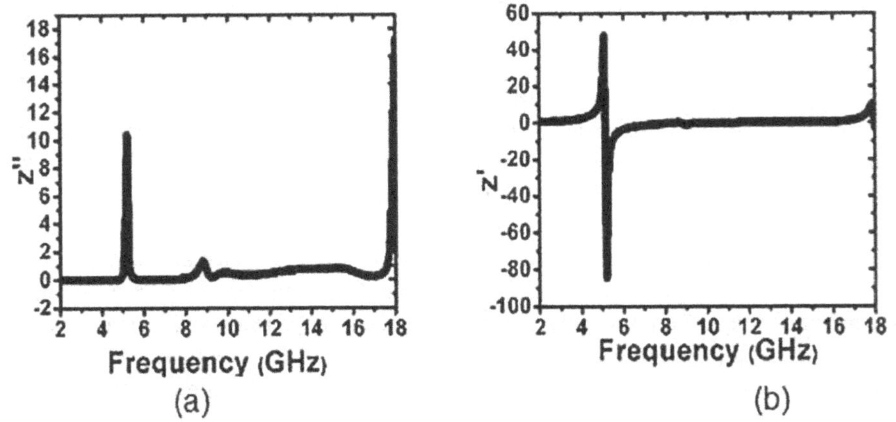

FIGURE 6.5 Frequency-dependent normalized impedance: (a) imaginary effective impedance and (b) real effective impedance.

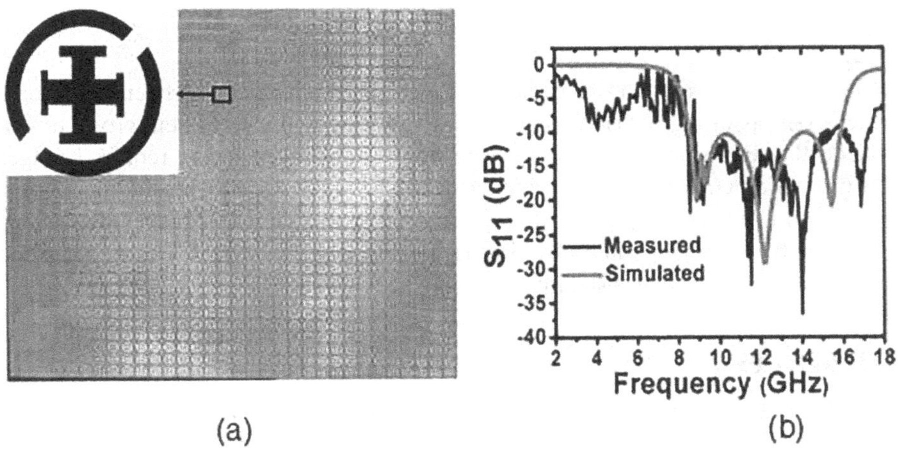

FIGURE 6.6 Evaluation of performance using experimental testing: (a) fabricated prototype of MMA and (b) frequency-dependent measured and simulated spectra in the range of 2 to 18 GHz.

TABLE 6.2
Properties of the Absorber Structure

Frequency (GHz)	μ′	μ″	ε′	ε″
8.9	1.31	–4.0	0.60	-3.5
9.4	1.8	–5.1	0.08	-4.5
12.3	1.96	–4.9	0.31	-4.6
15.6	0.31	–4.1	1.13	-2.3

6.4 FABRICATION AND PERFORMANCE EVALUATION OF ABSORBER

An array comprising of 32 × 32 unit cell structures using substrate layer (FR-4, 2.0 mm thick). The overall dimension is 256 mm × 256 mm of the fabricated sample. The proposed array structure is fabricated (Figure 6.6 (a)).

Further, the structure was created using printed circuit board technology. A one-port free-space microwave measurement setup was used to analyse the fabrication's performance in an anechoic chamber. The vector network analyser, model number E5063A from Keysight, is coupled to a broadband horn antenna as part of the measurement setup. A comprehensive description of the microwave measuring setup is described in [20]. Figure 6.6 (b) compares the results of RC-frequency spectrum of the absorber (simulation and measurement). Both these results agree well with each other are found to be in fairly good agreement. The fabricated sample exhibits –36.6 dB of peak RC value. Moreover, −10 dB absorption bandwidth of the fabricated sample is 8.26 GHz. However, due to finite structure dimensions and fabrication tolerances, fluctuations between results are noted.

6.5 CONCLUSION

In summary, a simple and broadband MM-inspired absorber has been presented, which consists of the hybrid geometry of a split ring and a Jerusalem cross. It was noticed that the broadband frequency response is received in the required range. Various parametric simulations are critically studied to analyse the structure's absorption mechanism. S-parameters were used to extract its frequency-dependent EM property. The fabricated absorber is flexible and thin, which can be used for radome structures and as a stealth.

REFERENCES

[1] R. Panwar and J.R. Lee, "Progress in frequency selective surface-based smart electromagnetic structures: A critical review," *Aerospace Science & Technology (Elsevier)*, vol. 66, pp. 216–234, 2017.

[2] A. Lai, K.M.K.H. Leong, and T. Itoh, "Infinite wavelength resonant antennas with monopolar radiation pattern based on periodic structures," *IEEE Transactions on Antennas and Propagation*, vol. 55, no. 3, pp. 868–876, March 2007.

[3] S.A. Cummer, B.I. Popa, D. Schurig, D.R. Smith, and J.B. Pendry, "Full wave simulations of electromagnetic cloaking structures," *Physical Review Letter (E)*, vol. 74, p. 036621, May 2006.

[4] H. Tao, N. Landy, C.M. Bingham, X. Zhang, R.D. Averit, and W.J. Padilla, "A metamaterial absorber for the terahertz regime: Design, fabrication and characterization," *Optics Express*, vol. 16, no. 10, pp. 7181–7188, May 2008.

[5] N. Zhang, P. Zhou, D. Cheng, X. Weng, J. Xie, and L. Deng, "Dual-band absorption of mid infrared metamaterial absorber based on distinct dielectric spacer layers," *Optics Letters*, vol. 38, no. 7, pp. 1125–1127, 1 April 2013.

[6] R. Mishra, R. Panwar, and D. Singh, "Equivalent circuit model for design of frequency selective, Terahertz-band, graphene based metamaterial absorber," *IEEE Magnetics Letters*, vol. 9, pp. 1–5, 2018.

[7] A. Fallahi, A. Yahaghi, H.-R. Benedickter, H. Abiri, M. Shahabadi, and C. Hafner, "Thin wideband radar absorbers," *IEEE Transactions on Antennas and Propagation*, vol. 58, pp. 4051–4058, 2010.
[8] M.H. Li, H.L. Yang, and X.W. Hou, "Perfect metamaterial absorber with dual bands," *Progress in Electromagnetics Research*, vol. 108, pp. 37–49, 2010.
[9] H. Xiong, J.-S. Hong, C.-M. Luo, and L.-L. Zhong, "An ultrathin and broadband metamaterial absorber using multilayer structures," *Journal of Applied Physics*, vol. 114, p. 064109, 2013.
[10] F. Costa, A. Monorchio, and G. Manara, "Theory, design and perspectives of electromagnetic wave absorbers", *IEEE Electromagnetic Compatibility Magazine*, vol. 5, pp. 67–74, 2016.
[11] C. Sabah, "Realization of polarization-angle-independent fishnet-based waveguide metamaterial comprised of octagon shaped resonators with sensor and absorber applications," *Journal of Materials Science Materials in Electronics*, vol. 27, no. 5, pp. 4777–4787, 2016.
[12] L.O. Nur and A. Munir, "Thin EM wave absorber composed of octagonal patch array and its characteristic measurement", in *3rd International Conference on Information and Communication Technology (ICoICT)*, IEEE Xplore, pp. 604–607, 27 May 2015. DOI: 10.1109/ICoICT.2015.7231494
[13] T.M. Kollatou, S.D. Assimonis, and C.S. Antonopoulos, "A family of ultra-thin, octagonal shaped microwave absorbers for EMC applications", *Materials Science Forum*, vol. 792, pp. 165–170, 2014.
[14] S. Ghosh, S. Bhattacharyya, D. Bhattacharyya, and K.V. Srivastava, "An ultrawideband ultrathin metamaterial absorber based on circular split rings", *IEEE Antennas and Wireless Propagation Letters*, vol. 14, pp. 1172–1175, 2015.
[15] S. Kaur and H.J. Kaur, "Comparative analysis of plasmonic metamaterial absorber for noble, alkaline earth and transition metals in visible region", *2019 6th International Conference on Computing for Sustainable Global Development (INDIACom)*, IEEE Xplore, pp. 513–516, 2019. ISSN 0973-7529; ISBN 978-93-80544-32-8.
[16] K. Hossain, T. Sabapathy, M. Jusoh, J.P. Soh, M.H. Jamaluddin, S.S. Al-Bawri, M.N. Osman, R.B. Ahmad, H.A. Rahim, M.N.M. Yasin, and N. Saluja, "Electrically tunable left-handed textile metamaterial for microwave applications", *Materials*, vol. 14, p. 1274, 2021. https://doi.org/10.3390/ma14051274
[17] M. Bakir, K. Delihacioglu, M. Karaaslan, F. Dincer, and C. Sabah, "U-shaped frequency selective surfaces for single-and dual-band applications together with absorber and sensor configurations", *IET Microwaves, Antennas & Propagation*, vol. 10, pp. 293–300, 2016.
[18] R. Panwar, S. Puthucheri, V. Agarwala, and D. Singh, "Fractal frequency selective surface embedded thin broadband microwave absorber coatings using heterogeneous composites", *IEEE Transactions on Microwave Theory and Techniques*, vol. 63, pp. 2438–2448, 2015.
[19] B.A. Numan and S.M. Sharawi, "Extraction of material parameters for metamaterials using a full-wave simulator", *IEEE Antennas and Propagation Magazine*, vol. 55, no. 5, pp. 202–211, 2013.
[20] R. Panwar and J.R. Lee, "Performance and non-destructive evaluation methods of airborne radome and stealth structures", *Measurement Science and Technology*, vol. 29, p. 062001, 2018.

7 Design Strategies and Challenges of Materials Sciences through Machine Learning

Kapil Mehta, Meenu Garg, Vandana Mohindru Sood, Himani Chugh, Isha Gupta

7.1 INTRODUCTION

Materials science research has progressed through various paradigms like experimental science, theory of mathematics, and simulation [1]. The starting paradigm investigation relies solely on intuitive observational experience, with no scientific basis for quantification. Physical stage models represented by mathematically based equations began a few decades ago, offering some theoretical studies, such as thermodynamics laws for materials research [2]. As a result of the decades-old discovery of computers, the third scientific paradigm emerges, permitting the simulation of complicated practically theory-based issues learned in the second stage paradigm [3]. The continuing artificial intelligence revolution has the potential to revolutionize civilization in ways that transcend science and technology. Machine learning is a fundamental component, for which progressively classy methods have been established, raising expectations that machines would able to surpass human beings in most jobs, including intelligent tasks, within a few decades.

Nonetheless, standard materials science research techniques, which largely rely on trial-and-error approaches, take more than 15 years or even far from vast research to implementation, and many conclusions that appeared to be erroneous have not been put into practice [4]. In the meantime, large volumes and higher dimensions of data or various characterizations of material techniques are being developed in the research of materials [5, 6].

The fundamental connections between artificial intelligence, computing, and big data science are emphasized in Figure 7.1.

The classical model creation in materials science research trusts heavily on physical mechanisms, such as the law of conversation and thermodynamics, followed by the mathematical formula derivation of parameter regression [7].

In contrast, using machine learning to investigate the symmetries and material data rules does not necessitate special principles but rather the training of flexible and non-linear models from the given data. Because of the advantages, machine

 DOI: 10.1201/9781003367154-7

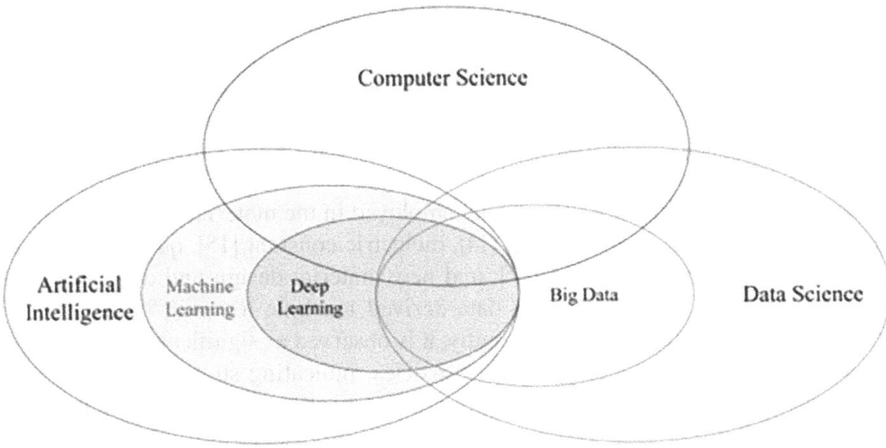

FIGURE 7.1 Basic concepts of artificial intelligence, computing, and data science.

learning has become a popular tool for predicting material attributes, selecting materials, and improving design [8].

7.2 MATERIALS AND METHODS

The reputation of big data and machine learning in materials sciences has been highlighted by programs of materials genome and multidimensional international attempts for developing a generic collaborating platform. Perhaps one of the most important goals is to go beyond the empirical methodologies that have dominated in the past:

1. Enable an integrated team approach in materials research by leading a cultural revolution.
2. Integrate experiment, computing, and theory, and provide advanced tools and methodologies to the materials community.
3. Consider digital data available.
4. Creation of first-class materials workforce prepared for employment in academics and industrial applications.

In today's world, AI-based machine learning is a major driving force for the industrial revolution and research [9, 10].

In the process of industry-based revolution and human societal progress, material innovation always plays a larger part [11]. Advanced alloys, semiconductor materials, polymer materials, and composite and superconductive materials, for example, have aided the novel energy, microelectronics, biotechnology, or spatial technology revolutions, which have ushered in an unparalleled information society. In the not-too-distant future, breakthrough materials science and technology will accelerate in an interconnected and intelligent direction, advancing the fourth industry-based revolution and green intelligent industry [12]. Because of the low computational cost, fast developing cycle, and great analysis of data and prediction capacity, machine learning is quickly becoming one of the most important AI approaches for material innovation.

Figure 7.2 illustrates the introduction of AI and machine learning in materials sciences.

Now, machine learning is extensively employed in the material prediction for the magnetocaloric effect [13], bandgap [14], dielectric constant [15], quantum chemistry [16], thermal characteristics [17], and new material design and discovery [18]. To summarize the current state of data-derived machine learning is in the early stages of material development. Currently, it is observed as significant growth in the machine learning and materials science articles, indicating study focus and scope are broadened. This study aims to highlight the machine learning major issues in

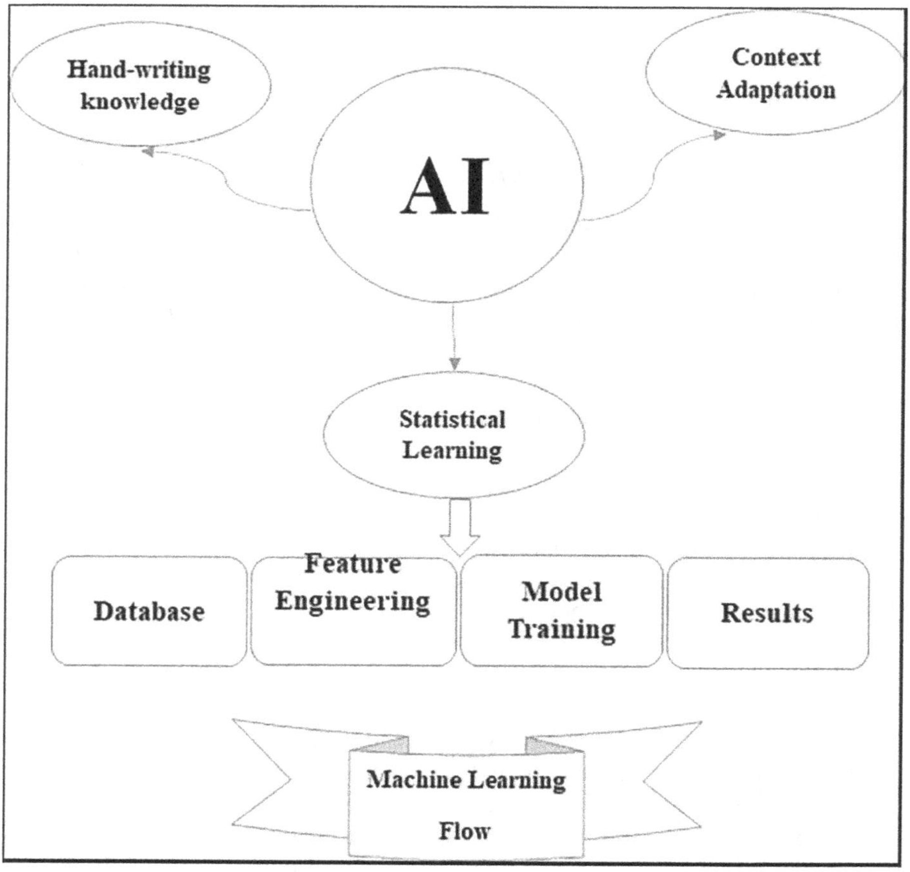

FIGURE 7.2 Introduction of artificial intelligence and machine learning in materials science.

materials science, with a particular objective on the research development of existing methodologies, as well as possible directives and perspectives for machine learning in materials innovation [19].

7.3 CHALLENGES OF MACHINE LEARNING IN MATERIALS SCIENCE

In materials science, machine learning is largely connected with supervised-type learning. The use of machine learning to solve design and development problems in materials science and related subjects is a rapidly emerging field that has seen remarkable abundance in the previous decade. Since 2014, the scientific research published in this sector has been steadily increasing, with the number of aids doubling every period [20]. While the early stages of this expansion were mostly focused on deepening one's grasp of machine learning model building, to evaluate the efficacy as well as efficiency of data-driven approaches to the development of materials. During this stage, research focused on answering fundamental questions such as "How do different statistical learning methods work?" "What are their potential strengths and weaknesses?" "How does one choose the best method for a given problem?" "What are some statistical learning best practices that one should follow for developing and validating a predictive model?" and so on [21]. The focus of research has switched to broad and major materials science concerns and has further evolved into a specialized research area [22]. Although the combination of materials science and machine learning research has yielded significant benefits, there are still some problems that must be addressed and thoroughly understood as elaborated in the following sections.

7.3.1 Inadequate Data

Not only is data the foundation of materials science; it is the initial hurdle in material applications. Materials science and technology, unlike most disciplines, have a diverse set of data categories. To put it another way, each category's data output is lower, and the feature dimension is less. The majority of studies gather research data through constrained practical circumstances, which takes a huge period in the interim [23].

7.3.2 Difficult to Identify and Feature Correlation

The most important aspect of material research is to investigate the link between the four aspects of materials that can be manipulated by experimental settings. Related to features and characteristics, the challenges are mentioned here.

7.3.2.1 Processing Difficulties

Processing is more challenging because of the greater number of features. Finally, as computational storage gets more complicated, classical learning methods become ineffective in achieving good outcomes.

7.3.2.2 Exploring Feature Interaction

Exploration of the link between multiple features or aspects is a challenge. The control value approach is utilized for discovering optimal metrics when studying the performances. However, establishing the association in numerous aspects is difficult.

7.3.2.3 Feature Classification

The classification of features is difficult because the relationship between the traits is still unknown, and their effects on materials science should be thoroughly examined.

7.3.2.4 Feature Recognition

For machine learning, the data that was initially obtained is frequently unidentifiable. To turn them into data features that can be recognized, a variety of altering procedures are required.

Materials science features must be able to acquire all critical data required for the differentiation between various nuclear or mineral environments. The amount of processing required is mostly determined by the algorithm. The extraction of features can be regarded model part for some techniques, such as deep learning.

7.3.3 Poor Universality of Algorithms

Machine learning demonstrates varied skills based on various algorithms; for example, the same algorithm may provide different results when dealing with different things, or different algorithms may produce different outcomes when dealing with the same object. However, this posed even more difficulties in terms of selecting and optimizing machine learning algorithms.

7.3.4 Poor Description

The deep network is a type of machine learning model, which has advanced quickly and demonstrated impressive capabilities in intelligent goods. As a result, in AI for materials science, the machine learning interpretability is also a study consideration [24].

7.3.5 Highly Specialized Research

Many researchers are combining materials science with machine intelligence, depending on a data-driven method for forecasting and analysing material properties, thanks to the development of AI technology. For machine learning, many programming languages can be utilized, such as Python, which has a simple yet powerful expansion package. However, in materials science, machine learning is currently progressing slowly due to a shortage of sophisticated theories of mathematics and programming expertise among materials scientists.

7.4 STRATEGIES FOR MACHINE LEARNING IN MATERIALS SCIENCE

Machine learning is a better approach for the design of explicit algorithms, which is difficult in the case of the detection of meaningful image elements. Several types of problem statements provide a large solutions variety. In materials science, many problems conform to this strategy which includes prediction of protein structure, design of the material, prediction of property, and model derivation. Here we describe five common strategies for machine learning in materials science.

7.4.1 Uninformative Priors

There are priorities that are not helpful. If no information about the process being modelled is available, the prior distribution assumptions about the comparative merits of distinct applicant functions would be a sensible choice. Uninformative priors are prior probability distributions that are not informative as shown in equation (7.1).

$$S(p) = -\int_{-\infty}^{\infty} p(x)\ln(p(x))dx \tag{7.1}$$

In the sense that it cannot be normalized to 1, a uniform prior that gives identical prior probability to infinite count of inappropriate prior functions.

7.4.2 Model Selection

Model selection is a typical version of this strategy, in which the process of learning is split twice. In the initial phase, a parameterized function (also known as a model) is chosen, thereby giving all other functions zero prior probability. Bayesian analysis can be used to derive a generic method for model selection. Let r be the smallest empirical risk that can be achieved inside a given batch of training data model, which is shown in equation (7.2).

$$2r + k\ln(n) \tag{7.2}$$

Under a wide range of circumstances, Schwartz established that the model with largest predicted subsequent probability is with the smallest value.

7.4.3 Prior Knowledge

Incorporating prior knowledge of process model into the prior distribution can help determine as substituting f, as shown in equation (7.3).

$$\partial f = f - f' -- \tag{7.3}$$

As an example, if f estimates the compound energy, so f denotes the composition-weighted average of the adjacent energy elements and ∂f is the energy estimate formation.

7.4.4 Hyperpriors

During space of possible predicted prior distributions, a probability distribution is an alternate method of selecting a specific prior distribution. A hyperprior is a probability distribution of this type. A hyperprior is formed as a probability distribution on various Gaussian variance values if a Gaussian distribution with zero means is chosen as prior.

7.4.5 Training Data

The training data could be derived from uncontrollable external phenomena observations, such as climate data that is defined in weather forecasting. In many circumstances, controlled experiments can be used to generate training data. Because generating training data can be costlier and/or time-consuming stage in process of learning, it is preferable for reducing the number of tests required to obtain a reasonable level of prediction error.

7.5 APPLICATIONS OF MACHINE LEARNING IN MATERIALS SCIENCE

In materials research, machine learning is gaining success, having a variety of accomplishments in alloys of high entropy, superconducting materials, steel materials, magnetic materials, and thermoelectric and catalytic materials. Materials science has become innovative after the machine learning application concepts in it. The applications and features of machine learning algorithms in materials sciences are shown in the Table 7.1.

The significant machine learning applications in metals, batteries, photovoltaic materials, and nanomaterials are elaborated in the following sections.

7.5.1 Metal Materials

Machine learning research has achieved some success in the realm of classic metal materials, like steel materials, in context to its application in new material research. There is a development of a new machine learning tool that can perform direct property prediction analysis, and properties-to-microstructure inverse estimation, dubbed Materials Genome Integration System Phase 1 (MIPHA). The goal was to assign material researchers with a unique way of data-driven materials design for speeding up the material discovery process. Few cast off a mixture techniques of machine learning, to forecast "time temperature transition" diagrams with relevant characteristics, such as alloying components, austenitizing temperature, and holding duration [34].

7.5.2 Battery Materials

The presentation of lithium-ion batteries is largely dependent on materials chosen and a design between materials, and the performance of the battery is difficult for the establishment of the design variable's complexity. Shandiz et al. have shown that the

TABLE 7.1
Applications and Features of Machine Learning Algorithms in Materials Sciences

Method	Category	Applicable Characteristic Features
ANN Technique [25]	Regression Technique	A big amount of data is required, and ANN's self-study and fault-tolerance skills are relatively powerful.
Linear and Logistic Regression [26]	Regression Technique	Fast modelling and strong interpretability are important traits that work with linearly conformable data which is unable to handle data with several characteristics.
Kernel Ridge and Support Vector Regression [27]	Regression Technique	Both can handle non-linear data, but speed is comparable in the case of big volumes.
Support Vector Classification [28]	Classification	SVC is a useful classification model for data with two classes.
Decision Tree (DT) [29]	Classification	DT can handle information having missed attributes that do not support online learning.
Random Forest [30]	Classification	RF can also prevent overfitting in case of little noise.
K-Means Clustering [31]	Clustering Technique	K-means is a traditional clustering approach sensitive to the original sources of data.
Hierarchical Cluster Analysis (HCA) [32]	Clustering Technique	HCA creates a hierarchy of clusters but requires a lot of computing power.
Hidden Markov Model (HMM) [33]	Clustering Technique	HMM is a popular stochastic signal model that deals with pattern recognition.

structure of the crystal model has a considerable effect, implying that crystal system analysis should be a useful tool for estimating cathode properties in batteries. The adjacent links between the crystal system and related properties of physical silica-based cathodes were established based on an evaluation of features in a statistical model [35].

7.5.3 Photovoltaic Materials

Photovoltaic materials, which were first utilized in photovoltaic power generation in 2009, have sparked a lot of research interest due to their remarkable optical and electrical properties or advantages of the easy mixture and low cost, and a wide range of raw materials [36]. Furthermore, the manufacture of photovoltaic solar cells produces a significant amount of toxicity, which limits the expansion of the noteworthy materials [37].

7.5.4 Nanomaterials

Theoretical research and application of nanomaterials require an understanding of the links between structures and properties. Traditional trial-and-error experimentation is time-consuming and expensive, while the DFT method necessitates the use of computational resources [38]. Machine learning approaches to promote the identification of new NMs with outstanding electrical, thermal, and optical properties by providing an accurate, rapid, validated, and high-throughput alternative [39]. The deep-learning neural-network approach can anticipate the sorption of many developing polar or ionizable organic pollutants, whereas existing adsorption models are typically created for neutral chemicals.

7.6 CONCLUDING REMARKS AND FUTURE PROSPECTS

Following the discussion in this chapter, machine learning has a significant capacity for mining novel materials in the data-driven world. Existing solutions and research efforts have been summarized and discussed in light of the machine learning challenges in materials. Although these relevant studies were carried out for the promotion of machine learning in materials science, unique procedural methodologies, efficient data pre-processing techniques, information mining, material structure, performance prediction, and novel prediction of functions and interaction between features were suggested in the future to improve the efficiency of work and research progress in materials science scientifically. As a result, it is thought that significant advancements are needed for meeting the practical applications. With so much at risk in the economics of firms and governments, we believe that the ongoing AI-based revolution come to an end, with many promising opportunities for materials science in near future.

REFERENCES

[1] B.O. Gregory. "Designing a new material world." *Science* 288 (2000): 993–998.

[2] Q. Luo, Y.L. Guo, B. Liu, Y.J. Feng, J.Y. Zhang, Q. Li, K.C. Chou. "Thermodynamics and kinetics of phase transformation in rare earth-magnesium alloys: A critical review." *J. Mater. Science and Technology* 44 (2020): 171–190.

[3] P.G. Boyd, Y. Lee, B. Smit. "Computational development of the nanoporous materials genome." *Nature Reviews Materials* 2 (2017): 17037.

[4] X. Wu, F.Y. Kang, W.H. Duan, J. Li. "Density functional theory calculations: A powerful tool to simulate and design high-performance energy storage and conversion materials." *Progress in Natural Science: Materials International* 29 (2019): 247–255.

[5] M.X. Zhang, C.L. Wang, A.L. Luo, Z.H. Liu, X.S. Zhang. "Molecular dynamics simulation on thermophysics of paraffin/EVA/graphene nanocomposites as phase change materials." *Applied Thermal Engineering* 166 (2020): 114639.

[6] J.L. Hou, M. Chen, Y.F. Zhou, L. Bian, F.Q. Dong, Y.H. Tang, Y.X. Nie, H.P. Zhang. "Regulating the effect of element doping on the CO_2 capture performance of kaolinite: A density functional theory study." *Applied Surface Science* 512 (2020): 145642.

[7] A. Kunwar, L.L. An, J.H. Liu, S.Y. Shang, P. Raback, H.T. Ma, X.G. Song, J. Mater. "A data-driven framework to predict the morphology of interfacial Cu6Sn5 IMC in SAC/Cu system during laser soldering." *Science and Technology* 50 (2020): 115–127.

[8] K. Momeni, Y.Z. Ji, Y.X. Wang, S. Paul, S. Neshani, D. Yilmaz, Y.K. Shin, D. Zhang, J.W. Jiang, H.S. Park, S. Sinnott, A. Duin, V. Crespi, L. Qing. "Multiscale computational understanding and growth of 2D materials: a review." *Chen, NPJ Computational Materials* 6 (2020): 22.

[9] S. Jonathan, R.G.M. Mário, B. Silvana, A.L.M. Miguel. "Recent advances and applications of machine learning in solid-state materials science." *NPJ Computational Materials* 5 (2019): 83.
[10] C. Stefan, T. Alexandre, E.S. Huziel, P. Igor, T.S. Kristof, K.R. Müller. "Machine learning of accurate energy-conserving molecular force fields." *Science Advances* 3 (2016): 1603015–1603022.
[11] M. Manik, C.D. Steven, L.K. Benjamin, G.B. Wolfgang. "Modeling and simulation of the thermodynamics of lithium-ion battery intercalation materials in the open-source software Cantera." *Electrochimica* Acta. 323 (2019): 134797.
[12] A.A. Adetokunbo, E. Koffi, J.B. Douglas. "On the formulation of the kinematics and thermodynamics for polycrystalline materials undergoing phase transformation." *International Journal of Plasticity* 123 (2019): 101–120.
[13] L. Enzo, A. Carlos, Z. Andrew, L. Andrew, L. Victor, S.G. Patrick. "Crystal nucleation in metallic alloys using x-ray radiography and machine learning." *Science Advances* 4 (2018): 4004–4014.
[14] D. Jia, H.T. Duan, S.P. Zhan, Y.L. Jin, B.X. Cheng, J. Li. "Design and development of lubricating material database and research on performance prediction method of machine learning." *Scientific Reports* 9 (2019): 20277–20288.
[15] Chaochao Gao, et al. "Innovative materials science via machine learning." *Advanced Functional Materials* 32.1 (2022): 2108044.
[16] Yann LeCun, Yoshua Bengio, Geoffrey Hinton. "Deep learning." *Nature* 521.7553 (2015): 436–444.
[17] Richard S. Sutton, Andrew G. Barto. *Reinforcement learning: An introduction*. MIT Press, 2018.
[18] S. Lloyd, C. Weedbrook. "Quantum generative adversarial learning." *Physical Review Letters* 121.4 (2018): 040502.
[19] Stephen L. Sass. *The substance of civilization: Materials and human history from the stone age to the age of silicon*. Arcade Publishing, 1998.
[20] Klaus Schwab. *The fourth industrial revolution*. Currency, 2017.
[21] Keith T. Butler, et al. "Machine learning for molecular and materials science." *Nature* 559.7715 (2018): 547–555.
[22] Rohit Batra, Le Song, and Rampi Ramprasad. "Emerging materials intelligence ecosystems propelled by machine learning." *Nature Reviews Materials* 6.8 (2021): 655–678.
[23] Bo Zhang, et al. "Machine learning technique for prediction of magnetocaloric effect in La (Fe, Si/Al) 13-based materials." *Chinese Physics B* 27.6 (2018): 067503.
[24] L. Weston and C. Stampfl. "Machine learning the band gap properties of kesterite I 2– II– IV– V 4 quaternary compounds for photovoltaics applications." *Physical Review Materials* 2.8 (2018): 085407.
[25] Chiho Kim, Ghanshyam Pilania, and Rampi Ramprasad. "Machine learning assisted predictions of intrinsic dielectric breakdown strength of ABX3 perovskites." *The Journal of Physical Chemistry C* 120.27 (2016): 14575–14580.
[26] Kristof T. Schütt, et al. "Quantum-chemical insights from deep tensor neural networks." *Nature Communications* 8.1 (2017): 1–8.
[27] Sherif Abdulkader Tawfik, et al. "Predicting thermal properties of crystals using machine learning." *Advanced Theory and Simulations* 3.2 (2020): 1900208.
[28] Jonathan Schmidt, et al. "Predicting the thermodynamic stability of solids combining density functional theory and machine learning." *Chemistry of Materials* 29.12 (2017): 5090–5103.
[29] Dane Morgan, Ryan Jacobs. "Opportunities and challenges for machine learning in materials science." *Annual Review of Materials Research* 50 (2020): 71–103.

[30] Vinayak Rai, et al. "Automated biometric personal identification-techniques and applications." In *2020 4th International Conference on Intelligent Computing and Control Systems (ICICCS)*. IEEE, 2020.

[31] Kapil Mehta, Yogesh Kumar, Aayushi Aayushi. "Enhancing time synchronization for home automation systems." *ECS Transactions* 107.1 (2022): 6197.

[32] K.A.P.I.L. Mehta, Yogesh Kumar. "Implementation of efficient clock synchronization using elastic timer technique in IoT." *Advances in Mathematics: Scientific Journal* 9.6 (2020): 4025–4030.

[33] Abhishek Bansal, Kapil Mehta, Sahil Arora. "Face recognition using PCA and LDA algorithm." In *2012 Second International Conference on Advanced Computing & Communication Technologies*. IEEE, 2012.

[34] Kapil Mehta, et al. "Machine learning based intelligent system for safeguarding specially abled people." In *2022 7th International Conference on Communication and Electronics Systems (ICCES)*. IEEE, 2022.

[35] Kapil Mehta, et al. "Enhancement of smart agriculture using internet of things." *ECS Transactions* 107.1 (2022): 7047.

[36] Vinayak Rai, et al. "Cloud computing in healthcare industries: Opportunities and challenges." *Recent Innovations in Computing* (2022): 695–707.

[37] Vandana Mohindru Sood, Kapil Mehta. "Autonomous UAV with obstacle management using AIoT: A case study on healthcare application" (Computing, 2022), "AIoT Technologies and Applications for Smart Environments", Chap. 14, pp. 251–273, http://doi.org/10.1049/PBPC057E_ch14, IET Digital Library, https://digitallibrary.theiet.org/content/books/10.1049/pbpc057e_ch14.

[38] K. Mehta, V.M. Sood, J. Singh, D. Sharma, P. Chhabra, "Securing cyber infrastructure of IoT-based networks using AI and ML," In *2022 Seventh International Conference on Parallel, Distributed and Grid Computing (PDGC)*, Solan, Himachal Pradesh, India, 2022, pp. 378–383, http://doi.org/10.1109/PDGC56933.2022.10053326.

[39] Kapil Mehta, Vandana Mohindru Sood. "Agile software development in the digital world–trends and challenges." *Agile Software Development: Trends, Challenges, and Applications* (2023): 1–22.

8 Wideband Antenna Design for Biomedical Applications with Scope of Metamaterial for Improved SAR

Poonam Jindal, Aarti Bansal, Shivani Malhotra

8.1 INTRODUCTION

There has been a significant growth in the design of wearable antennas targeting various biomedical and healthcare applications like brain tumour detection, skin cancer detection, breast cancer detection, and telemetry applications [1–3]. Wearable antennas are further categorized to be utilized in two different types: on-body and off-body wearable antenna designs. These antennas are targeted for two frequency bands: ISM band and Medical Implant Communications Service (MICS) band operating at frequency range of 2.4–2.48 GHz and 402–405 MHz, respectively [4–5]. These antennas, however, face different design challenges when mounted on human body surface. Firstly, these antennas must be compact in size [3] so as to be easily placed on different part of human body. For this, miniaturized antennas proposed in recent literature includes fractal antennas [6], PIFA [7], meandered antennas [3, 8], shorted pin antennas [9–10], and spiral-shaped antennas [11–12]. These antennas tend to decrease the resonant frequency by increasing the electrical length of the antenna.

The next challenge faced by this antenna is to be easily compatible with the dielectric properties of the human body. As the human body is lossy in nature, the performance of the antenna suffers when placed on the human body surface. Hence, the antenna must be optimized to exhibit reasonable performance while considering dielectric losses of the human body. Additionally, when the human body is exposed in front of radio frequency radiations that exceeds the specific absorption rate (SAR) limits [13], it affects the human tissue seriously. This is because the power absorbed by the biological tissue is in direct proportion to the electric field intensity (E):

$$P_{abs} = \int \sigma \left| E \right|^2 dV \tag{8.1}$$

DOI: 10.1201/9781003367154-8

where σ is a conductivity of the human tissue [14–15]. In this research work, a wearable antenna is proposed for on-body communication, which will help in real-time health monitoring applications. To design the antenna, different slots are etched to obtain the miniaturization, leading to the final dimensions of 31 × 17.5 × 0.8 mm^3. Further, to check the performance of the proposed wearable antenna for on-body applications, it was simulated and optimized first in free space and then by mounting on the human surface using CST Studio Suite 2018. Additionally, to validate its utility on human body applications while considering safety of human tissues, SAR analysis was carried out to obtain the maximum power limit of the antenna proposed.

The description of the subsequent sections in this chapter is given further. Section 2 describes the design configuration of the proposed wearable antenna. In Section 3, the designed antenna is simulated and optimized by mounting it on human body model. Section 4 presents the performance of designed antenna in terms of gain, efficiency, surface current distribution, and most importantly SAR computation. Finally, the conclusion and future scope is discussed in the Section 5.

8.2 ANTENNA DESIGN CONFIGURATION

Figure 8.1 shows the final geometry of the designed miniaturized antenna for wearable applications. The chosen substrate for the designed antenna is FR4 (ϵ_r = 4.4 and $tan\delta$ = 0.02) having 0.8 mm of thickness. The conventional patch antenna with two narrow slots is etched to achieve miniaturization. To reduce the size of antenna, slots

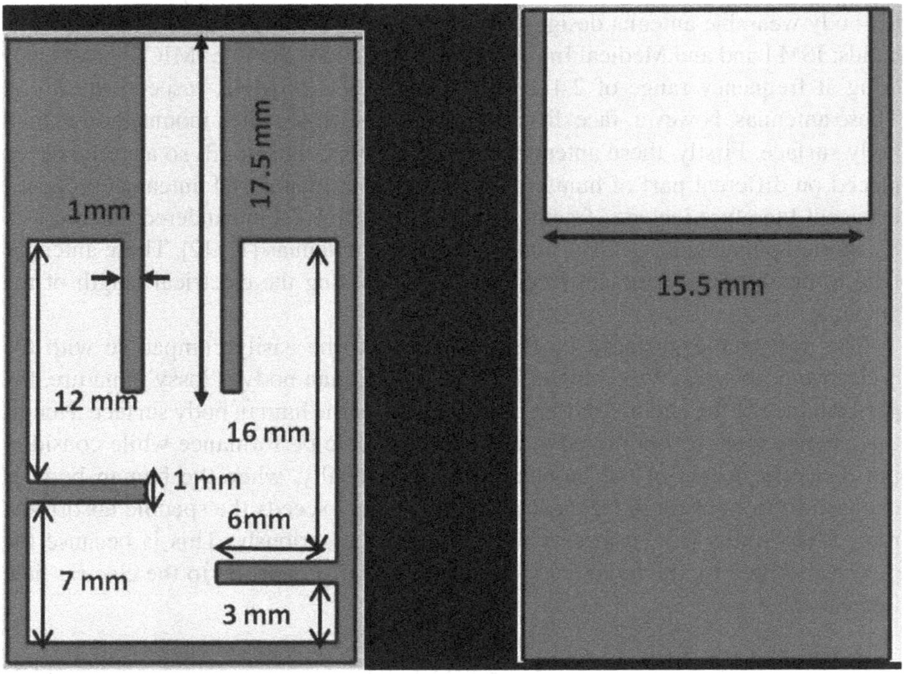

FIGURE 8.1 Geometry of proposed antenna: (a) top patch, (b) bottom ground.

are etched in the conventional microstrip patch, resulting in increase in its electrical length. The antenna is designed, simulated, and optimized using CST studio suite. The length and width of the narrow slots along with patch dimensions are parametrically varied to optimize the resonant frequency of the designed antenna. Further, the ground dimensions are also optimized to attain wide bandwidth of the designed antenna.

The antenna is excited using inset feeding technique, and the optimized dimensions of the microstrip feed line are 17.5 mm and 4 mm, respectively. Also, defected ground structure is etched at the back of the antenna having a partial ground with dimensions of 15.5 mm × 10 mm. The antenna is designed in two steps. Initially, the antenna is designed for off-body communication in which antenna is designed and optimized for ISM band resonating at 2.4 GHz. Further, to validate its utility in biomedical applications, in the second step, the antenna is mounted on the human body model as explained in the next section.

8.3 ANTENNA DESIGN ON THE HUMAN BODY MODEL

To make the antenna compatible to the human body surface, the antenna is mounted over a human model, considering its different layers: muscle, bone, fat, and skin, as shown in Figure 8.2 (a–b). Also, the dielectric property and thickness of each of the tissue layer is shown in Table 8.1 [16–18]. Also, the SAR value of the antenna loaded on human body is then calculated so as to consider human safety considerations.

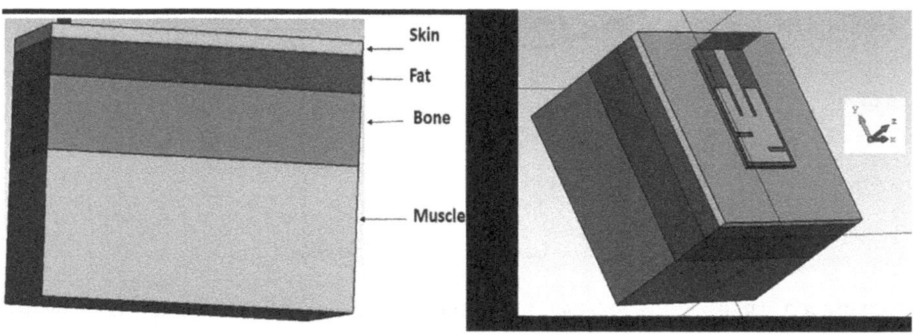

FIGURE 8.2 (a) Four-layered human model comprising skin, fat, bone, and muscle. (b) Antenna mounted on a human model.

TABLE 8.1
Dielectric Property and Thickness of Human Tissues at the ISM Band

Layer of Human Tissue	Dielectric Constant	Thickness (mm)	Density (kg/m³)	Conductivity (S/m)
Skin	5.27	2	1001	0.11
Fat	52.67	5	900	1.77
Bone	37.95	10	1008	1.49
Muscle	53.574	20	1040	0.82

8.4 SIMULATION RESULTS

The proposed antenna is evaluated in terms of various parameters, such as reflection coefficient, gain, and radiation efficiency at the resonating frequency of 2.4 GHz to validate its performance for wearable applications. The S11 parameters of the designed antenna are evaluated in both free space and on the human model and are plotted in Figure 8.3. It is observed that the antenna, when placed on human surface, resonates well in the frequency range from 1.51 GHz to 2.67 GHz, which is well within the desired ISM band.

The gain and efficiency of the proposed antenna for free space compared with that on the human body are shown in Figure 8.4 and Figure 8.5, respectively. The designed antenna exhibits a realized gain of 1.21 dB in free space compared to –9.6 dB when mounted on the human model. Further, the radiation efficiency is also observed to degrade from 78% in free space to about 6% when mounted on the human model due to losses. The degradation in the performance of antenna when mounted on the human body is obvious, owing to its high dielectric properties and lossy nature.

The surface current density is shown in Figure 8.6. The proposed antenna is observed to resonate well at the resonant frequency of 2.45 GHz lying in the ISM

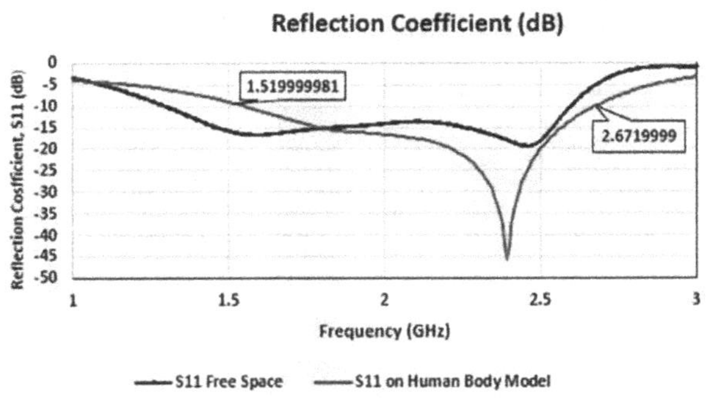

FIGURE 8.3 Reflection coefficient, S11 (dB) for free space and on the human body model.

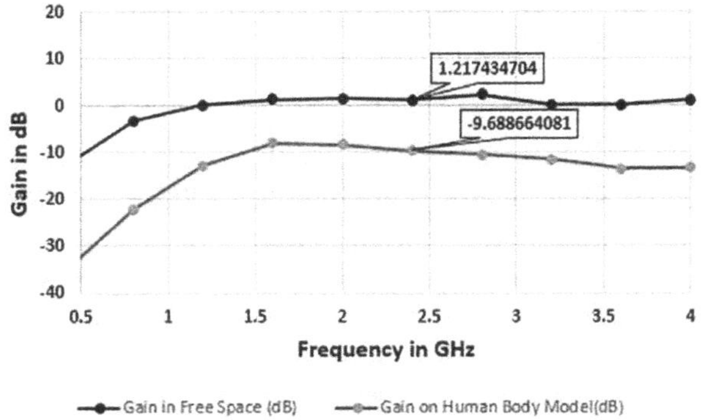

FIGURE 8.4 Simulated radiation gain for the proposed antenna.

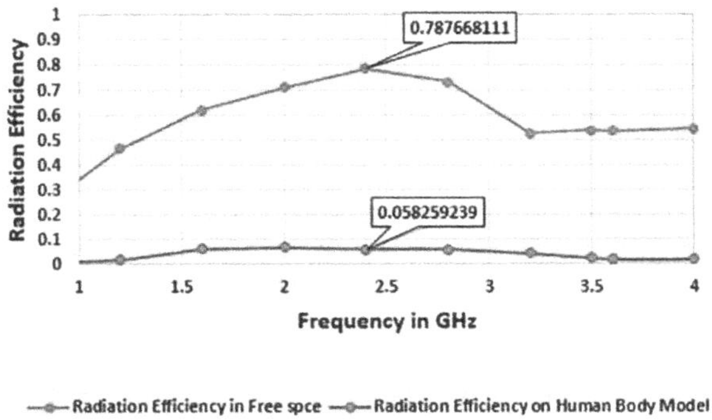

FIGURE 8.5 Simulated radiation efficiency results for the proposed antenna.

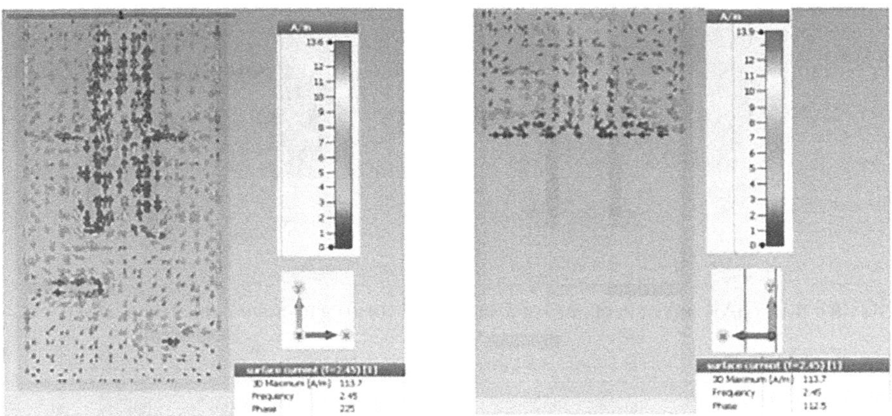

FIGURE 8.6 Surface current distribution of the proposed antenna on (a) top patch and (b) bottom ground plane.

band. The maximum current is observed to be concentrated around the inset feed validating that the antenna is excited significantly at the required resonant frequency.

8.4.1 Far Field Radiation Patterns

The far field radiation pattern (3D and 2D) exhibited by the proposed antenna is shown in Figure 8.7.

8.4.2 Specific Absorption Rate (SAR) Analysis

Further, to check the performance of the designed antenna on human body surface, its specific absorption rate (SAR) results for 1 g/10 g averaged tissue must be within the safety limits specified by FCC and ICNIRP standards. Thus, the impact of SAR for the proposed antenna on the four-layered human model is calculated at reference

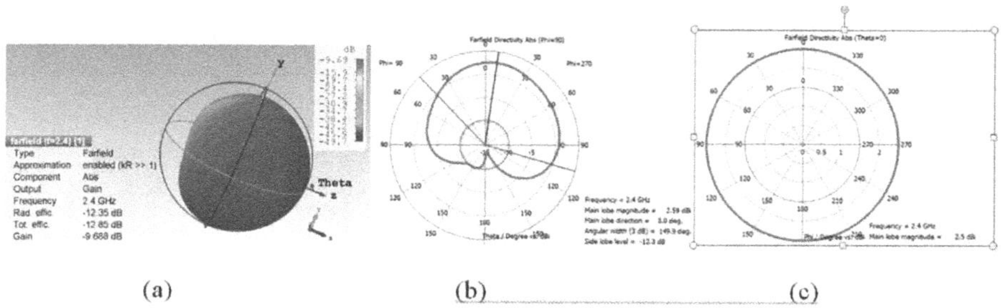

FIGURE 8.7 (a) 3D radiation pattern and 2D polar plots in (b) elevation and (c) azimuth planes.

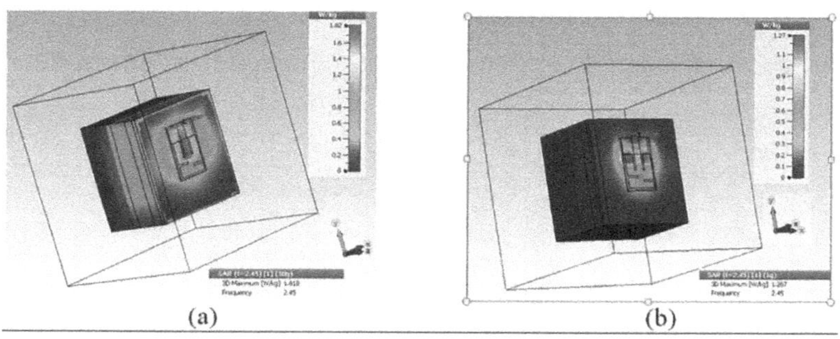

FIGURE 8.8 SAR analysis on (a) 1 g average and (b) 10 g average cubic tissue at 0.05 W.

TABLE 8.2
Comparison of Performance of the Proposed Antenna with Recent Antennas

References	Resonant Frequency (GHz)	Volume (mm³)	SAR (1 g-averaged)	Peak gain (dB)	Bandwidth ($\|S_{11}\| \leq -10$ dB)
2	2.44	27.5×21×1.6	–	–27.46	12.57%
15	2.45	22×16×1.27	2.15×10^{-3}	–19.5	1.6%
16	2.34	14×14×1.27	482	–19	12%
17	2.45	27×9×1.27	–	–20	>6%
18	2.4	19×30×1.6	–	–13	>8%
This work	2.45	31 x 17.5 x 0.8	1.26 (1 g) 1.81(10 g)	–9.7	47%

power of 0.05 W. From Figure 8.8(a–b), the SAR is observed to be 1.27 for 1 g averaged cubic tissue and 1.81 for 10 g averaged cubic tissue, which are well within the specified limits specified in IEEE C95.1 standard.

Table 8.2 summarizes the proposed antenna performance compared to that of recently developed antennas for biomedical applications. The proposed antenna is seen to exhibit significant gain with small size and wide bandwidth of 47% ranging from 1.53 GHz to 2.66 GHz.

8.5 CONCLUSION AND FUTURE SCOPE

In this chapter, the wideband antenna is designed for biomedical applications resonating in the ISM band. The designed antenna is compact having dimensions of $31 \times 17.5 \times 0.8$ mm^3 and a peak gain of −9.6 dB. Two narrow slots are etched in the patch to miniaturize its size. The bandwidth of the antenna must be large so as to prevent frequency detuning. Hence, defected ground structure is utilized at the back of antenna to increase its bandwidth up to 47%. The designed antenna exhibits reasonable gain and an efficiency of 78% and 6% in free space and when mounted on the human model, respectively.

Also, the SAR parameter for 1 g/10 g averaged cubic tissue is found to be well within the safety limits and are found to be 1.27 and 1.8, respectively, at 0.05 W of reference power. The radiation characteristics of the antenna can be further enhanced by integrating metamaterial structures with the antenna. Additionally, the metamaterial also helps in further reducing the SAR for higher reference power. Also, the introduced metasurface as a superstrate or high impedance surface helps in cancellation of the electric field due to reflection from the underlying human surface [19–21]. Thus, the antenna has better performance in terms of various parameters, such as reduced size, low SAR, and wider bandwidth, making it a preferable choice for consideration in biomedical applications.

8.6 ABBREVIATIONS

SAR	specific absorption rate
ISM	Industrial Scientific and Medical
PIFA	planar inverted F antenna
FCC	Federal Communications Commission
ICNIRP	International Commission on Non-Ionizing Radiation Protection

REFERENCES

1. Kiourti A, Nikita KS. A review of implantable patch antennas for biomedical telemetry. *IEEE Antennas Prop Mag*. 2012;54(3): 210–228.
2. Sukhija S, Sarin RK. Low-profile patch antennas for biomedical and wireless applications. *J Comput Electron*. 2017;16(2): 354–368.
3. Cho Y, Yoo H. Miniaturised dual-band implantable antenna for wireless biotelemetry. *Electron Lett*. 2016;52(12): 1005–1007.
4. Ahlawat S, Srivastava G, Kumar G. Design of skin implantable radiator with capacitive and CSRR loadings for ISM band applications. *Int J Inf Technol*. 2019: 1–10.
5. Liu C, Guo YX, Xiao S. Compact dual-band antenna for implantable devices. *IEEE Antennas Wirel Propag Lett*. 2012;11: 1508–1511.
6. Bhaskar S, Siddiqui M, Singhal S, Bansal A. Miniaturized circularly polarized vicsek-cross-shaped slot antenna for UHF-RFID reader handset applications. *Int J Radio Freq Identif*. 2022;6(22): 515–523.

7. Hossain MI, Faruque MI, Islam MT. Investigation of hand impact on PIFA performances and SAR in human head. *J Appl Res Technol.* 2015;13(4): 447–453.
8. Karacolak T, Cooper R, Topsakal E. Electrical properties of rat skin and design of implantable antennas for medical wireless telemetry. *IEEE Trans Antennas Propag.* 2009;57(9): 2806–2812.
9. Soontornpipit P, Furse CM, Chung YC. Design of implantable microstrip antenna for communication with medical implants. *IEEE Trans Microw Theory Techn.* 2004;52(8): 1944–1951.
10. Lee S, Seo W, Ito K, Choi J. Design of an implanted compact antenna for an artificial cardiac pacemaker system. *IEICE Electron Express.* 2011;8(24): 2112–2117.
11. Kim J, Rahmat-Samii Y. Implantable antennas inside a human body: simulations, designs, and characterizations. *IEEE Trans Microw Theory Techn.* 2004;52(8): 1934–1943.
12. Lee JH, Seo DW. Compact and tissue-insensitive implantable antenna on magneto-dielectric substrate for wireless biotelemetry. *J Electromagn Waves Appl.* 2019;33(18): 2339–2461.
13. Zhang Y, Liu C, Zhang K, Cao H, Liu X. Design and in-vivo testing of a low-cost miniaturized capsule system for body temperature monitoring. *Int J RF Microw Comput-Aided Eng.* 2019;29(8): e21793.
14. Kovar S, Spano I, Gatto G, Valouch J, Adamek M. SAR evaluation of wireless antenna on implanted cardiac pacemaker. *J Electromagn Waves Appl.* 2017;31(6): 627–635.
15. Yeap K, Voon C, Hiraguri T, Nisar H. A compact dual-band implantable antenna for medical telemetry. *Microw Opt Technol Lett.* 2019;61(9): 2105–2109.
16. Usluer M, Cetindere B, Basaran SC. Compact implantable antenna design for MICS and ISM band biotelemetry applications. *Microw Opt Technol Lett.* 2020;62(4): 1581–1587.
17. Lei W, Chu H, Guo Y. Miniaturized differentially fed dual-band implantable antenna: Design, realization, and in vitro test. *Radio Sci.* 2015;50(10): 959–967.
18. Gozasht F, Mohan AS. Miniaturized slot PIFA antenna for tripleband implantable biomedical applications. *2013 IEEE MTT-S International Microwave Workshop Series on RF and Wireless Technologies for Biomedical and Healthcare Applications (IMWS-BIO)*, pp. 1–3, 2013, December. IEEE.
19. Kaur S, Kaur HJ. Comparative analysis of plasmonic metamaterial absorber for noble, alkaline earth and transition metals in visible region. *2019 6th International Conference on Computing for Sustainable Global Development (INDIACom)*, pp. 513–516, 2019, March. IEEE.
20. Hossain K, Sabapathy T, Jusoh M, Soh JP, Jamaluddin MH, Al-Bawri SS, Osman MN, Ahmad RB, Rahim HA, Yasin MNM, Saluja N. Electrically tunable left-handed textile metamaterial for microwave applications. *Materials.* 2021;14: 1274. https://doi.org/10.3390/ma14051274.
21. Shaik S, Yalavarthi SK, Bandi A, Dasari V, Konuku SR. Metamaterial wearable antenna design and analysis for wireless applications. *2022 International Conference on Communication, Computing and Internet of Things (IC3IoT)*, pp. 1–5. IEEE Xplore.

9 Comparative Study of Different Cushion Materials Used to Manufacture Vehicle Seats Using FEM

Shubham Sharma, Jagjit Singh, Sachin Kalsi

9.1 INTRODUCTION

A human body, while driving, sitting in a bus, performing agricultural operations, and so on, experiences a vibrational effect that induces fatigue, discomfort, injury, and other adverse effects. The reason behind the vibrations induced in human body is due to certain factors—the seat cushion substance, posture, amplitude and frequency of the vibration, and so on. The most important component impacting a person's comfort is considered to be their direct touch with seat and cushion materials. To reduce vibration, excellent cushion materials are used. The fabric of the seat and back cushions directly affects seat and back comfort, so choosing high-quality cushions is essential for enhancing comfort in vehicles. Because of its stiffness and dampening qualities, seat comfort is one of the most essential aspects. Numerous studies have been conducted on human comfort, seat ergonomics, and seat materials. Tewari & Prasad [1] worked on four distinct seat pans and backrests with varying profile curvatures, as well as three different backrest inclination angles, used in the experiments. Mehta & Tewari [2] used indentation force-deflection indenter to load the seat's cushion material under dynamic load in order to make the setup realistic and mimic the actual loading circumstances that a tractor operator would experience. De Looze et al. [3] associated the sitting comfort and discomfort with an objective data such as pressure measurements, electromyography (EMG), and analysis of sitting posture experimentally. Mohideen [4] proposed a study that centred on the postures people adopt while seated in automobile seats. Wang et al. [5] undertook a study to demonstrate the influence of various seated postures and seat geometry on the mechanical energy absorption characteristics of human participants when they are subjected to z-axis vibrations. Kolich [6] conducted a study to elaborate on the limitations of the process used to improve vehicle seat comfort. Seradio et al. [7] did a study to determine the connection between the seat cushion material characteristics and how

DOI: 10.1201/9781003367154-9

uncomfortable it is for people to ride in cars. Y & B [8] present a study that uses FEM to perform a natural frequency analysis of an automobile seating system in order to address vibration issues. Kiran [9] proposed a research to analyse the transmission of vibrations from the air-inflated cushion on top of the polyurethane foam seat to the driver. Kumar et al. [10] assessed the feet-to-head and seat-to-head transmissibility; harmonic analysis was done on a human subject while they were standing and seated. Fairley & Griffin [11] discovered a technique to assess the transmissibility of a seat without placing a person on it by observing the dynamics of the seat. Howarth & Griffin [12] conducted two lab-based investigations to determine if discomfort is felt by male human individuals who are subjected to vertical whole body vibration while sitting. Donati [13] did research to improve the seat design to reduce the seat's transmissibility for the operator's improved riding comfort. Duke & Goss [14] used non-linear stiffness and an on-off damper in studies to lower the seat's transmissibility. Jain K. K. [15] conducted laboratory tests on a common tractor seat with a focus on the dimensions of the seat to predict the vibration effect. Toward & Griffin [16] conducted a study on a human subject's vertical apparent mass that is seated on a rigid surface with a backrest that can be adjusted in angle was measured. Mehta & Tewari [17] conducted research on nine seat cushion materials that are commercially accessible and vary in thickness, density, and composition. Toward & Griffin [18] conducted study to find the car seat transmissibility using 80 people in frequency range of 0.6 to 20 Hz with acceleration magnitudes of 0.5, 1.0, and 1.5 m/s^2 to see how the apparent mass of the occupant affects the seat. In Dewangan et al. [19], three different cushion materials were used to test the apparent mass response of the human body with and without first-person support at three levels of vertical acceleration (0.23, 0.50, and 0.75 m/s^2) throughout a frequency range of 0.5 to 20 Hz (flat and contoured polyurethane foam and airs cushion). Mandapuram [20] did experimental work on a seat to forecast the power absorption of vehicle vibration under various operating conditions. Nawayseh [21] investigated the impact of vibration transmission from a car seat to a human subject under various seating settings. The objective of this study was to compare the effects of vibration on car drivers utilizing several types of cushions, featuring vehicle seat cushions made of coir-based foam and polyurethane foam.

9.2 METHODOLOGY

The effectiveness of various cushion materials in reducing the vibration effect on the car seat was compared. Various mechanical attributes go into making cushions with high shock-absorber qualities. Polyurethane foam, synthetic foam, and coir rubber foam are the three types of cushion materials used in the current study. The cushion's dimensions are 600 mm wide, 500 mm long, and 60 mm thick, and ANSYS 18.1 software was used to perform the structural analysis. From an experimental study, the material properties of various cushions were taken into consideration, as shown in Table 9.1.

One of the surfaces of the cushion material specimen was subjected to a load of up to 760 N, which is about 76 kg of an Indian human male subject—that is, 95th percentile anthropometric data of Indian male population, as shown in the Figure 9.1(c).

TABLE 9.1
Material Properties of Cushion Materials

Sr. No.	Foam Material	Density (kg/m³)	Young's Modulus (MPa)	Poisson's Ratio
1	Polyurethane foam	12.66	25	0.3
2	Synthetic rubber foam	69.72	20	0.3
3	Coir-based cushion	40.18	22	0.3

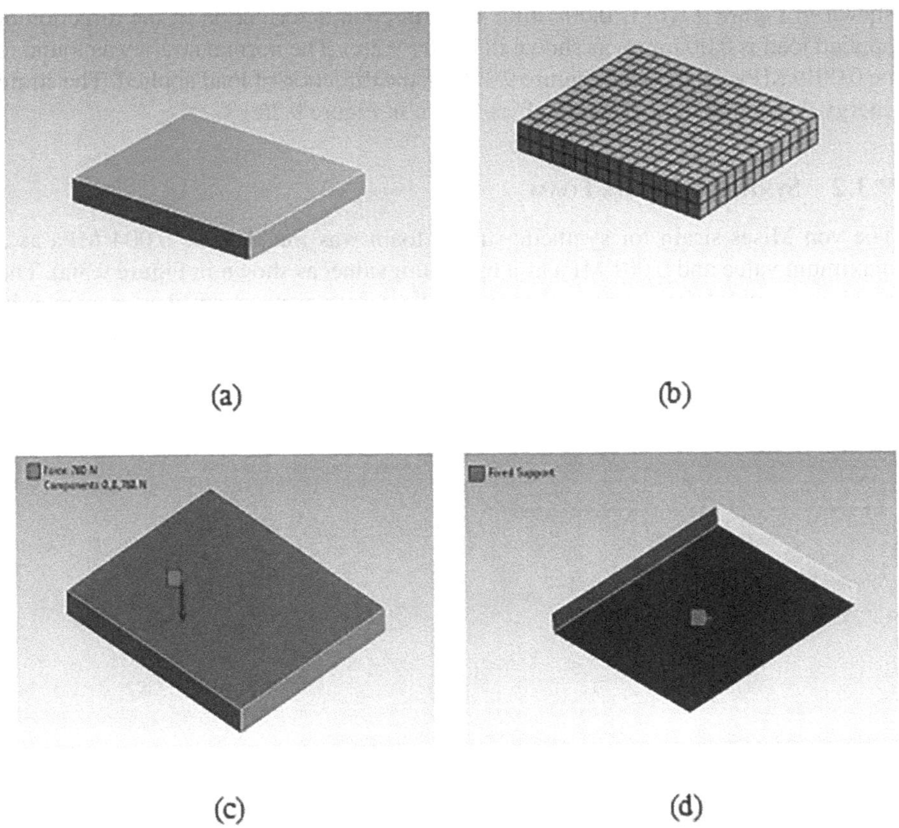

FIGURE 9.1 (a) CAD model of specimen; (b) meshing; (c) load of 760 N; (d) fixed support.

Another surface of the cushion material specimen was fixed with boundary condition as fixed support. The mesh for the CAD model has been applied with a Hexa mesh element with the element size of 0.1 mm that includes 2,655 elements and 3,655 nodes, as shown in Figure 9.1(b). The meshing quality parameters—Jacobian ratio, aspect ratio, and skewness—were found to be within the permissible limits.

9.3 RESULTS AND DISCUSSION

The structural analysis using FEM has been performed on three types of cushion materials to check the different structural properties of a material for Indian human subject with a mass of 76kg.

9.3.1 Polyurethane Foam

The von Mises strain for the polyurethane foam was found to be 0.004 MPa as a maximum value and 0.0013 MPa as a minimum value that has been shown in Figure 9.2(a). The maximum total deformation was found at the corner parts of cushion as shown in Figure 9.2(b) is 0.008 mm. The direction deformation in the direction of applied load is 0.007 mm, as shown in Figure 9.2(c). The normal strain was found to be 0.0019 MPa, as shown in Figure 9.2(d), in the direction of load applied. The strain energy was found to be 0.0074 MJ, as shown in Figure 9.2(e).

9.3.2 Synthetic Rubber Foam

The von Mises strain for synthetic rubber foam was found to be 0.004 MPa as a maximum value and 0.001 MPa as a minimum value, as shown in Figure 9.3(a). The maximum total deformation was found at the corner parts of cushion, as shown in Figure 9.3(b) is 0.01 mm. The direction deformation in the direction of applied load is 0.008 mm, as shown in Figure 9.3(c). The normal strain was found to be 0.0019

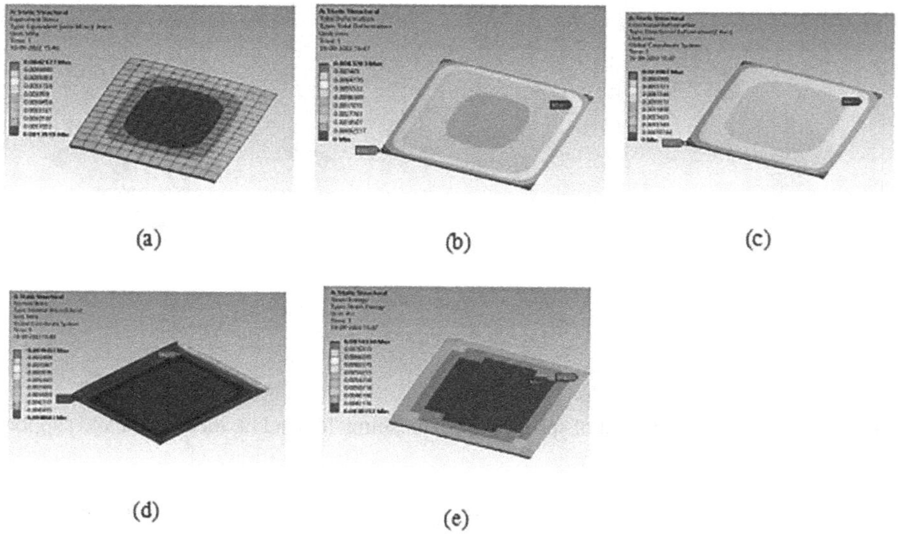

FIGURE 9.2 Results for polyurethane foam: (a) von Mises strain; (b) total deformation; (c) directional deformation; (d) normal strain; (e) strain energy.

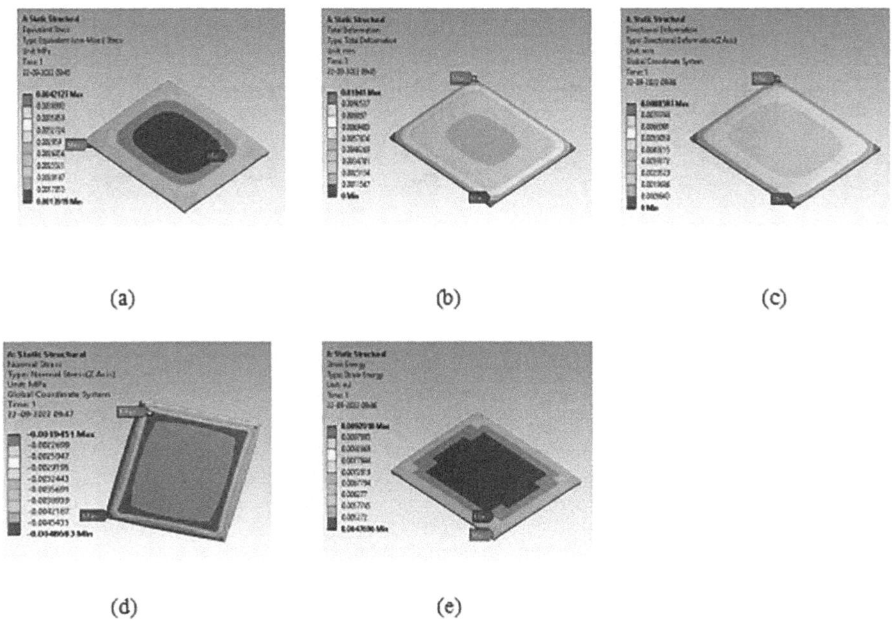

FIGURE 9.3 Results for synthetic rubber foam: (a) von Mises strain; (b) total deformation; (c) directional deformation; (d) normal strain; (e) strain energy.

MPa, as shown in Figure 6.3(d), in the direction of load applied. The strain energy was found to be 0.0092 MJ, as shown in Figure 9.3(e).

9.3.3 Coir-Based Foam

The von Mises strain for synthetic rubber foam was found to be 0.004 MPa as a maximum value and 0.001 MPa as a minimum value that has been shown in Figure 9.4(a). The maximum total deformation was found at the corner parts of cushion, as shown in Figure 9.4(b) is 0.009 mm. The direction deformation in the direction of applied load is 0.008 mm, as shown in Figure 9.4(c). The normal strain was found to be 0.0019 MPa, as shown in Figure 9.4(d), in the direction of load applied. The strain energy was found to be 0.0084 MJ, as shown in Figure 9.4(e).

From the results, the value of maximum von Mises strain was found to be same for all types of cushion materials, the value of deformation was found to be more for synthetic rubber foam material, and the value of normal strain was found to be same for all types of cushion material. Synthetic rubber foam shows higher value of strain energy and Polyurethane foam with the lowest. It shows that during static conditions, all types of cushion materials show significantly the same mechanical behaviour, and in the future scope, these cushion materials need to be tested under dynamic conditions.

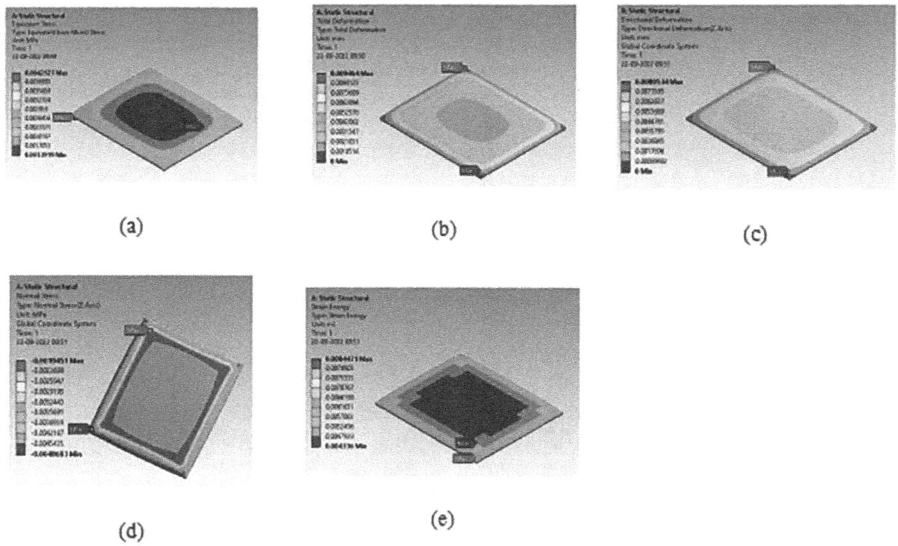

FIGURE 9.4 Results for coir-based foam: (a) von Mises strain; (b) total deformation; (c) directional deformation; (d) normal strain; (e) strain energy.

9.4 CONCLUSIONS

The 95th percentile sample measurements for the male population of India and a 76 kg Indian human subject are utilized in the current study to test several cushion materials commonly used in the manufacture of vehicle seats under static settings. From the results, the value of maximum von Mises strain was found to be same for all types of cushion materials, the value of deformation was found to be more for synthetic rubber foam material, and the value of normal strain was found to be same for all types of cushion materials. Synthetic rubber foam shows higher value of strain energy, and polyurethane foam with the lowest. It shows that during static conditions, all types of cushion materials show significantly same mechanical behaviour, and in the future scope, these cushion materials need to be tested under dynamic conditions.

REFERENCES

[1] V. K. Tewari and N. Prasad, "Optimum Seat Pan and Back-Rest Parameters for a Comfortable Tractor Seat," *Ergonomics*, vol. 43, no. 2, pp. 167–186, 2000, doi:10.1080/001401300184549.

[2] C. R. Mehta and V. K. Tewari, "IT—Information Technology: Real Time Characteristics of Tractor Seat Cushion Materials," *J. Agric. Eng. Res.*, vol. 80, no. 3, pp. 235–243, Nov. 2001, doi:10.1006/JAER.2001.0737.

[3] M. P. De Looze, L. F. M. Kuijt-Evers, and J. Van Dieën, "Sitting Comfort and Discomfort and the Relationships with Objective Measures," *Ergonomics*, vol. 46, no. 10, pp. 985–997, 2003, doi:10.1080/0014013031000121977.

[4] A. Mohideen, Reduction of Whole-Body Vibration transmitted to test drivers performing complete vehicle durability testing at Hällered Proving Ground. Master thesis, Department of Applied Mechanics, Chalmers University of Technology, Gothenburg, Sweden, 2018.
[5] W. Wang, S. Rakheja, and P. É. Boileau, "The Role of Seat Geometry and Posture on the Mechanical Energy Absorption Characteristics of Seated Occupants under Vertical Vibration," *Int. J. Ind. Ergon.*, vol. 36, no. 2, pp. 171–184, 2006, doi:10.1016/j.ergon.2005.09.006.
[6] M. Kolich, "A Conceptual Framework Proposed to Formalize the Scientific Investigation of Automobile Seat Comfort," *Appl. Ergon.*, vol. 39, no. 1, pp. 15–27, 2008, doi:10.1016/j.apergo.2007.01.003.
[7] P. Servadio, A. Marsili, and N. P. Belfiore, "Analysis of Driving Seat Vibrations in High Forward Speed Tractors," *Biosyst. Eng.*, vol. 97, no. 2, pp. 171–180, 2007, doi:10.1016/j.biosystemseng.2007.03.004.
[8] S. U. Y and K. S. B, "Natural Frequency Analysis of Automotive Seating System by Using FEM Software," *Int. J. Mech. Eng. Robot. Vib.*, vol. 1, no. 2, pp. 93–98, 2003.
[9] P. K. Kiran, "Effect of Air Cushion Seating in Passenger Car on Different Road Conditions," *Indian J. Sci. Technol.*, vol. 13, no. 10, pp. 1224–1231, 2020, doi:10.17485/ijst/2020/v13i10/150099.
[10] R. Kumar, S. Kalsi, and I. Singh, "Vibration Effect on Human Subject in Different Postures using 4-Layered CAD Model," *Int. J. Innov. Technol. Explor. Eng.*, vol. 9, no. 7, pp. 168–174, 2020, doi:10.35940/ijitee.g5104.059720.
[11] T. E. Fairley and M. J. Griffin, "A Test Method for the Prediction of Seat Transmissibility," *SAE Tech. Pap.*, Feb. 1986, doi:10.4271/860046.
[12] H. V. C. Howarth and M. J. Griffin, "Subjective Reaction to Vertical Mechanical Shocks of Various Waveforms," *J. Sound Vib.*, vol. 147, no. 3, pp. 395–408, Jun. 1991, doi:10.1016/0022-460X(91)90488-6.
[13] P. Donati, "Survey of Technical Preventative Measures to Reduce Whole-Body Vibration Effects When Designing Mobile Machinery," *Undefined*, vol. 253, no. 1, pp. 169–183, May 2002, doi:10.1006/JSVI.2001.4254.
[14] M. Duke and G. Goss, "Investigation of Tractor Driver Seat Performance with Non-linear Stiffness and On-off Damper," *Biosyst. Eng.*, vol. 96, no. 4, pp. 477–486, Apr. 2007, doi:10.1016/J.BIOSYSTEMSENG.2007.01.005.
[15] A. K. Shrivastava, C. R. Mehta, and K. K. Jain, "Analysis of Selected Tractor Seats for Seating Dimensions in Laboratory," *Agric. Eng. Int. CIGR J.*, vol. X, no. January, pp. 1–14, 2008 [Online], www.researchgate.net/publication/264166792.
[16] M. G. R. Toward and M. J. Griffin, "Apparent Mass of the Human Body in the Vertical Direction: Inter-Subject Variability," *J. Sound Vib.*, vol. 330, no. 4, pp. 827–841, 2011, doi:10.1016/j.jsv.2010.08.041.
[17] C. R. Mehta and V. K. Tewari, "Damping Characteristics of Seat Cushion Materials for Tractor Ride Comfort," *J. Terramechanics*, vol. 47, no. 6, pp. 401–406, 2010, doi:10.1016/j.jterra.2009.11.001.
[18] M. G. R. Toward and M. J. Griffin, "The Transmission of Vertical Vibration Through Seats: Influence of the Characteristics of the Human Body," *J. Sound Vib.*, vol. 330, no. 26, pp. 6526–6543, 2011, doi:10.1016/j.jsv.2011.07.033.
[19] K. N. Dewangan, S. Rakheja, P. Marcotte, and A. Shahmir, "Effects of Elastic Seats on Seated Body Apparent Mass Responses to Vertical Whole Body Vibration," *Undefined*, vol. 58, no. 7, pp. 1175–1190, Jul. 2015, doi:10.1080/00140139.2015.1052852.

[20] S. Mandapuram, S. Rakheja, P. É. Boileau, and W. Bin Shangguan, "Energy Absorption of Seated Body Exposed to Single and Three-axis Whole Body Vibration," *J. Low Freq. Noise Vib. Act. Control.*, vol. 34, no. 1, pp. 21–38, 2015, doi:10.1260/0263-0923.34.1.21.
[21] N. Nawayseh, "Effect of the Seating Condition on the Transmission of Vibration Through the Seat Pan and Backrest," *Undefined*, vol. 45, pp. 82–90, Feb. 2015, doi:10.1016/J.ERGON.2014.12.005.

10 Research and Analysis of Selection Criteria and Usage of Sustainable Materials in Interior Design

Indu Aggarwal, Atul Dutta

10.1 INTRODUCTION

Sustainability is the effort to maintain the ecosystem by avoiding the depletion of natural resources globally. It is the moral responsibility of all sectors of the society to play a significant role in the protection, preservation, and restoration of the ecosystem (Rashdan & Ashour, 2017). The industrial revolution led to rapid growth in society, manufacturing of everything from soap to steel was done in factories, and industries using machines created a new social order enhancing human lifestyle (Fazio et al., 2008). This increased the flow of raw materials from natural sources to factories and industries. Also, procuring raw materials from the sources increased embodied energy consumption and contributed to several environmental, climatic, and social problems. Due to the free flow of materials, depletion and exploitation of renewable resources took place in addition to their life cycle impacts (Fiksel, 2006).

The abundant usage of natural resources and energies resulted in degrading the ecosystem causing harm to environment. In recent decades, the dependency on energy supply for enhanced lifestyle has increased the demand for energy consumption in the form of electricity as most things like lifts and computers are operated with its help. Due to so much of energy supply and consumption, environmental problems arise, such as global warming, diminishing resources and biodiversity, pollution, acidic precipitations, loss of the ozone layer, wastages, forest loss, health hazards, water scarcity, and many other issues. These mentioned concerns will impart a negative impact on the environment, and innovative solutions need to be seriously considered to protect it from degrading and safeguard our own survival (Dincer, 2000).

10.1.1 Environmental Issues and Their Implications

- Change in the climate: Global warming is the result of changes in the patterns of climate. The drastic shifts in weather conditions, like increase or

DOI: 10.1201/9781003367154-10

decrease in moisture, humidity, and precipitation and sudden changes in climate due to various factors over different timescales, refers to climate change and may be due the impacts of global warming. The prominent reason for this global warming is the concentration of greenhouse gases, such as carbon dioxide, due to expeditious industrialization and the production of pollutants originating in human activities. Global warming conditions, altering rain scenarios, changes in the glaciers and oceans, and increasing frequencies and intensities of cyclones/floods all contribute to climate vulnerability. The implication of these can be now seen in the damages to water resources, agriculture, forests, and the ecosystem (Hasnat et al., 2018).

- Diminishing resources: Natural resources, like forests, fossil fuels, and minerals, have been rapidly used in comparison to their replacement by nature, which has created an unsustainable situation. These natural resources and fossil fuels take millions of years to replenish. Their rapid usage will lead to a situation when future generations may not get them as they will be depleted. Biodiversity includes all living organisms from all ecosystems. Extraction of fossil fuels and deforestation for human activities, like road construction, have lead to biodiversity loss, making the ecosystem vulnerable or leading to extinction (Butt et al., 2013).
- Waste: Due to industrialization, enhanced lifestyles and population growth have led to increase in municipal and industrial waste on the earth in the form of landfills. These wastes release pollutants and harmful gases into the soil and atmosphere. Some wastes are biodegradable, but some are non-biodegradable, such as plastic, which remains here for a long time. All kinds of waste are occupying places, creating mounds of debris in both land and sea and imposing great danger to human beings and wildlife (Artiola, 2019).
- Health hazards, allergies, and stress: Other than industrial exposures, humans are also majorly exposed within the indoor environment. The release of chemicals from finished surfaces and furniture exposes occupants to allergies, asthma, fatigue, headache, and rashes, and this is referred to as sick building syndrome. The levels of air pollutants are much higher in the indoor environment. This syndrome is caused by a lack of natural ventilation and increased use of mechanical ventilation systems to circulate fresh air inside buildings. Further, increased use of synthetic materials in construction and finishing of buildings has contributed to rise in sick building syndrome (Redlich et al., 1997).
- Population growth: The population has rapidly increased in last few decades in this developing world. This has led to change in the consumption patterns of all natural resources and rise in energy demand, resulting in poor air/water quality, degraded environment, and global

warming and changes in climatic conditions. As a result, one of the key drivers of environmental damage is population growth (Ray & Ray, 2011).

- Water scarcity: Water is used for drinking, sanitation, and making products. Water is also used in industrial works and for generation of electricity. Groundwater availability for agriculture and livestock is 84%, and 16% is used for residential purposes, industry consumption, and generation of electricity. Water availability is declining and scarcity is increasing, leading to health problems (Ray & Ray, 2011).
- The building industry: This plays a definite role in climate change, greenhouse gases and carbon dioxide emissions majorly contribute to environmental problems. Construction needs heavy machines and materials, which use natural resources and negatively affect the ecosystem (Spence & Mulligan, 1995). Per the records, the United States industrial sector is the third highest greenhouse gas emitter and accounts for 6% of the total emissions, 32% of nitrogen oxide, and 37% of particulate matter from all mobile sectors. In the same way, Canada's construction industry approximately produces 5.5% of nitrogen oxide and 18% of total particulate matter emissions to environment. The report of the UK Green Building Council presents that the construction industry alone is responsible for 50% of the total greenhouse gas emissions. These alarming data should make us consider thoughtful solutions to minimize all these environmental issues. Interior designers are part of construction industry and can contribute hugely to safeguard the ecosystem by providing innovative methods of construction and choice of materials to achieve sustainability (Wang et al., 2018).

The overview of the environmental issues and their affects are listed in Table 10.1.

TABLE 10.1
Overview of the Environmental Issues

Environmental Issues	Causes	Affected Zones	Implications
Change in the climate; diminishing resources; waste; health hazards, allergies, and stress; population growth; water scarcity; construction industry	Industrial revolution, rapid urbanization, and mass production of all material things which enhance lifestyle	Soil, water, air, and all species, including humans	Deforestation leads to loss of natural habitats and biodiversity, degrading ecosystems, decrease in agricultural yields, water scarcity, and diseases in humans

10.2 METHODOLOGY

The purpose of the research is to discuss sustainable building materials that support safeguarding the environment. The methodology followed is based on literature study from various authentic sources and consists of discussions on the role of interior designers in sustainability, criteria followed for the selection of sustainable materials, types of sustainable materials, application of various sustainable materials, and analysis and assessment of sustainable materials.

10.2.1 Sustainability and Interior Design

The conventional approach of an interior designer is to be concerned with functional, aesthetic, and economic criteria while designing any interior space for a client (Ching & Binggeli, 2018). But various researches have indicated that designers should engage in design criteria not only in the conventional context but also in the environmentally oriented context (Celadyn, 2018). Previous conventional approaches ignored the energy saving and emission reductions of materials, and finishes may have a negative effect on the psychological and physical wellbeing of the occupants and on the overall environment. In recent decades, designers have changed their approach, with design considerations more oriented on establishing a healthy and sustainable environment for occupants to live and work in. The architects and designers have a major influence in the selection of materials in built environments (Dutta & Kumar, 2022b). Society as a whole, like designers and clients, are beginning to understand their moral responsibilities regarding protection of environment as everything is interconnected with each other. Now designers and occupants both have their charm of seeking design of interior spaces, which are more environmentally sustainable (Hayles, 2015). The advances in technology play a key role in projects success in terms of sustainability (Figure 10.1). Various technologies that support temperature changes, wastewater reusage, energy saving, and power cooling systems with photovoltaic mechanisms are available for the construction industry to achieve sustainable buildings, termed as green buildings (Patil et al., 2022). This new concept of green buildings is providing many benefits to people by providing the ability to save water, reduce energy consumption and construction/demolition wastes, and establish a platform for attaining healthy environment (Yousif et al., 2022).

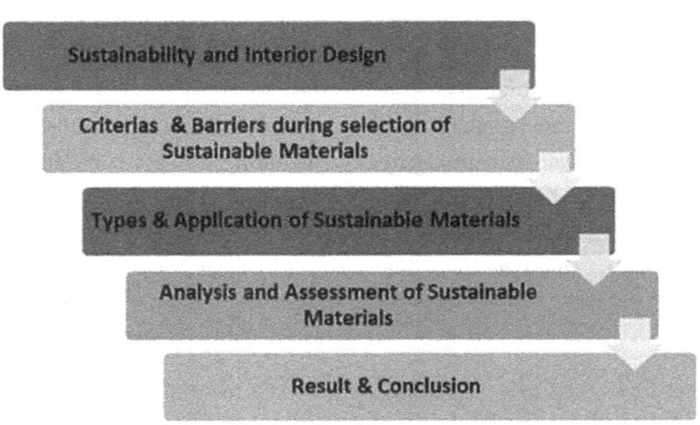

FIGURE 10.1 Conceptualization of methodology for interior environment sustainability.

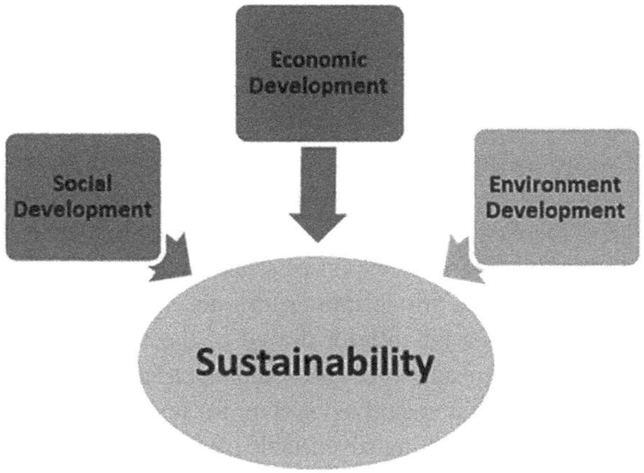

FIGURE 10.2 Three aspects of sustainability.

Various professionals have interpreted sustainability in relation to their own professional zones. Environmentalists connect sustainability with natural resources and systems. Economists take care of sustainable economic development by putting emphasis on lifestyle or standard of living. Sociologists tend to emphasize the social and cultural aspects of life. The integration of these three dimensions results in fully achieving sustainability (Kamg & Guerin, 2008). Sustainability has social, economic, and environmental aspects, and it is defined in many ways. One of the definitions given by the WCED (World Commission of Environment and Development, 1987) is that sustainability is an approach to "meet current demands without jeopardizing future generations' abilities to fulfill their own requirements." GDRC (Global Development Research Center, 2008) explains it as "maintaining a delicate balance between the human need to improve lifestyles and feeling of wellbeing on one hand and preserving natural resources and ecosystem, on which we and future generations depend." The United States Environmental Protection Agency (EPA, 2012) explains that sustainability is determined by the interplay of three pillars: ecological sustainability, economic efficiency, and societal sustainability shown in Figure 10.2.

These definitions give us the eye-opening explanation that the interiors can achieve sustainability when materials and all systems integrate together to benefit both occupants and environment and provide positive impact on social, financial, and ecological systems throughout the structure's lifespan (E. Lee et al., 2013).

10.2.2 Criteria and Barrier's for Selection of Sustainable Materials

Materials have a significant environmental influence. The growing need for a healthier, more energy-efficient environment has driven interior designers, architects, and other facility managers to pay close attention to material choices. The material selection in interior environment involves various factors like cost and energy efficiency, whereas aesthetics and maintenance requirements make the whole process challenging for designers to reach to decisions while working on design solutions for any project. The interiors involve multiple materials suitable for each element, which adds on to challenges for designers while following sustainable-oriented approach for selecting materials (Gilani et al., 2022). In comparison to residential projects, commercial projects are altered and redesigned every few years within a decade's time. So such projects create a large amount of waste and put unnecessary pressure on natural resources. This results in the usage of embodied energy on various components of interiors, like furniture, fittings, and fixtures, increasing the cost of operational energy over the lifespan of a building and clarifying the enormous expenses associated with the use of unsustainable resources (Hayles, 2015). There are certain criteria for selecting sustainable materials engaged in providing solutions for interior design of the space. Sustainable materials disseminate sustainable design solutions which support a safe and healthy environment and provide platform for a better organized system for resources and energy consumption. The criteria adopted for selecting sustainable materials are as follows:

1. Reuse: The usage of new materials should be kept to a minimum to encourage the reuse of existing materials.
2. Recycle: Materials should be recycled.
3. Renewable: Materials selected should be renewable and certified by local authorities.
4. Manufacturers: The manufacturer who uses sustainable processes and techniques in production must be identified.
5. Wastage: During construction, installation, and packaging of materials, the wastage that occurs should be at a minimum to prevent pollution and environmental damage.
6. Durability and flexibility: Materials should be durable and flexible in nature.
7. Local materials: The consumption of embodied energy should be as low as possible during processing and shipment. Therefore, one should use as many local products as possible.
8. Indoor air quality: The materials which emit radiation, toxic gases, and harmful chemicals should be avoided to provide healthy indoor air quality.
9. Maintenance and replacement: The materials should not demand an extraordinary amount of maintenance and replacement.
10. Energy and water consumption: The materials chosen should be energy-efficient, which means they reduce the rate of energy and water consumption (Ching & Binggeli, 2018).

The integration of these criteria can establish a high performance benchmark for interior design solutions (Rashdan & Ashour, 2017). The formulation of appropriate policies and frameworks with understanding causative factors certainly ensure the execution processes in the desired domains (Dutta & Kumar, 2022a). According to various surveys and research undertaken worldwide, many common grounds are noticed as barriers in the selection and promotion of sustainable materials. The common barriers in the selection of furnishings, finishes, furniture, and other equipment for interior environments are lack of information and knowledge, clients' preferences and needs, cost of materials, limited availability and lack of information about authentic suppliers, time limitation for completing projects, and in-depth knowledge of the impact of usage of unsustainable materials on environment. The survey also showed that interior designers and professionals are inclined towards the value of sustainability and accept their moral responsibility to provide environmentally sustainable interior spaces to clients. It is also the moral obligation of product designers and manufacturers to design, produce, and continually supply products that are made in an environmentally responsible way (Hayles, 2015). The successful sustainable design project can be conducted and achieved by utmost necessary involvement of various people, including the owner, the community, architects, interior designers, landscape designers, various types of engineers, and plumbing and electrical contractors, throughout the various stages of the project (Y. S. Lee, 2014).

10.2.3 Types and Application of Sustainable Materials in the Interior Environment

10.2.3.1 Types of Sustainable Materials

Materials are the physical aspect of sustainable interior design practice (Ayalp, 2013). The choice of design solutions and careful selection of materials permit users to control the environment, satisfy their needs, and maximize resource efficiency. Users can accomplish these objectives by reducing the quantity and size of materials and finishes and investing in longevity and durability, which reduce the need for change. Flexibility means spaces or products that perform multiple purposes, and reusing, recycling, and employing maximum usage of biodegradable products are a must (Mátė, 2007). Sustainable materials are broadly classified as natural, converted, and artificial materials. During the process of interior design practice, some materials are used as base materials, and some are required as finishing materials. Materials are also classified per their application on interior components, like windows, surfaces, flooring, walls, and ceilings. Sustainable materials are classified as follows:

1. Natural materials: These materials are available in nature, with the advantage of remaining stable in naturalistic form by employing interventions to their outer layers, such as stone or wood.
2. Modified materials: Some natural materials are modified into some new materials, called modified materials. Tiles and bricks are such materials.
3. Artificial materials: These are materials are obtained by manufacturing processes and are not available in nature. Plastics, pigments, and glasses are examples of artificial materials (Alfuraty, 2020).

10.2.3.2 Applications of Sustainable Materials

Floors: Floors are finished with hard flooring, such as natural stones, like marble and granite; wood; finishes that can be quickly replenished, like cork, bamboo, and linoleum; recycled rubber; tiles and chips created from both pre- and post-consumption elements, which makes them sustainable; and also concrete, which can be finished in situ. Rugs and carpets could be created with recycled materials. Carpets could be weaved from wool, cotton, linen, bamboo, straw, or yarn, with underlying material created from recycled materials.

Walls and ceilings: These could be finished with volatile or non-volatile organic paints, such as water-based and clay paints. Recycled tiles made from glass, ceramic, porcelain, and from pre- and post-consumed industrial waste can be used to finish the walls and ceilings. Nature-based clay plastering permits walls to soak and transfer the wetness as required so that earth-based plasters are safest from health point of view to finish walls. Wallpaper made from cork grass and other plant fibres (renewable) needs to be utilized and glued with eco-friendly adhesives. Wall panelling must be done with panels manufactured with eco-friendly materials which are non-toxic, recyclable, and renewable. A faux stone manufactured from recycled water and post-consumer waste can be used in interiors.

Surface materials: To maintain the quality of air for both indoor and outdoor environment, the surfaces should be finished with products which do not release pollutants and harmful toxic gases to save the occupants and labours from sick building syndrome. So the selection of green and sustainable materials makes a huge difference on environment and human health.

Furnishings: Fabrics are used for furnishings made from plant, animal, and synthetically created fibres. The sustainable fibres created from plants include organically generated cotton/linen, bamboo, century plant, stinger, cannabis, and soya. The sustainable fibres created from animals include wool, pashmina, silk, pacos, and camel hair/leather. Synthetically created fibres from petroleum products can also ne green. Both pre- and post-consumption elements like plastic bottles and industrial waste are also used to spun synthetic fibres, reducing the trash in landfills. The fabrics obtained from recycled waste are not only durable but more sustainable as they can be recycled into new raw materials on the end of their lifespan. These synthetic fabrics are easy to maintain.

Window treatments: The function of the window is to provide light and ventilation, and by providing some kind of treatment to windows, we can control their amount. Window treatment consists insulating and light-blocking, thus controlling the amount and intensity of heat and light. Window treatment must be chosen from the products manufactured from green and sustainable fabrics, such as natural grass, bamboo, wood, and 100% recyclable and renewable composites (Hayles, 2015).

10.3 RESULTS AND DISCUSSIONS: ANALYSIS AND ASSESSMENT OF SUSTAINABLE MATERIALS

The building materials which are employed in interior design practices are evaluated and assessed individually and as a part of the whole design assessment. The formal certification of materials helps interior designers to fairly evaluate environmental benefits and substantiate them to the clients. Certifications bear an authorized emblem and include validation and documentation. The life cycle assessment (LCA) of a material is a procedure which includes levels like excavation, manufacturing, transportation, building, operation and management, demolition and disposal, and recycling and reusing, giving holistic analysis of the material's effects on the environment. Many organizations grant certifications for the LCA of materials, such as Ecoinvent (Swiss LCA database), BRE Eco-profiles (UK), and Eco Indicator (NL). Nature Plus and Eco Label are the two famous authentic labels for green products; the former deals with good-quality components for built environments, construction supplies, and household furnishes, and the latter has a flower symbol used in European products, guiding customers in a simple and accurate way (Asdrubali et al., 2012).

Some more assessment tools which motivate professionals to ensure environmental sustainability are BREEAM (England), SBTOOL (international), LEED (USA),

Eco-Profile (Norway), Promise (Finland), Green Mark Buildings (Singapore), Green Star (Australia), and CASBEE (Japan). Amongst these, BREEAM and LEED are the most widely used assessment tools worldwide. Building Research Establishment Environmental Assessment Method (BREERAM) was established by Building Research Establishment and covers nine topics for evaluating building sustainability: governance, health and welfare, power, transportation, water, resources, waste, land use, and pollution. Leadership in Energy and Environmental Design (LEED, USA) evaluates buildings under six categories: site treatment with sustainability, water- and power-efficient use, resource choice, interior environment, quality and innovation, and design method. Nowadays, various software programs are also available to evaluate materials' environmental potential. Programs like Athena and BEES are assessment tools which support designers to great extent in evaluating and spotting materials with low or high environmental impacts (Ayalp, 2013).

10.4 CONCLUSION AND FUTURE SCOPE

Interior designers and other related professionals should follow the principles of sustainability. They should put all efforts to their design solutions to achieve energy efficiency, reduce emissions and waste, and eventually aim to zero-energy interiors that are in real sense creative to both client and environment.

This chapter explored the serious environmental problems which should be addressed promptly to prevent the collapse of ecosystems, depletion of fossil fuels, and destruction of biodiversity and environment in all respects. This chapter also highlighted the ways how interior construction can play vital role to solve environmental issues to a great extent. The criteria presented and discussed in this chapter are evident in motivating the designers to first reuse materials before recycling them. The life cycle assessment of materials is the key for selecting sustainable products. The selection of sustainable materials which are certified by authentic local government agencies can support the interior designers and contractors in providing innovative design solutions. This chapter has reviewed the various types of sustainable materials which can be employed in different surfaces of the interior environments and also in furniture, fittings, and fixtures. The various types of recommended finishes, such as non-volatile organic compounds, are also discussed. Sustainable guidelines should be developed for interior designers to integrate various interior elements with sustainable materials and construction technology to secure the health, safety, and welfare of those living in those built environments.

This research can be taken further, involving the identification of further avenues and sustainable technologies and applying net-zero concepts. Sustainability is multifaceted, and there is a vast scope of research with interdisciplinary collaborations, such as high performance, recycled materials, smart materials, and so on, to address sustainable developments for our future generations in a comprehensive manner.

REFERENCES

Alfuraty, A. B. (2020). Sustainable Environment in Interior Design: Design by Choosing Sustainable Materials. *IOP Conference Series: Materials Science and Engineering*, *881*(1), 012035. https://doi.org/10.1088/1757-899X/881/1/012035

Artiola, J. F. (2019). Industrial Waste and Municipal Solid Waste Treatment and Disposal. In *Environmental and Pollution Science* (Third Edition, pp. 377–391). Academic Press. https://doi.org/10.1016/B978-0-12-814719-1.00021-5

Asdrubali, F., Schiavoni, S., & Horoshenkov, K. V. (2012). A Review of Sustainable Materials for Acoustic Applications. *Building Acoustics, 19*(4), 283–312. https://doi.org/10.1260/1351-010X.19.4.283

Ayalp, N. (2013). Multidimensional Approach to Sustainable Interior Design Practice. *International Journal of Energy and Environment, 4*(7), 143–151. http://www.naun.org/main/NAUN/energyenvironment/d0116-569.pdf

Butt, N., Beyer, H. L., Bennett, J. R., Biggs, D., Maggini, R., Mills, M., Renwick, A. R., Seabrook, L. M., & Possingham, H. P. (2013). Conservation. Biodiversity Risks from Fossil Fuel Extraction. *Science (New York, N.Y.), 342*(6157), 425–426. https://doi.org/10.1126/SCIENCE.1237261

Celadyn, M. (2018). Environmental Activation of Inner Space Components in Sustainable Interior Design. *Sustainability, 10*(6), 1945. https://doi.org/10.3390/SU10061945

Ching, F. D. K., & Binggeli, C. (2018). *Interior Design Illustrated* (Fourth Edition). John Wiley & Sons, Inc. https://www.wiley.com/en-in/Interior+Design+Illustrated%2C+4th+Edition-p-9781119377207

Dincer, I. (2000). Renewable Energy and Sustainable Development: A Crucial Review. *Renewable and Sustainable Energy Reviews, 4*(2), 157–175. https://doi.org/10.1016/S1364-0321(99)00011-8

Dutta, A., & Kumar, A. (2022a). Factors Influencing the Policy Formulations and Strategies for Disaster Risk Reduction. *ECS Transactions, 107*(1), 8675–8683. https://doi.org/https://doi.org/10.1149/10701.8675ecst

Dutta, A., & Kumar, A. (2022b). The Imperative Relationship between Architecture, Urban Design and Development and Disaster Management. *ECS Transactions, 107*(1), 8657–8666. https://doi.org/10.1149/10701.8657ECST/XML

Fazio, M. W., Moffett, M., Wodehouse, L., & Moffett, M. (2008). *A World History of Architecture*, p. 592. https://books.google.com/books/about/A_World_History_of_Architecture.html?id=DLMyAQAAIAAJ

Fiksel, J. (2006). A Framework for Sustainable Materials Management. *JOM, 58*(8), 15–22. https://doi.org/10.1007/S11837-006-0047-3

Gilani, G., Hosseini, S. M. A., Pons-Valladares, O., & de la Fuente, A. (2022). An Enhanced Multi-Criteria Decision-Making Approach Oriented to Sustainability Analysis of Building Facades: A Case Study of Barcelona. *Journal of Building Engineering, 54*, 1–16. https://doi.org/10.1016/J.JOBE.2022.104630

Hasnat, G. N. T., Kabir, M. A., & Hossain, M. A. (2018). Major Environmental Issues and Problems of South Asia, Particularly Bangladesh. In C. M. Hussain (Ed.), *Handbook of Environmental Materials Management* (pp. 1–40). Springer. https://doi.org/10.1007/978-3-319-58538-3_7-1

Hayles, C. S. (2015). Environmentally Sustainable Interior Design: A Snapshot of Current Supply of and Demand for Green, Sustainable or Fair Trade Products for Interior Design Practice. *International Journal of Sustainable Built Environment, 4*(1), 100–108. https://doi.org/10.1016/J.IJSBE.2015.03.006

Kamg, M., & Guerin, D. A. (2008). The Characteristics of Interior Designers Who Practice Environmentally Sustainable Interior Design. *Environment and Behavior, 41*(2), 170–184. https://doi.org/10.1177/0013916508317333

Lee, E., Allen, A., & Kim, B. (2013). Interior Design Practitioner Motivations for Specifying Sustainable Materials: Applying the Theory of Planned Behavior to Residential Design. *Journal of Interior Design, 38*(4), 1–16. https://doi.org/10.1111/JOID.12017

Lee, Y. S. (2014). Sustainable Design Re-examined: Integrated Approach to Knowledge Creation for Sustainable Interior Design. *International Journal of Art & Design Education*, *33*(1), 157–174. https://doi.org/10.1111/J.1476-8070.2014.01772.X

Mâté, K. (2007). Using Materials for Sustainability in Interior Architecture and Design. *Journal of Green Building*, 2(4), 23–38. https://doi.org/10.3992/JGB.2.4.23

Patil, M., Boraste, S., & Minde, P. (2022). A Comprehensive Review on Emerging Trends in Smart Green Building Technologies and Sustainable Materials. *Materials Today: Proceedings*, *65*, 1813–1822. https://doi.org/10.1016/J.MATPR.2022.04.866

Rashdan, W., & Ashour, A. F. (2017). Criteria for Sustainable Interior Design Solutions. *WIT Transactions on Ecology and The Environment*, *223*, 311–322. https://doi.org/10.2495/SC170271

Ray, S., & Ray, I. A. (2011). Impact of Population Growth on Environmental Degradation: Case of India. *Journal of Economics and Sustainable Development*, 2(8), 72–77. www.iiste.org

Redlich, C. A., Sparer, J., & Cullen, M. R. (1997). Sick-building Syndrome. *Lancet (London, England)*, *349*(9057), 1013–1016. https://doi.org/10.1016/S0140-6736(96)07220-0

Spence, R., & Mulligan, H. (1995). Sustainable Development and the Construction Industry. *Habitat International*, *19*(3), 279–292. https://doi.org/10.1016/0197-3975(94)00071-9

Wang, B. Z., Zhu, Z. H., Yang, E., Chen, Z., & Wang, X. H. (2018). Assessment and Management of Air Emissions and Environmental Impacts from the Construction Industry. *Journal of Environmental Planning and Management*, *61*(14), 2421–2444. https://doi.org/10.1080/09640568.2017.1399110

Yousif, N., Awwad, I., & Mabad, G. A. M. F. (2022). An Analytical Field Study on the Principles of Sustainability in the Interior Design of Modern Egyptian Buildings. *Journal of Architecture, Arts & Humanistic Sciences*, *7*(36), 96–106. https://doi.org/10.21608/MJAF.2021.57256.2163

11 Prediction of Mechanical Characteristics of Single-Walled Carbon Nanotube

Shahbaz P., Amruthamol N. A.,
Sumit Sharma, Rajesh Kumar

11.1 INTRODUCTION

Nanoscience, a brand-new field of study in materials science, has grown in prominence. Hundreds of millions of dollars have already been spent in an effort to reveal the mysteries of nanomaterials. Even yet, it was all worthwhile since these useful nanoscale materials offer a wide range of fascinating, unique, and distinctive features. Water entrapped inside the carbon nanotubes (CNTs) has been discovered to exist in the fourth state of matter, one that is not a gas, liquid, or solid (Vivanco-Benavides et al., 2022a). On a molecular level, CNTs are a hundred times stronger than steel. Being just one-sixth the weight of steel and having a high aspect ratio makes them ideal as a filler material that can enhance mechanical properties (Ho et al., 2021). These are the major motivating factor behind this investigation.

An overview of current developments in materials computing has been discussed in Tanaka et al., 2019, which is mentioned the potential impact of learning algorithms on domains including mechanics, electronics, photonics, and dynamic system models. A study of learning algorithms used to solve various nanofluids related issues have been explored in Bahiraei et al., 2019. The literature (Bandaru et al., 2017; Kim et al., 2020; Lu et al., 2017; Voyles, 2017) demonstrates its promise as a tool for prediction and optimization. In Kim et al., 2020, both small and large databases are addressed for the identification of innovative materials. In Lu et al., 2017, discussed materials optimization using data mining support.

The developments in the nexus of data science and computations at the atomic level were covered as presented by Bandaru et al., 2017, with an emphasis on investigations of inorganic solid-state materials. Additionally, they noted important problems, including the restricted software availability needed to employ these methodologies and openings for atomistic modelling research on materials' informatics. A high-resolution electron microscope (HREM) has been used in Tulaphol et al., 2016, to study the distortion characteristics of single-walled and multi-walled

DOI: 10.1201/9781003367154-11

CNTs (SWCNTs and MWCNTs) under compression. Utilizing the carbon atoms' many-body potential set to a reasonable value provided a more quantitative explanation for the data. According to Liew et al., 2004, there was buckling behaviour in SWCNTs and MWCNTs under axial compression in a theoretically based simulation, and they found that the buckling stress is at its peak at a certain desired diameter for SWCNTs in order to evaluate the modulus of elasticity of free-standing at room-temperature vibrations of CNTs in TEM. In Vijayaraghavan et al., 2014, performed thermal vibration analysis and reported that the value of modulus of elasticity is 1.25 TPa. Under axial compression, Zhang et al., 2009, studied the buckling behaviour of SWCNTs and MWCNTs using theoretically based modelling. They discovered that the buckling load for SWCNTs can reach its optimum value at a certain dimension. Vivanco-Benavides et al., 2022b, looked into the impact of flaws on the buckling behaviour of CNTs. They discovered that, in comparison to the Stone-Wales (SW) defect, point imperfections substantially reduce the elastic properties.

Today, it is possible to depict physical properties without doing actual tests. Due to computer simulations and data analytics for material modelling and designing, this occurred. Different simulation approaches have been provided in Nordlund, 2019, such as density functional theory (DFT), molecular dynamics (MD), and so on. As a result, theoretical models may be employed as a practical counterpart for time-consuming for estimating the properties of the material at the nanoscale. Despite this, a complete method to forecast its mechanical properties at the nanoscale is still a gap (Vivanco-Benavides et al., 2022b). So the use of an artificial intelligence (AI) algorithm appears to be required in the prediction of SWCNT mechanical properties.

In several ways, researchers have already implemented AI algorithm or ML algorithm to understand different characteristics of CNT. In Frydrych et al., 2021, the reliability of the estimation of their features has been improved by using data from sequential experimental characterization of CNT using supervised algorithms. However, in Maulana Kusdhany & Lyth, 2021, unsupervised algorithms have been utilized to manage massive volumes of data solely based on MD simulations. In comparison to experimental characterization, it is clear that machine learning approaches are quicker and more economical.

There are still insufficient benchmarks available for molecular dynamics (MD) verification, despite significant breakthroughs in the experimental study of CNTs. Another problem is that most present research only looks at aspects of MD modelling. Even still, there is no full model yet to predict its mechanical characteristics at the nanoscale. Here, we are attempting to create a thorough model that will predict SWCNT mechanical properties.

The main contribution of this study are as follows:

- Comprehensive model of the prediction of mechanical characteristics of SWCNTs.
- Evaluating the performance metrics of the models has been examined, and the most appropriate model has been recommended.

11.2 MOLECULAR STRUCTURES AND PREDICTION METHODOLOGY

11.2.1 Structure of SWCNTs (Single-Walled Carbon Nanotubes)

A SWCNT can be viewed as a rolled graphene sheet (Mohammadpour et al., 2011). A typical SWCNT structure is illustrated in Figure 11.1(a). A detailed explanation of the relationships determining the geometry of SWCNT as well as a visual representation has been presented in Figure 11.1(b), where j_1 and j_2 are the lattice vectors. The chiral vector π is defined in terms of integers (n, m) and the basis vectors j_1 and j_2 of the honeycomb lattice as presented in equation (11.1). When m = 0, it is called "zigzag"; when n = m, it is called "armchair"; and generally, when n > m, it is called "chiral," and these are shown in Figure 11.1(c).

$$\Pi = n \times j_1 + m \times j_2 \tag{11.1}$$

11.2.2 Ridge Regression

Linear regression seems to be the most fundamental supervised learning-based approach. Linear regression is used to forecast a dependent variable related to a specific independent variable.

$$y = h_0 + h_1w_1 + h_2w_2 + \ldots + h_nw_n \tag{11.2}$$

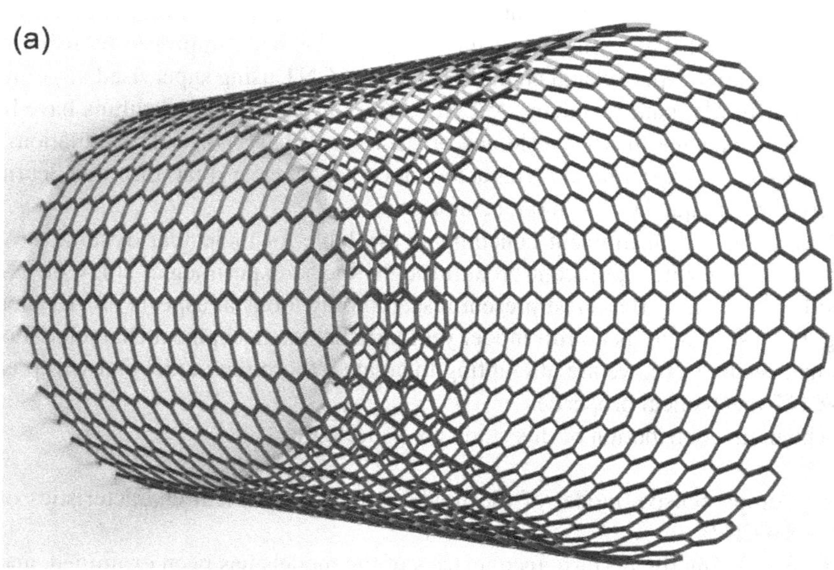

FIGURE 11.1 (a) Molecular structure of section of CNT (Ruoff et al., 2003); (b) Geometry of SWCNT (Mohammadpour et al., 2011); (c) A typical molecular structure of armchair and zigzag (He et al., 2013)

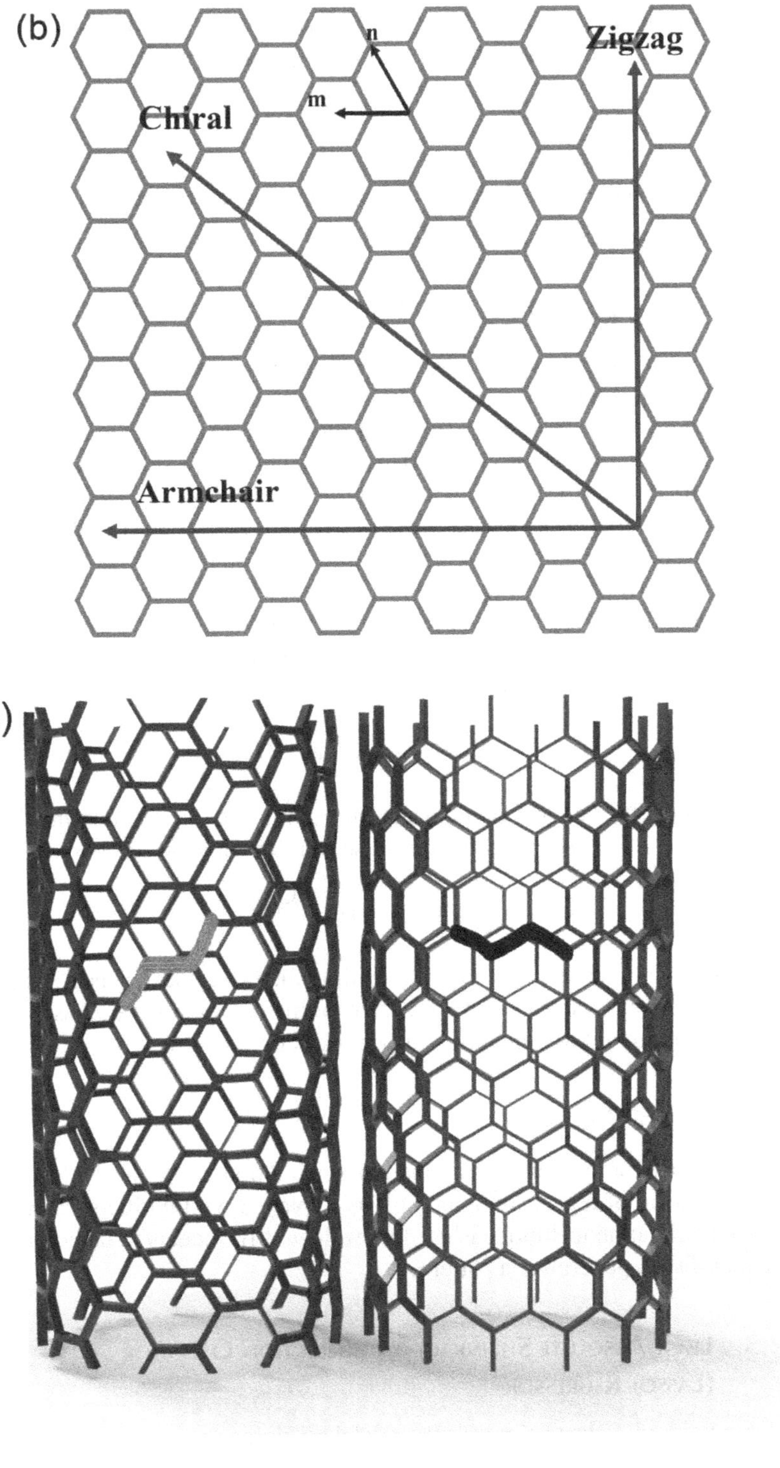

FIGURE 11.1 (Continued)

where y is the dependent variable and w is the independent variable. Ridge regression is a kind of linear regression that incorporates regularization. It is defined as follows:

$$A(h) = MSE(h) + \beta \sum_{i=1}^{n} h_i^2 \tag{11.3}$$

where $\beta \sum_{i=1}^{n} h_i^2$ is the regularization. β represents the degree of regularization. Regression employing the L2-norm is known as ridge regression (square regulation).

11.2.3 Kernel Ridge Regressor

The kernel technique is used in ridge regression, the outcome is kernel ridge regression (KRR), a non-linear variation of ridge regression in which the kernel controls the kind of non-linearity. It is important to keep in mind that the technique only has to be derived and implemented once. Thereafter, applying with an alternative kernel would essentially produce a new non-linear version of ridge regression. Kernel trick is a technique which uses kernel functions to operate inside a high-dimensional, implicit feature space without ever calculating the positions of data there. Instead, it computes all inner products seen between pictures among all couples of data there in feature space. In KRR, a kernel basis function is applied to every component (every component) inside the training dataset, (M_k), as well as connected toward a characteristic of just a question component M, G_{KRR} (M), by

$$G_{kRR}(m) = \sum_{i=1}^{N_m} \beta_k k(M, M_k) \tag{11.4}$$

where, N_m being the quantity of molecules in the training set, K becoming the kernel, and βk being the regression coefficients discovered by linear regression:

$$\beta = (k + \lambda I)^{-1} G^{(ref.)} \tag{11.5}$$

If the training set's noise is assumed to be minimal, the regularize term λ, multiplied by the unit matrix I, is often tiny. We refer to this common method as single-state fitting since it only allows the fitting of one molecular feature at a time.

11.2.4 RidgeCV Regressor

RidgeCV stands for the ridge regression cross-validation technique. Ridge regression is a specific kind of regression that is often employed with multicollinear datasets. This cross-validation technique is used to evaluate ML models to detect overfitting, or the inability to generalize a pattern.

11.2.5 Least Absolute Shrinkage and Selection Operator (Lasso) Regressor

The lasso method seeks to simplify the model by giving the model coefficients more weight. The lasso method specifically seeks to cut down on polynomial terms to overcome the overfitting issues. It minimizes the residual underneath the condition that the parameters l1-norm are constrained by a fixed t: $\sum_{j=1}^{q} \left|\delta j\right| \leq t$. This same expression is describing in (6).

$$\min_{\alpha_j} \sum_{i=1}^{N} \left(y_i - \sum_{j=1}^{m} x_{ij}\alpha_j \right)^2$$

$$subjected\, to \sum_{j=1}^{m} \left|\alpha_j\right| \leq t \tag{11.6}$$

where t is a tuning parameter that is not negative. When the aforementioned equation (11.6) is resolved using a soft threshold, lasso concurrently performs shrinkage and automated attribute selection. Based on the threshold, lasso will zero out some variables. One by one, a few of the coefficients provided by lasso will be nonzero if t is first set to 0 and subsequently raised. The coefficients returned by lasso-pure, on the other hand, will all have nonzero values.

11.2.6 LARS Lasso

A novel model selection approach called least angle regression can be thought of as a faster vector-based version of phase wise. For the whole set of solutions equal to OLS, where m is the number of covariates, only m steps are necessary. This is how LARS operates. It begins by setting each coefficient α to zero before identifying the covariate that is most strongly connected with the answer. Then, until another covariate has a same amount of correlation with the present residual, it moves as far in the direction of this covariate as it can. When a third variable has an equal or greater correlation with the residual than this one, LARS continues in the direction that is equiangular to the predictor variables. Till the kth covariate, or α_k, is included in the model, the procedure is iterated. When k = m, the OLS model is produced. The goal is to choose k to produce a model that is easier to understand and more universal.

11.2.7 Pipeline Polynomial Regressor

The several sequential phases that make up a machine learning pipeline include everything from extracting data and preprocessing to model training and testing. In a manual operation, everything must be handled manually, including preprocessing, testing, and training. Researchers have developed a new technique called pipeline machine learning to prevent this, which comprises all processes from preprocessing to training and testing. Fitting polynomial algorithms with a linear model is a straightforward method. In the linear model, the degree of input variables that is called polynomial regression is increased. The mathematical expression of polynomial regression as follows:

$$y_{pred} = \beta_1 X + \beta_2 X^2 + \beta_3 X^3 + \ldots + \alpha \tag{11.7}$$

where β and α are polynomial features.

11.2.8 Decision Tree Regressor

One of the most useful learning models is the decision tree. This choice was made using the analysis of the training dataset. Starting from the tree root, the decision tree algorithm divides the training data into features based on the greatest gain (G), equation (11.9) illustrates this. This procedure is repeated on every child node up until all the leaves are pure, so the overfitting problem can be easily corrected. By limiting the tree's ability to move beyond a certain depth (maximum depth), we can try to cut down the tree. One of the main advantages is that, even when working with non-linear data, there is no need to change features.

$$Gain\left(A_i, x_j\right) = I\left(A_i\right) - \frac{N_{left}}{N_i} I\left(A_{left}\right) - \frac{N_{right}}{N_j} I\left(A_{right}\right) \tag{11.8}$$

11.2.9 Dataset

There are 818 data samples collected from "Deep learning framework for carbon nanotubes: Mechanical properties and modeling strategies" [18], including training, testing, and validating data. The feature set consists of chirality indices n and m, initial diameter, initial length, modulus of elasticity, and Poisson's ratio to find out UTS and the strain at fracture. Parameters, minimum values, and max values of dataset are well given in Table 11.1.

11.2.10 Performance Indices

In this study, five performance indices were utilized to evaluate the effectiveness of the model built using the training data. The performance indices are as follows:

11.2.10.1 Mean Squared Error (MSE)

The mean squared error between the projected value and the actual value. Every data point's true value is measured vertically from its corresponding predicted value on

TABLE 11.1
Dataset Overview

Parameters	Minimum Value	Maximum Value
n	3	51
m	0	29
Initial diameter	0.36	3.916
Initial length	1.899	19.836
Young's modulus	903.846	1016.954
Poisson's ratio	0.047	0.27
Strain at fracture	0.13218	0.21101
Ultimate tensile strength	76.03	108.86

the fit line, and the difference is squared. The last step is to sum together the entire squared numbers and divide the result by the total number of points.

$$MSE = \frac{1}{n}\sum_{n=1}^{n}\left[D_p - D_a\right]^2 \tag{11.9}$$

11.2.10.2 Root Mean Square Error (RMSE)

It is the square root of MSE.

$$RMSE = \sqrt{\frac{1}{n}\sum_{n=1}^{n}\left[D_p - D_a\right]^2} \tag{11.10}$$

11.2.10.3 Mean Absolute Error (MAE)

The average size of the absolute errors between the projected value and the actual value is measured using the MAE metric. The MAE is frequently referred to as the mean absolute deviation (MAD).

$$MAE = \frac{1}{n}\sum_{n=1}^{n}\left[D_p - D_a\right] \tag{11.11}$$

11.2.10.4 Mean Absolute Percentage Error (MAPE)

It is equal to the average absolute percent error for each time period divided by the true value.

$$MAPE = \frac{1}{n}\sum_{i=1}^{n}\frac{x_{act(i)} - x_{pred(i)}}{x_{act(i)}} \tag{11.12}$$

11.2.10.5 R^2 Score

What percentage of the variation in the dependent variable can be predicted from the independent variable(s) is called R^2 score. Therefore, if it is really 100%, the two variables are fully connected—that is, there is absolutely no variation.

$$R^2 = 1 - \frac{RSS}{TSS} \tag{11.13}$$

where RSS is the sum of square of residuals and TSS is the total sum of squares.

11.3 RESULTS AND DISCUSSION

11.3.1 Correlation Plot

A listing of the correlation coefficients among all attributes is known as a correlation matrix. The correlation between any two attributes is shown in each table column.

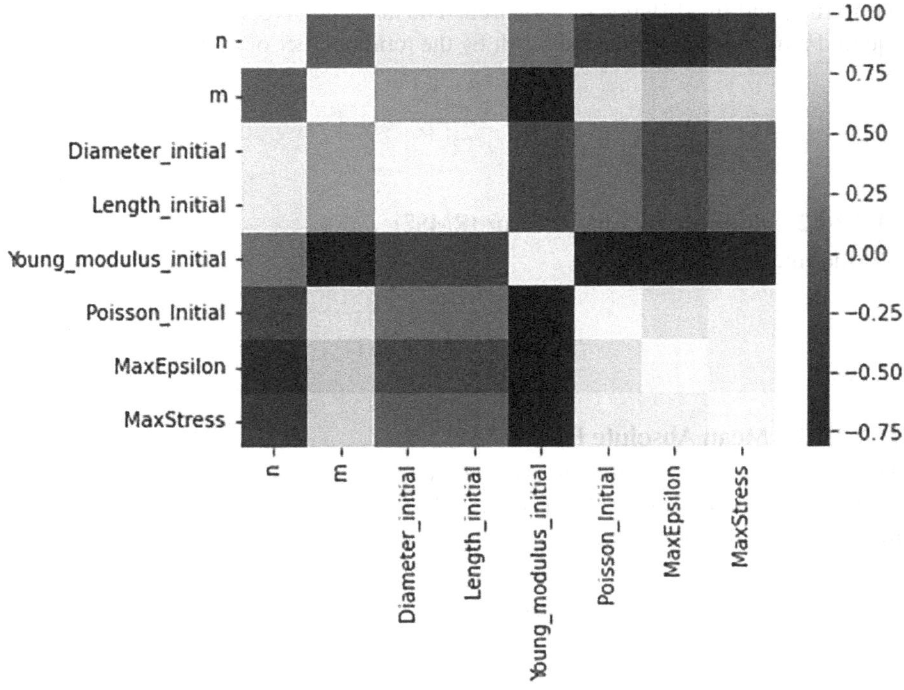

FIGURE 11.2 Correlation plot.

The correlations among all the factors are shown in Figure 11.2. In order to develop a model among the attributes and the targets, we need to know whether there is a connection among all the parameters.

11.3.2 Prediction Analysis

11.3.2.1 Maximum Strain at Fracture Point

The actual and predicted value of maximum strain at the fracture point is successfully plotted. The predicted value and actual value are not having much variation as shown in Figure 11.3(a).

11.3.2.2 UTS

The actual and predicted value of UTS is successfully plotted. The predicted value and actual value are not having much variation as shown in Figure 11.3(b).

11.3.2.3 Significance of Performance Indices

Analysing Figure 11.5 and Figure 11.6, we can clearly see the differences in the plots of actual and predicted values using different methods. These differences are due to errors in prediction. Utilizing five statistical parameters, which are shown in Table 11.2, the errors seen between actual and predicted values were calculated

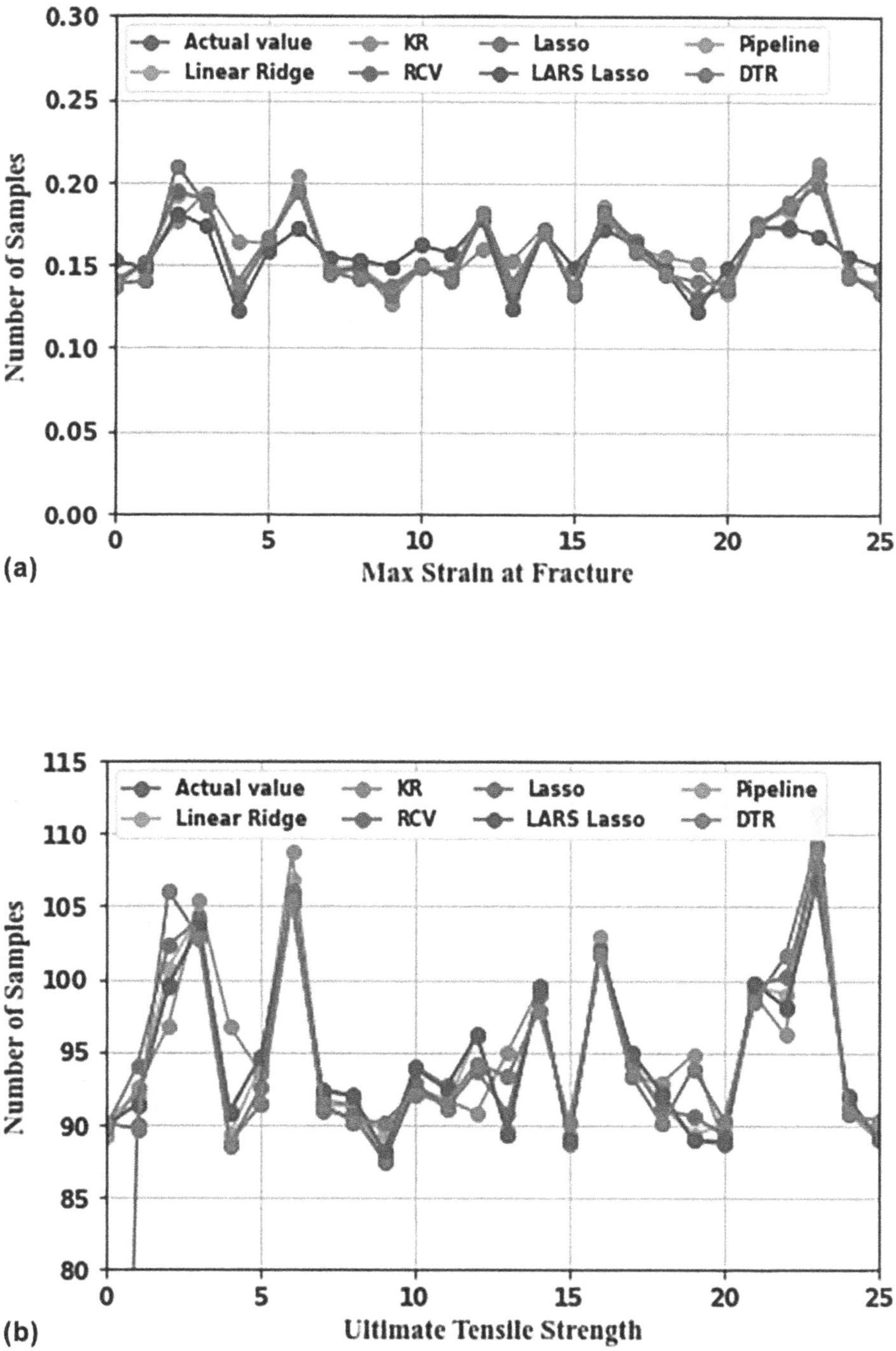

FIGURE 11.3 (a) Actual and predicted values of maximum strain at fracture point. (b) Actual and predicted values of UTS.

TABLE 11.2
Performance Analysis of Proposed Models

Model	MSE	RMSE	MAE	R^2 Score	MAPE
Ridge	2.41758	1.55486	0.68110	0.88426	0.02105
Kernel ridge	3.27121	1.80865	0.88781	0.81221	0.02810
RidgeCV	0.81339	0.90188	0.49938	0.94566	0.01740
Lasso	4.07794	2.01939	0.97630	0.68553	0.04302
LARS lasso	3.84766	1.96154	0.91836	0.69275	0.04241
Pipeline polynomial	0.92306	0.96076	0.28346	0.96167	0.00914
Decision tree	0.92306	0.96076	0.28346	0.96167	0.00914

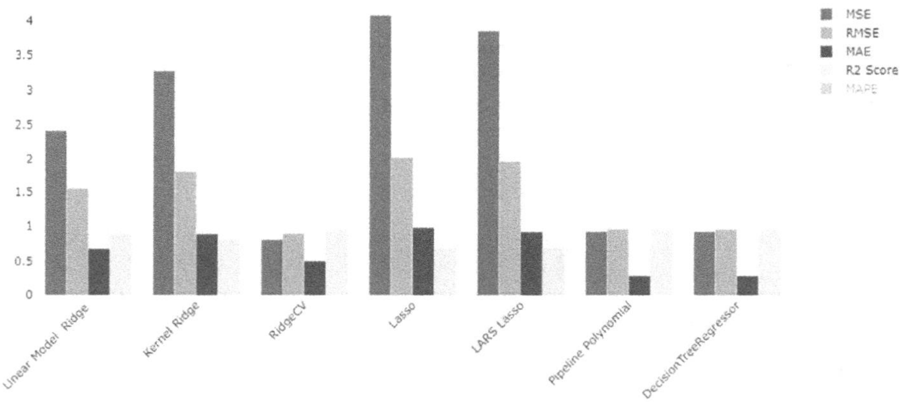

FIGURE 11.4 Comparing performance indices.

for the present study. Here, pipeline polynomial and decision tree regressor have a higher R^2 score compared to all other methods. RidgeCV also has a good R^2 score. Performance indices of all models are plotted in Figure 11.4. A model is said to be more accurate if the R^2 Score is closer to 1 or if MAE, MSE, RMSE, and MAPE values are as low as possible given that these variables indicate error. By examining Table 11.2 and the bar chart given in Figure 11.4, we can conclude that pipeline polynomial and decision tree are the most effective methods for this inquiry.

11.4 CONCLUSION

In this study, seven machine learning regressor techniques were used to accurately predict maximum strain at fracture point and ultimate tensile strength of single-walled carbon nanotube. Comparisons are also made between the performances of regressor models with varied features. The correlation plot clearly shows that chiral unit vectors have a strong link with these mechanical characteristics. Plots were made in between expected and actual values for each mechanical property individually.

The pipeline polynomial and decision tree regressors are the most effective models obtained from this study.

11.5 FUTURE SCOPE

In this study, the following works could be conducted in the future.

- Forecasting of mechanical characteristics of MWCNT could be found out.
- Foreseeing of mechanical characteristics of SWCNT having diameter more than 4 nm could be found out.
- Vibrational and thermal characteristics could be evaluated and predicted.

REFERENCES

Bahiraei, M., Heshmatian, S., & Moayedi, H. (2019). Artificial intelligence in the field of nanofluids: A review on applications and potential future directions. *Powder Technology*, *353*, 276–301. https://doi.org/10.1016/j.powtec.2019.05.034

Bandaru, S., Ng, A. H. C., & Deb, K. (2017). Data mining methods for knowledge discovery in multi-objective optimization: Part A—survey. *Expert Systems with Applications*, *70*, 139–159. https://doi.org/10.1016/j.eswa.2016.10.015

Frydrych, K., Karimi, K., Pecelerowicz, M., Alvarez, R., Dominguez-Gutiérrez, F. J., Rovaris, F., & Papanikolaou, S. (2021). Materials informatics for mechanical deformation: A review of applications and challenges. *Materials*, *14*(19), 1–31. https://doi.org/10.3390/ma14195764

He, H., Pham-Huy, L. A., Dramou, P., Xiao, D., Zuo, P., & Pham-Huy, C. (2013). Carbon nanotubes: Applications in pharmacy and medicine. *BioMed Research International*, *2013*. https://doi.org/10.1155/2013/578290

Ho, N. X., Le, T. T., & Le, M. V. (2021). Development of artificial intelligence based model for the prediction of Young's modulus of polymer/carbon-nanotubes composites. *Mechanics of Advanced Materials and Structures*, *29*(27), 5965–5978. https://doi.org/10.1080/15376494.2021.1969709

Kim, E., Jensen, Z., Van Grootel, A., Huang, K., Staib, M., Mysore, S., Chang, H. S., Strubell, E., McCallum, A., Jegelka, S., & Olivetti, E. (2020). Inorganic materials synthesis planning with literature-trained neural networks. *Journal of Chemical Information and Modeling*, *60*(3), 1194–1201. https://doi.org/10.1021/acs.jcim.9b00995

Liew, K. M., Wong, C. H., He, X. Q., Tan, M. J., & Meguid, S. A. (2004). Nanomechanics of single and multiwalled carbon nanotubes. *Physical Review B—Condensed Matter and Materials Physics*, *69*(11), 1–8. https://doi.org/10.1103/PhysRevB.69.115429

Lu, W., Xiao, R., Yang, J., Li, H., & Zhang, W. (2017). Data mining-aided materials discovery and optimization. *Journal of Materiomics*, *3*(3), 191–201. https://doi.org/10.1016/j.jmat.2017.08.003

Maulana Kusdhany, M. I., & Lyth, S. M. (2021). New insights into hydrogen uptake on porous carbon materials via explainable machine learning. *Carbon*, *179*, 190–201. https://doi.org/10.1016/j.carbon.2021.04.036

Mohammadpour, E., Awang, M., & Abdullah, M. Z. (2011). Predicting the Young's modulus of single-walled carbon nanotubes using finite element modeling. *Journal of Applied Sciences*, *11*(9), 1653–1657. https://doi.org/10.3923/jas.2011.1653.1657

Nordlund, K. (2019). Historical review of computer simulation of radiation effects in materials. *Journal of Nuclear Materials*, *520*, 273–295. https://doi.org/10.1016/j.jnucmat.2019.04.028

Ruoff, R. S., Qian, D., & Liu, W. K. (2003). Mechanical properties of carbon nanotubes: Theoretical predictions and experimental measurements. *Comptes Rendus Physique*, *4*(9), 993–1008. https://doi.org/10.1016/j.crhy.2003.08.001

Tanaka, G., Yamane, T., Héroux, J. B., Nakane, R., Kanazawa, N., Takeda, S., Numata, H., Nakano, D., & Hirose, A. (2019). Recent advances in physical reservoir computing: A review. *Neural Networks*, *115*, 100–123. https://doi.org/10.1016/j.neunet.2019.03.005

Tulaphol, S., Bunsan, S., Kanchanatip, E., Miao, H. Y., Grisdanurak, N., & Den, W. (2016). Influence of chlorine substitution on adsorption of gaseous chlorinated phenolics on multi-walled carbon nanotubes embedded in SiO2. *International Journal of Environmental Science and Technology*, *13*(6), 1465–1474. https://doi.org/10.1007/s13762-016-0984-5

Vijayaraghavan, V., Garg, A., Wong, C. H., & Tai, K. (2014). Estimation of mechanical properties of nanomaterials using artificial intelligence methods. *Applied Physics A: Materials Science and Processing*, *116*(3), 1099–1107. https://doi.org/10.1007/s00339-013-8192-3

Vivanco-Benavides, L. E., Martínez-González, C. L., Mercado-Zúñiga, C., & Torres-Torres, C. (2022a). Machine learning and materials informatics approaches in the analysis of physical properties of carbon nanotubes: A review. *Computational Materials Science*, *201*(September 2021). https://doi.org/10.1016/j.commatsci.2021.110939

Vivanco-Benavides, L. E., Martínez-González, C. L., Mercado-Zúñiga, C., & Torres-Torres, C. (2022b). Machine learning and materials informatics approaches in the analysis of physical properties of carbon nanotubes: A review. *Computational Materials Science*, *201*(June 2021). https://doi.org/10.1016/j.commatsci.2021.110939

Voyles, P. M. (2017). Informatics and data science in materials microscopy. *Current Opinion in Solid State and Materials Science*, *21*(3), 141–158. https://doi.org/10.1016/j.cossms.2016.10.001

Zhang, Y. Y., Wang, C. M., & Tan, V. B. C. (2009). Buckling of carbon nanotubes at high temperatures. *Nanotechnology*, *20*(21). https://doi.org/10.1088/0957-4484/20/21/215702

12 A Review on Facial Expression Recognition Application, Techniques, Challenges, and Tools Used for Video Datasets

Shinnu Jangra, Gurjinder Singh, Archana Mantri, Anjali, Prashant Gupta

12.1 INTRODUCTION

The computational identification of human facial expressions (FE) is a very interesting and difficult topic to investigate. Humans communicate with one another mostly through speech, but they can also express their emotions and highlight points of speech with body language. Emotions can be expressed physically, vocally, and visually. An increasing body of research demonstrates that emotional intelligence includes emotional skills. Humans frequently express their emotions through their FEs.

Non-verbal signs are communicated from facial expressions and human emotions, and these signs are crucial for social and mutual communication. Natural human behavioural unit interfaces can benefit greatly from automatic FE detection, which can further be applied in behaviour studies and clinical applications [1]. It is caused by its potential implementation to natural HCI (human-computer interaction), HEA (human emotion analysis), image indexing, interactive video, and retrieval. Facial expression recognition (FER) and analysis is presently a very ongoing research subject in the field of pattern recognition, computer vision, and artificial intelligence [2]. This chapter describes the features extraction method and classifications used for FER and also explains the software used for the emotion recognition with different types of APIs and programming algorithms. At the end, also mentioned are the challenges faced by the authors in the research area of FER of the dataset used with the classification.

The remaining part of this chapter is structured as follows: Section 12.3 describes the literature review in the field of FER approach. This section explains the classification methods of FER, tools and techniques used by the researchers, dataset and models designed by the author for extracting the emotions of humans. Section 12.3 explains the tools and techniques required for the FER with its features. Section 12.4

DOI: 10.1201/9781003367154-12

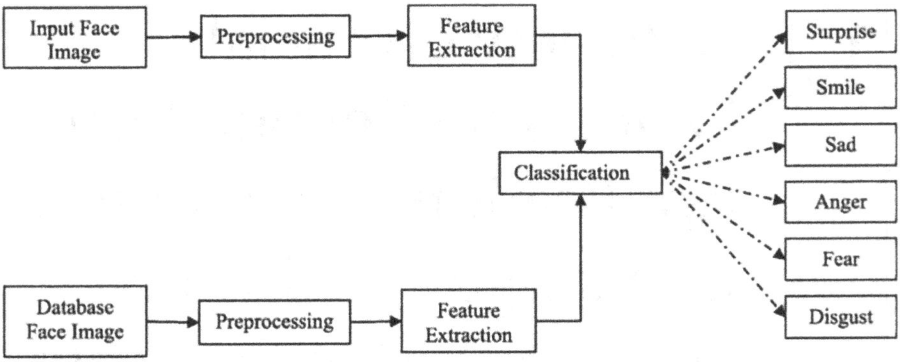

FIGURE 12.1 Classification of the facial expression recognition (FER) model [3].

describes the FER APIs, and Section 12.5 explains the opportunities and challenges faced by the researchers in facial recognition and classification techniques.

The basic classification of the FER model is shown in Figure 12.1.

12.2 RELATED WORK

This study focuses on emotion expression identification from facial photographs and videos and thoroughly reviews articles. The main goal of the study is to present the almost all accepted techniques and tools for reading and detecting human FEs, published in the recent year. This chapter describes the extraction and classification methods used for FER.

The significant work in the field of FER undertaken by various researchers is listed in Table 12.1.

12.3 TOOLS AND TECHNIQUES AND APIS USED FOR FER

The main goal of the study is to present the almost all accepted techniques and tools for reading and detecting human FEs, published in recent years [12]. This chapter describes the features extraction and classification methods used for FER.

Table 12.2 and Table 12.3 reviews the softwares used for analysis and APIs in the field of FER, respectively.

12.4 CHALLENGES AND OPPORTUNITIES

Even though FEs are largely recognized by humans without delay or effort, accurate expression recognition by technology is still difficult [13]. Finding the best approach is a major challenge for extraction, selection, classification, and processing of features, especially when input dataset variability is present [14]. These are some challenges:

1. Complex pictures, object tracking, background textures, and uncontrolled brightness have a potential impact on recognition that is undesirable.
2. Sources of facial variations are irrelevant.

TABLE 12.1
Literature Review in the Field of FER

Author/Year	Classification Methods for FER	Techniques/Tools	Type of Dataset Used/Name	Key Features	Designed Models/Techniques
Zhao, X., & Zhang, S. (2016) [2]	Hidden Markov-model (HMM), artificial neural network (ANN), BN network, K nearest neighbour (KNN), SVM, SRC classification	Active shape model (ASM), SIFT transform model, local binary patterns (LBP), Gabor wavelet representation (GWR)	Static and dynamic images dataset JAFFE database and Cohn-Kanade	Static image recognition used two methods: geometric feature-based methods (extracting these features generally requires an exact feature-point detection approach) and appearance-based methods. Dynamic image recognition also used two methods: optical flow detection and feature-point tracking.	Designed an FER model for human facial feature extraction for emotional analysis and facial expression classification
Cohen, I., Sebe, N., Garg, A., Chen, L. S., and Huang, T. S. (2017) [4]	Naive Bayes classifiers, hidden Markov models (HMMs)	Gaussian tree-augmented Naive (TAN) (The relationships between the features are modelled as it efficiently learns from data to create a tree structure, and the resulting tree structure increases the likelihood function with certainty.)	Two video input datasets used: (1) subjects displaying FE dataset and (2) Cohn-Kanade database	Bayesian network classifiers (BNCs) for categorizing expressions from video datasets. Using HMM technique, identifying the facial expression using a video sequence. Authors investigate and contrast both human-dependent and human-independent expression recognition techniques.	Two types of methods used for FER: (1) dynamic and (2) static classification. Piecewise Bezier volume deformation (PBVD) tracker. Motion units (MUs) used for face trackers.

(Continued)

TABLE 12.1 ***(Continued)***
Literature Review in the Field of FER

Author/Year	Classification Methods for FER	Techniques/Tools	Type of Dataset Used/ Name	Key Features	Designed Models/ Techniques
Stöckli, S., Schulte-Mecklenbeck, M., Borer, S., & Samson, A. C. (2018) [5]	Facial Action Coding System (FACS), electromyography	iMotions's software (used EEG, GSR, EMG, ECG, eye tracking for recording of dataset)	The Warsaw Set of Emotional Facial Expression Pictures, Amsterdam Dynamic Facial Expression Set (ADFES), and Radboud Faces Database (RaFD)	A research used 2,110 participants' facial expressions and used three international datasets for the FER analysis. Using the iMotions software, the accuracy of facial expressions were measured. Using action units, a total 46 emotions were measured. Also used were video frames with 24 frames per second. FER analyses the emotions such as sadness, anger, contempt, disgust, happiness, fear, and surprise.	Two algorithms used for FER: AFFDEX and FACET (formerly the Computer Expression Recognition Toolbox [CERT] algorithm)
Huang, Y., Chen, F., Lv, S., and Wang, X. (2019) [6]	Two main approaches used: conventional and deep-learning-based approaches. Facial landmarks (FLs), facial action units (AUs), kNN, SVM, probabilistic neural network (PNN)	2D and 3D images. Notable elements: image dimension, shooting environment, labelling approach, and elicitation method.	JAFFE, CK+ (image preprocessing, feature extraction, and expression classification)	In this study, four FER techniques were described, related aspects of the database that may affect the selection and extraction of FER techniques, as well as the 17 most often utilized FER datasets.	Psychology, computer vision, and pattern recognition have all given the research of FER a lot of attention. FER has many uses in a variety of fields, such as human-computer interaction, virtual reality, augmented reality, enhanced driver assistance systems, education, and entertainment. (Active shape model [ASM])

Author/Year	Classification Methods for FER	Techniques/Tools	Type of Dataset Used/ Name	Key Features	Designed Models/ Techniques
Samadiani, N., Huang, G., Cai, B., Luo, W., Chi, C. H., Xiang, Y., and He, J. (2019) [7]	Support vector machine (SVM), K nearest neighbour (K-NN), hidden Markov model (HMM)	Electrocardiogram (ECG), electromyography (EMG), electroencephalograph (EEG), and electrooculography (EOG), eye-tracker	AFEW dataset, CK+ dataset, MMI dataset	Three classes were used for FER: detailed-face, non-visual, and target-focused sensors. Also discussed is human interaction for FER and challenges facing in this research area.	Universal manifold model (UMM), deep neural network
Kulke, L., Feyerabend, D., & Schacht, A. (2020) [8]	Facial electromyography (EMG)	iMotions with EMG, Affectiva Affdex software	Video dataset	In this study, they compared two software for facial expression recognition and measured happy, angry, and neutral faces. A total of 20 users were used in the dataset.	Comparison study between EmoVu, FaceReader, FECAT, and Affectiva Affdex for reliable detection of emotions
Revina, I. M., & Emmanuel, W. S. (2021) [3]	Texture feature methods (TFM) used for texturing, local binary pattern (LBP), SVM, HMM, Online Sequential Extreme Learning Machine (OSELM),	This survey only uses complexity algorithms for the classification of emotions from different-different regions.	JAFFE, CK dataset (happiness, sadness, anger, disgust, surprise, and fear), AR face database, OWN dataset etc.	This paper explains the different types of FER techniques and finds out the best three stages: extraction, preprocessing, and classification. It also compared different types of algorithms based on the complexity and number of expressions.	Histogram equalization model used for variations like contrast and lighting.

(Continued)

TABLE 12.1 ***(Continued)***
Literature Review in the Field of FER

Author/Year	Classification Methods for FER	Techniques/Tools	Type of Dataset Used/ Name	Key Features	Designed Models/ Techniques
Adem, K. I., Ramakrishnan, K., Govardhanan, K., & Kannan, R. (2022) [9]	Neural network (NN), deep learning	Machine learning, affect recognition,	AffectNet and RAF-DB dataset	This paper explores the different types of datasets used for FER and also finds out the best technique for FER using the AffectNet dataset with neural network.	Convolutional neural network models were used for accuracy, and it was determined that the Visual Geometry Group 16-layer (VGG16) model is the best model.
Wolf, K. (2022) [10]	Electromyography for video data, Facial Action Coding System (FACS), EMFACS	Sophisticated High-Speed Object Recognition Method (SHORE system) with C++ software libraries	Videotaped face images	In this study, researchers have looked for certain patterns of expression in experimental investigations that measure facial expressions.	Three main methods were used for measuring facial expressions. Google Glass was used for real-life data.
Canal, F. Z., Müller, T. R., Matias, J. C., Scotton, (2022) [11]	Human-computer interaction, neural network, convolutional neural networks	Used artificial intelligence and neural experiments.	Video and facial images collection	This study reviewed a total of 51 papers on facial expression selected from different databases. Results of classification and neural network algorithms were compared.	Neural-network-based architectures used for emotional expressions capturing

TABLE 12.2
Software Used for Facial Expression Recognition and Analysis

S. No.	Name of Software	Dataset Used	How to Measure Human Expression	APIs and Tools Needed
1.	FaceReader FER	Valence, arousal, head orientation, heart rate, and heart rate variability and images and videos dataset	FaceReader is a facial analysis programme that has been taught to categorize various facial emotions. Three processes are used in this FaceReader to categorize facial expressions. Face finding, face modelling, and face classification.	Deep face model used with deep learning.
2.	iMotions FEA	image/video/audio, website, screen, face and scene recording, GlassesSync, 360 image (VR), 360 Video (VR)	It can be applied with all the biometric devices and capture live visualization of the data stream. It can measure human expressions, like act, think, and feel, using eye tracking, facial expression analysis, EEG, heart rate, and skin response.	Inbuilt APIs Tobii, Eyetech, Emotiv, and Emotient webcam. R studio is used for the processing.
3.	Amazon's (AWS) Rekognition FER	Input data requires text, images, and videos for analysis.	Import the input metadata in S3 Amazon for object, scenes, and activities detection. Create a lambda function for the Rekognition and moderation. It is based on deep learning with APIs, Console, and Command line.	Requires machine learning and pip installation of AWS-Shell, Lambda Function APIs
4.	3Divi Face SDK	Works with image and video streaming dataset	Facial recognition that works quickly and precisely. Works on masked faces. Analyses face features and gathers demographic information about people passing by to produce analytical reports.	Face Machine Server (FMS) and multiple Face Machine Client (FMC), GraphQL APIs
5.	Viso FER	It works for video and audio data.	In order to document participant behaviours and interactions and to instruct, train, and enhance the abilities of students and professionals, Viso is the simple-to-use solution.	Does not require any specific APIs

TABLE 12.3
Facial Expression Recognition APIs

S. No.	Name of APIs	Recognition Features of Software	Key Features	Cost	Accuracy
1.	Microsoft Computer Vision API	Best for image processing contents like colour schemes, label visual functions, recognizing items and brands, finding faces in images, and identifying image classes.	High-level programming algorithms for image processing and information return are provided via the Microsoft Computer Vision Facial and Image Recognition API.	Basic plan is free and ranges $19 to $ 199.	96%
2.	Lambda Labs API	Lambda API is best for face detection, features identification, and gender identification.	This is recognized and classified by gender, and additionally, it has a few options for placing the lips, nose, and eyes.	Free for 1 month and $149 to $1,449 per month.	99%
3.	Inferdo	Inferdo API best for face detection and age estimations.	It works with machine learning. Human faces are located on images using Inferdo face detection. The API is effective in identifying faces, genders, and ages, as well as identifying facial traits.	Basic plan is free and $10 to $1,000 per month.	100%
4.	Face++	This is used for extracting facial expressions, facial identification, and comparison from images.	Megvii's Face++ Face Detection API uses deep learning models (DLMs) to recognize and analyse people. These algorithms have been trained on millions of images in order to deliver the best level of face recognition accuracy.	Face++ is free.	99%
5.	EyeRecognize	This best for face detection from biometrics.	The nose, mouth, and eyes coordinates are provided by the EyeRecognize Face Detection API. Additionally, it functions with biometric attributes, such as race, gender, and age.	Basic plan is free and $29 to US $249 per month.	99%

S. No.	Name of APIs	Recognition Features of Software	Key Features	Cost	Accuracy
6.	Kairos	Kairos used for finding faces and extracting features. It is used in image recognition.	The API identifies faces in photos, videos, and real-time streaming while defining gender and age. The use of this API helps to safeguard client communications and stop identity fraud.	Paid $19 to $499 per month	62%
7.	Animetrics	Best for deep learning face recognition, revisualization option, and SetPose.	It treats details about facial features, such as the lips, chin, ears, nose, and eyebrows, as coordinates on the image. Additionally, it specifies gender and plots the faces' orientation along three axes.	Basic plan is free and $49 to $999 per month.	100%
8.	Macgyver	It is used for facial recognition and comparison, and it is used with machine learning, TensorFlow, and deep neural networks.	It can determine whether there are faces in the image, compare two faces, and return the coordinates (x, y) of the individuals in the photo.	It has a free version, but certain types of use require a fee.	74%
9.	BetaFace	It is used for facial recognition and transforming.	Faces are simple to find, examine, identify, and compare. With various facial references and more than 40 facial traits, the technology can also identify gender, age, and ethnicity.	Basic plan is free and $245 to $1,595 per month.	81%
10.	Luxand.cloud	It is mainly used for detection and comparison of human faces.	Age, gender, and emotion are all noted in the video, and it also recognizes previously labelled individuals.	Free for basic plan and $19 to $499 per month.	Not defined

12.5 CONCLUSION

A thorough explanation of how expressions and emotions are controlled and how the activities of human FEs can be described could advance our understanding of social science's fundamental principles. Unanswered concerns in the domain of emotion research may be fully addressed by new research in the area of FER display of emotion.

This study focuses on emotion expression and identification from facial photographs and videos and reviews many articles. The main goal of the study is to present the almost all accepted techniques and tools for reading and detecting human FEs, published in recent years. This chapter describes the features extraction and classification methods used for FER techniques and also explains the software used for the emotion recognition with different types of APIs and programming algorithms. At the end, also mentioned are the challenges faced by the authors in the area of FER approaches and analysis of dataset used with classification.

REFERENCES

1. Chibelushi, C. C., & Bourel, F. (2003). Facial expression recognition: A brief tutorial overview. *CVonline: On-Line Compendium of Computer Vision*, *9*.
2. Zhao, X., & Zhang, S. (2016). A review on facial expression recognition: Feature extraction and classification. *IETE Technical Review*, *33*(5), 505–517.
3. Revina, I. M., & Emmanuel, W. S. (2021). A survey on human face expression recognition techniques. *Journal of King Saud University-Computer and Information Sciences*, *33*(6), 619–628.
4. Cohen, I., Sebe, N., Garg, A., Chen, L. S., & Huang, T. S. (2003). Facial expression recognition from video sequences: Temporal and static modeling. *Computer Vision and Image Understanding*, *91*(1–2), 160–187.
5. Stöckli, S., Schulte-Mecklenbeck, M., Borer, S., & Samson, A. C. (2018). Facial expression analysis with AFFDEX and FACET: A validation study. *Behavior Research Methods*, *50*(4), 1446–1460.
6. Huang, Y., Chen, F., Lv, S., & Wang, X. (2019). Facial expression recognition: A survey. *Symmetry*, *11*(10), 1189.
7. Samadiani, N., Huang, G., Cai, B., Luo, W., Chi, C. H., Xiang, Y., & He, J. (2019). A review on automatic facial expression recognition systems assisted by multimodal sensor data. *Sensors*, *19*(8), 1863.
8. Kulke, L., Feyerabend, D., & Schacht, A. (2020). A comparison of the Affectiva iMotions Facial Expression Analysis Software with EMG for identifying facial expressions of emotion. *Frontiers in Psychology*, *11*, 329.
9. Adem, K. I., Ramakrishnan, K., Govardhanan, K., & Kannan, R. (2022). Taking facial expression recognition outside the lab and into the wild by using challenging datasets and improved performance metrics. *F1000Research*, *11*(349), 349.
10. Wolf, K. (2022). Measuring facial expression of emotion. *Dialogues in Clinical Neuroscience*, *17*(4), 457–462.
11. Canal, F. Z., Müller, T. R., Matias, J. C., Scotton, G. G., de Sa Junior, A. R., Pozzebon, E., & Sobieranski, A. C. (2022). A survey on facial emotion recognition techniques: A state-of-the-art literature review. *Information Sciences*, *582*, 593–617.

12. Zempelin, S., Sejunaite, K., Lanza, C., & Riepe, M. W. (2021). Emotion induction in young and old persons on watching movie segments: Facial expressions reflect subjective ratings. *PLoS ONE*, *16*(6), e0253378.
13. Jiang, L., Yin, D., & Liu, D. (2019). Can joy buy you money? The impact of the strength, duration, and phases of an entrepreneur's peak displayed joy on funding performance. *Academy of Management Journal*, *62*(6), 1848–1871.
14. Stöckli, S., Schulte-Mecklenbeck, M., Borer, S., & Samson, A. C. (2018). Facial expression analysis with AFFDEX and FACET: A validation study. *Behavior Research Methods*, *50*(4), 1446–1460.

13 Multilevel Inverter System Topologies for Photovoltaic Grid Integration

Parul Gaur

13.1 INTRODUCTION

Usage of renewable energy sources (RESs) has been increased rapidly in the past few years. Due to environmental protection issues RESs, such as photovoltaic (PV) or wind power systems, are widely being used for the generation of electric power. According to Energy Policy Review India 2020, grid connected renewable electricity capacity has reached 84 GW in November 2019, with 32 GW coming from the solar PV and remaining from the wind and small hydro. More usage of RESs like solar energy and wind energy results in energy security, notable environmental changes, and economic benefits. Solar power is a technique in which the sun's energy is converted into the electrical energy, either directly using the PV panels or indirectly with the help of concentrating solar energy. Research is going on the power inverters based on PV cells [1–4], and these inverters are known as the solar power inverters. These solar power inverters form the very essential part of the solar electric power system as they convert the variable DC output of the PV panels into the AC current. This generated AC electricity can be utilized for the residential purposes to operate various appliances [5–7]. MLIs (multilevel inverters) are playing a very significant role for these solar power inverters. Integration of a MLI with PV sources is more advantageous both economically and environmentally. However, PV integration to the MLI has posed some technical challenges to the researchers in the smooth operation of MLI. The major challenge associated with PV integration is to maintain the power from the output at constant level. However, this power keeps on changing because of external conditions, such as temperature and irradiation. Thus, PV integration to the MLIs is nowadays an important issue to optimize the RESs utilization and to enhance the system performance.

The scenario of global energy is now witnessing a rapid surge in the energy requirements. The utilization of RESs results in reduction in pollution, emission and more environmental friendly. Among the existing RESs, solar energy demand has grown rapidly by 25% per annum over the past twenty years [8–10]. PV integration with the grid is gaining attention nowadays in most of the countries. A MLI can be

DOI: 10.1201/9781003367154-13

integrated with RESs such as PV. Integration with RESs results in lesser emission and more utilization of the RESs. Figure 13.1 represents a solar PV system tied with a residential grid. Solar energy is acquired from the PV system and then it is converted into the useful electrical energy, which is beneficial for industrial as well as residential purposes [11–14]. With the help of PV effects, solar energy is converted into the electrical energy using solar panels in the PV system and the DC current is changed to AC current using the inverters. The energy from the PV panels is DC current. Power inverters are used to convert this DC current into AC current as the electrical grid and power electronics equipment use the AC form. After this, AC electricity is supplied to domestic appliances or the utility grid.

Research has been conducted in MLIs integrated with PV to address the parameters, such as cost, efficiency, and reliability [15–19]. The reliability and lesser cost of MLIs make them as a strong candidate for the efficient, reliable, and robust solar power converters. The solar-based MLI architecture provides the active and reactive power support, improves the system stability by providing the smart grid capabilities, and leads to overall system performance enhancement, better stability, and reliability of the complete system [20–21]. Apart from regulating real power flow, solar-based MLI structure provides the better voltage regulation, frequency regulation, compensation of harmonics, reduction in total harmonic distortion and damping of transients, and so on.

Out of available MLI structures, the CHB structure is generally used nowadays for integration [22–25]. PV-based CHB inverters have many commercialized applications, such as large variable-speed drives and reactive power compensator devices. These structures result in better utilization of power and efficiency. A maximum power point tracking (MPPT) algorithm can be deployed in the PV-based inverters for maximization of power received from the solar panels [26–30]. The CHB

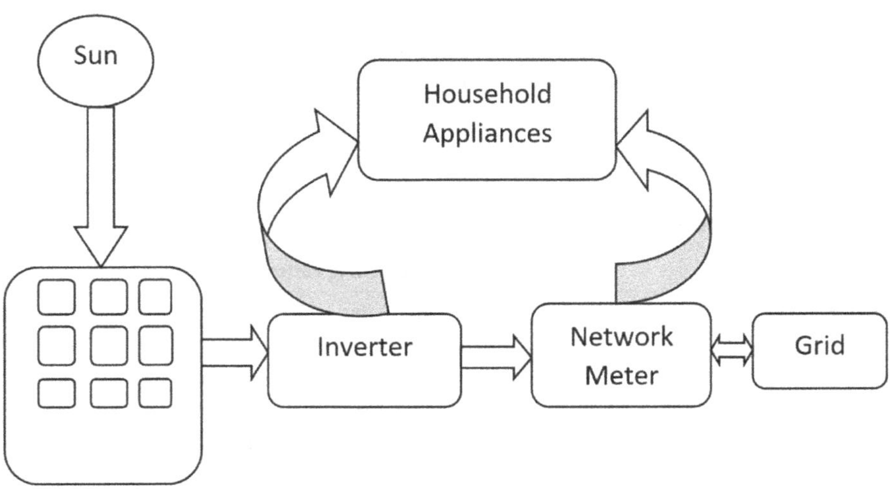

FIGURE 13.1 A residential grid-tied solar PV system.

topology of MLIs makes possible to use independent voltage control and tracking of maximum power point in every PV panel, which further leads to maximization of the efficiency of overall system.

This research paper is designed into five sections. The first section presents the introduction to RESs, MLIs, topologies, and integration with the PV grid. MLI system with PV integration is presented in the second section. Output voltage characteristics of the proposed MLI system are depicted in the third section. Finally, the fourth section is the conclusion.

13.2 PROPOSED MULTILEVEL INVERTER SYSTEM WITH PHOTOVOLTAIC INTEGRATION

A MLI based on CHB topology consists of series of H-bridges having a discrete DC source as input. As the output is taken in series, these DC sources must be separated from each other. Each H-bridge produces staircase waveform in the form of three levels: $+V_{dc}$, 0, and $-V_{dc}$. The output AC voltage is the sum of the voltages produce by the H-bridges. The general outline of the proposed system is depicted in Figure 13.2. It consists of a PV source as input DC, CHB MLI, SHE-PWM technique, and the optimization algorithms. MLI is fed with the input using PV source. The CHB MLI comprises a series of three H-bridges associated in series to obtain the seven-level output voltage. The output is further fed to the AC load. Switching circuits provide the switching pulses to the switches of MLI. The switching pattern of the proposed inverter is shown in Table 13.1.

The seven-level cascaded H-bridge consists of three H-bridges: H_1, H_2, and H_3. There are four IGBTs per H-bridge; therefore, the total number of switches per phase is 12. Hence, total number of IGBTs switches is 36 for all the three phases in the seven-level CHB inverter. These H-bridges of each phase are connected with PV sources.

PV panels are used for generation of input DC voltage and are configured with the optimized parameters as represented in Table 13.2. The voltage pulse generator ramps the voltage up from 0 to 40 and allows to result in VI curves. Output is taken via two scopes. Figure 13.3 depicts the Simulink model of single-phase seven-level MLI, while Simulink model of PV panels is represented in Figure 13.4.

TABLE 13.1
Switching Pattern for Multilevel Inverter System

	Switching Pattern											
Output Voltage (Vo)	**Sa**	**Sb**	**Sc**	**Sd**	**Se**	**Sf**	**Sg**	**Sh**	**Si**	**Sj**	**Sk**	**Sl**
+V	1	1	0	0	0	1	0	1	0	1	0	1
+2V	1	1	0	0	1	1	0	0	0	1	0	1
+3V	1	1	0	0	1	1	0	0	1	1	0	0
0	0	1	0	1	0	1	0	1	0	1	0	1
–V	0	0	1	1	0	1	0	1	0	1	0	1
–2V	0	0	1	1	0	0	1	1	0	1	0	1
–3V	0	0	1	1	0	0	1	1	0	0	1	1

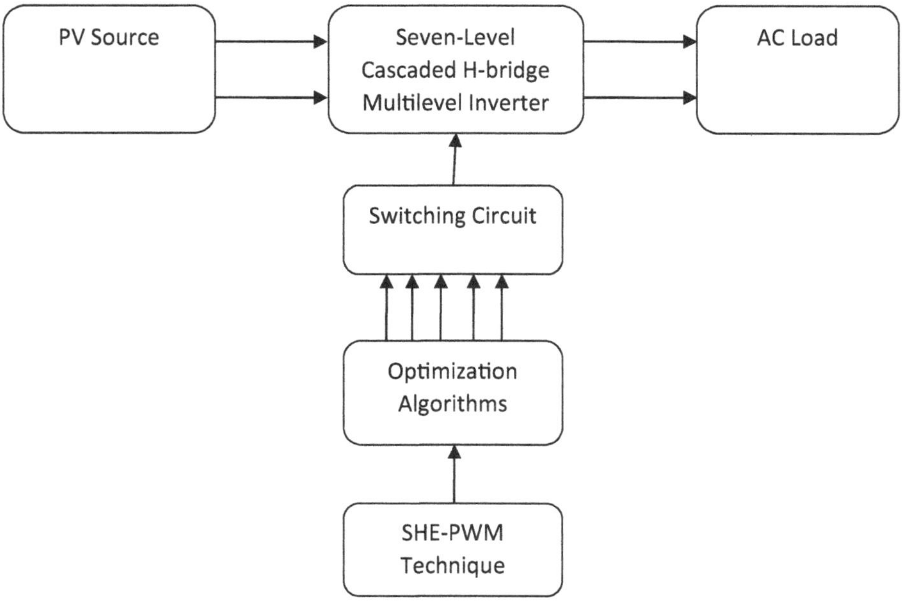

FIGURE 13.2 Block diagram of the proposed multilevel CHB inverter.

TABLE 13.2
Optimized Parameters of a Solar Cell

Parameters	Values
Diode saturation current (I_s)	2.7098e^{-10} A
Solar-generated current (I_{ph})	9.0012 A
Series resistance (R_s)	0.0057 Ω
Parallel resistance (R_p)	27 M Ω
First-order temperature coefficient	0.002 K^{-1}
Energy gap (EG)	1.2570 eV
Series resistor temperature exponent	2.305e^{-07}
Parallel resistor temperature exponent	0.3071

13.3 RESULTS: OUTPUT VOLTAGE CHARACTERISTICS OF PV INTEGRATED MULTILEVEL INVERTER SYSTEM

After simulations in Xilinx and MATLAB, output voltage levels are obtained in the form of stairs. For the proposed seven-level MLI, obtained THD is low, as shown in Figure 13.5. Table 13.3 demonstrates the THD (%) of seven-level MLI having PV source as input. From the simulated results, it is observed that with the help of designed seven-level MLI, minimum THD and harmonics are obtained in the output voltage.

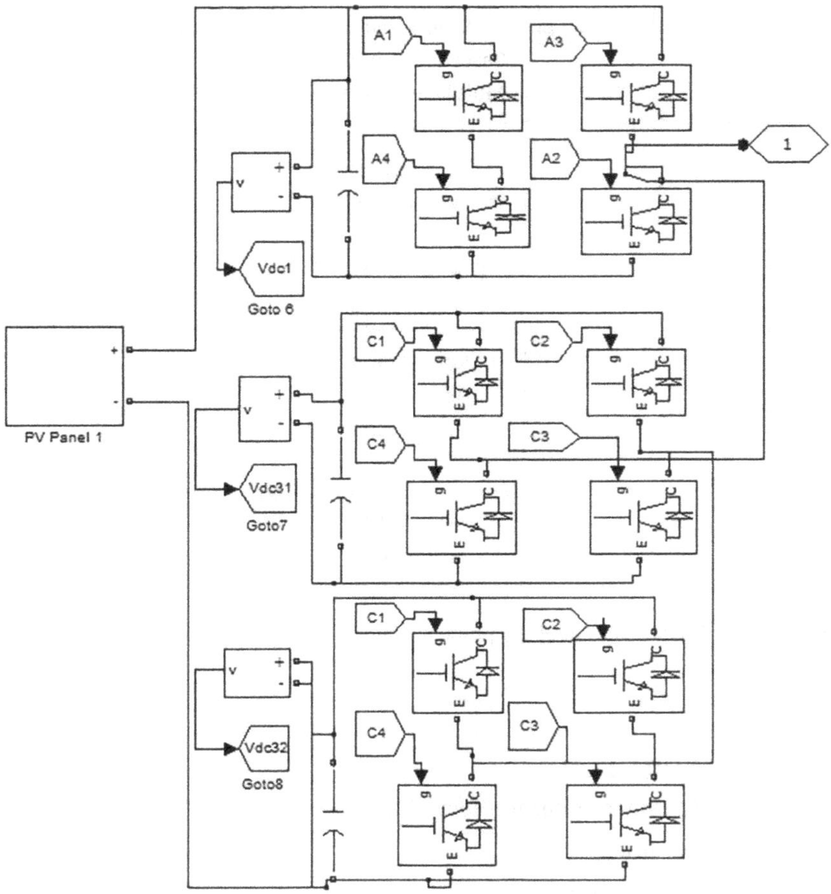

FIGURE 13.3 Simulink model of single-phase seven-level MLI.

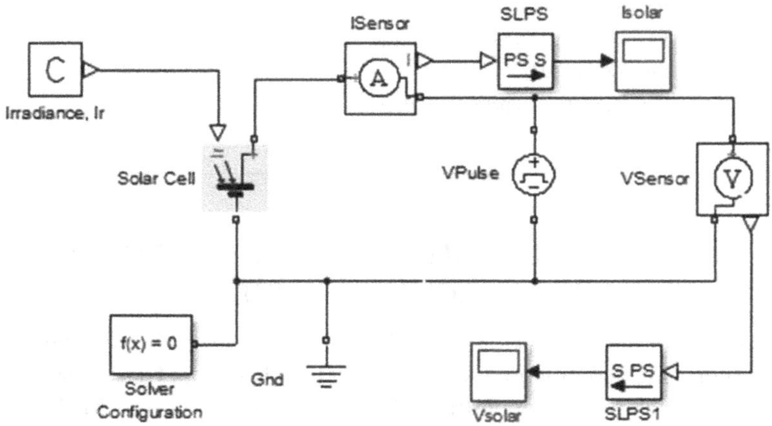

FIGURE 13.4 Simulink model of PV panels.

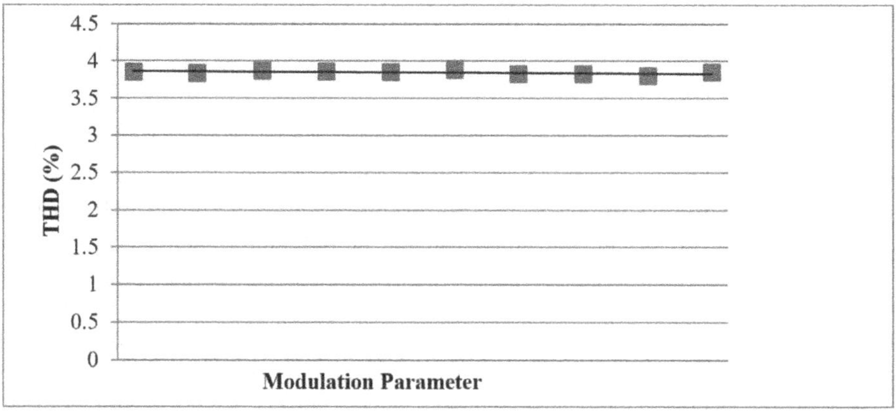

FIGURE 13.5 THD (%) in seven-level MLI with PV integration.

TABLE 13.3
Obtained THD in Seven-Level MLI with Input as PV Source

Modulation Parameter (Ma)	THD in Seven-Level MLI (%)
0.10	3.85
0.15	3.84
0.20	3.84
0.25	3.86
0.30	3.87
0.35	3.88
0.40	3.86
0.45	3.87
0.50	3.85
0.55	3.89
0.60	3.88
0.65	3.82
0.70	3.83
0.75	3.85
0.80	3.83
0.85	3.85
0.90	3.80
0.95	3.86
1.00	3.85

13.4 CONCLUSION

The implementation of CHB topology based seven-level inverter system with minimum THD in the output waveform has been presented. PV source acts as input for the designed system. The renewable source (PV), on connecting with the proposed seven-level CHB topology, results in 3.85% THD on average, which is within the acceptable limits per IEEE standard. The modulation technique used for the firing of switches of seven-level CHB inverter is SHE-PWM technique. For high-power applications, fundamental switching frequency method, such as selective harmonic elimination (SHE), is needed to keep the losses below the acceptable values. The output sinusoidal voltage is obtained in the form of levels—seven levels in this case—and is having lesser THD. The performance of the proposed MLI system is measured in terms of harmonics. Observing the obtained output waveform and MLI performance, harmonics content is drastically reduced, and high-quality output waveform is obtained. This feature facilitates the use of the proposed MLI system in high-power industrial applications. In fact, grid integration of MLIs with the RESs such as PV is very prominent nowadays due to the advantages offered by the MLIs, such as lower electromagnetic interferences, lesser DV/DT stresses, better output voltage, and lesser harmonic contents. Further, one of the applications of the implemented MLI system is also in controlling of the speed of induction motor. As a future scope, the proposed MLI system can be used in other renewable sources, such as variable-speed wind turbine (WT) systems.

REFERENCES

[1] G. A. Kamal, A. A. Maurya, A. Y. Dhawale, A. Mahajan, P. Ghuse and P. Ganvir, "Solar power based inverter," *International Journal of Innovative Research in Technology*, vol. 6, pp. 186–188, April 2020.

[2] A. Harish, A. Ramanan, P. K. Das, G. Giridhar, K. Balaraman and T. Prabu, "Formulation of efficiency of inverters for solar photovoltaic power plants-Indian case study", *International Conference on Power Energy, Environment and Intelligent Control*, Greater Noida, February 2020.

[3] M. Calais, V. G. Agelidis and M. Meinhardt, "Multilevel converters for single phase grid connected photovoltaic systems: An overview," *Solar Energy*, vol. 66, no. 5, pp. 325–335, August 1999.

[4] A. B. Nezhad, A. Namadmalan and A. Rahdarian, "Cascaded H-bridge multilevel inverters with discrete variation of DC sources," *International Journal of Electronics*, vol. 106, no. 10, pp. 1480–1497, 2019.

[5] H. Taghizadeh and M. T. Hagh, "Harmonic elimination of cascaded multilevel inverters with non equal DC sources using PSO," *IEEE Transactions on Industrial Electronics*, vol. 57, no. 11, pp. 3678–3684, November 2010.

[6] M. Marcos, E. R. Cadaval, M. A. G. Martinez and M. I. M. Montero, "Cooperative operation of inverters for grid connected photovoltaic generation systems," *Electric Power Systems Research*, vol. 96, pp. 47–55, March 2013.

[7] en.wikipedia.org/wiki/Photovoltaic_system

[8] M. Islam and S. Mekhilef, "An improved transformerless grid connected photovoltaic inverter with reduced leakage current," *Energy Conversion and Management*, vol. 88, pp. 854–862, December 2014.

[9] D. S. Vanaja, A. A. Stonier, G. Mani and S. Murugesan, "Investigation and validation of solar photovoltaic fed modular multilevel inverter for marine water pumping applications," *Electrical Engineering*, vol. 104, pp. 1163–1178, August 2021.

[10] H. Bouaouaou, D. Lalili and N. Boudjerda, "Model predictive control and ANN based MPPT for a multilevel grid connected photovoltaic inverters," *Electrical Engineering*, vol. 104, pp. 1229–1246, August 2021.

[11] C. Boonmee, Y. Kumsuwan and N. Watjanatepin, "Cascaded H-bridge multilevel inverter for single-phase grid-connected PV system with low power on PV string," *2022 25th International Conference on Electrical Machines and Systems (ICEMS)*, Chiang Mai, Thailand, 2022, pp. 1–4.

[12] K. Geetha, "Implementation of five level multilevel inverter with reduced leakage current," *2022 IEEE International Conference on Distributed Computing and Electrical Circuits and Electronics (ICDCECE)*, Ballari, India, 2022, pp. 1–6.

[13] A. Mittal, K. Janardhan and A. Ojha, "Multilevel inverter based grid connected solar photovoltaic system with power flow control," *2021 International Conference on Sustainable Energy and Future Electric Transportation (SEFET)*, Hyderabad, India, 2021.

[14] T. Rampradesh, B. Elangovan and G. Anbumani, "Design and implementation of a novel Z source multilevel inverter for photovoltaic applications," *2022 Second International Conference on Advances in Electrical, Computing, Communication and Sustainable Technologies (ICAECT)*, Bhilai, India, 2022.

[15] M. P., K. Annamalai, S. Dhara and V. T. Somasekhar, "A quasi-z-source-based space-vector-modulated cascaded four-level inverter for photovoltaic applications," *IEEE Journal of Emerging and Selected Topics in Power Electronics*, vol. 10, no. 4, pp. 4749–4762, August 2022.

[16] S. Jakkula, N. Jayaram, S. V. K. Pulavarthi, Y. R. Shankar and J. Rajesh, "A generalized high gain multilevel inverter for small scale solar photovoltaic applications," *IEEE Access*, vol. 10, pp. 25175–25189, 2022.

[17] F. B. Grigoletto, "Multilevel common-ground transformerless inverter for photovoltaic applications," *IEEE Journal of Emerging and Selected Topics in Power Electronics*, vol. 9, no. 1, pp. 831–842, February 2021.

[18] M. Sarebanzadeh, M. A. Hosseinzadeh, C. Garcia, E. Babaei, S. Islam and J. Rodriguez, "Reduced switch multilevel inverter topologies for renewable energy sources," *IEEE Access*, vol. 9, pp. 120580–120595, 2021.

[19] S. R. Khasim, D. C, S. Padmanaban, J. B. Holm-Nielsen and M. Mitolo, "A novel asymmetrical 21-level inverter for solar PV energy system with reduced switch count," *IEEE Access*, vol. 9, pp. 11761–11775, 2021.

[20] H. K. Jahan and M. Abapour, "Switched-capacitor-based multilevel inverter for grid-connected photovoltaic application," *IEEE Transactions on Power Electronics*, vol. 36, no. 9, pp. 10317–10329, September 2021.

[21] M. A. Rezaei, M. Nayeripour, J. Hu, S. S. Band, A. Mosavi and M.-H. Khooban, "A new hybrid cascaded switched-capacitor reduced switch multilevel inverter for renewable sources and domestic loads," *IEEE Access*, vol. 10, pp. 14157–14183, 2022.

[22] N. M. Chakkamath Mukundan, V. Kallaveetil, S. S. Kumar and J. Pychadathil, "An improved H-bridge multilevel inverter-based multiobjective photovoltaic power conversion system," *IEEE Transactions on Industry Applications*, vol. 57, no. 6, pp. 6339–6349, November–December 2021.

[23] M. D. Siddique et al., "Single-phase boost switched-capacitor-based multilevel inverter topology with reduced switching devices," *IEEE Journal of Emerging and Selected Topics in Power Electronics*, vol. 10, no. 4, pp. 4336–4346, August 2022.

[24] P. M. Lingom, J. Song-Manguelle, D. L. Mon-Nzongo, R. C. C. Flesch and T. Jin, "Analysis and control of PV cascaded H-bridge multilevel inverter with failed cells and changing meteorological conditions," *IEEE Transactions on Power Electronics*, vol. 36, no. 2, pp. 1777–1789, February 2021.

[25] S. Padmanaban, C. Dhanamjayulu and B. Khan, "Artificial neural network and Newton raphson (ANN-NR) algorithm based selective harmonic elimination in cascaded multilevel inverter for PV applications," *IEEE Access*, vol. 9, pp. 75058–75070, 2021.

[26] M. J. Sathik, N. Sandeep, D. J. Almakhles and U. R. Yaragatti, "A five-level boosting inverter for PV application," *IEEE Journal of Emerging and Selected Topics in Power Electronics*, vol. 9, no. 4, pp. 5016–5025, August 2021.

[27] M. A. Hosseinzadeh, M. Sarebanzadeh, E. Babaei, M. Rivera and P. Wheeler, "A switched-DC source sub-module multilevel inverter topology for renewable energy source applications," *IEEE Access*, vol. 9, pp. 135964–135982, 2021.

[28] K. Kalyan, A. O. Sir and R. S. Sir, "Applications of multilevel inverter for grid integration of renewable energy sources," *2021 7th International Conference on Electrical Energy Systems (ICEES)*, Chennai, India, 2021.

[29] M. S. O. Yeganeh, M. Rahmani, N. Mijatovic, T. Dragicevic, F. Blaabjerg and P. Davari, "Voltage control scheme for multilevel interfacing PV application: Real-time MRAC-based approach," *2022 24th European Conference on Power Electronics and Applications (EPE'22 ECCE Europe)*, Hanover, Germany, 2022.

[30] P. K. Chamarthi, A. Al-Durra, T. H. M. EL-Fouly and K. A. Jaafari, "A novel three-phase transformerless cascaded multilevel inverter topology for grid-connected solar PV applications," *IEEE Transactions on Industry Applications*, vol. 57, no. 3, pp. 2285–2297, May–June 2021.

14 Design and Analysis of Different Levels of Modular Multilevel Converter

Parishrut Singh Charak, Dr Parul Gaur

14.1 INTRODUCTION

Multilevel converter topologies have grown in popularity and are gradually replacing two-level converters due to their significant benefits in terms of reducing the size of harmonic filters, increasing power capability, and reducing component voltage stress [1]. Professor Marquardt first proposed the modular multilevel converter (MMC) in a German patent in 2001 [2], which has since been expanded to serve various purposes like medium-voltage motor drives and power quality improvement [3]. MMC has gained significant popularity in high- and medium-voltage applications due to its several advantages. It is scalable and modular and can produce a nearly sinusoidal output voltage [8] and also can eliminate the need for bulky inductors and transformers due to its superior harmonic performance [9]. The popular and traditional topologies of multilevel converters that have been marketed recently, replacing conventional converters in numerous industrial applications, include cascaded full bridge, flying capacitor (FC), neutral point clamped (NPC), and active neutral point clamped (ANPC) [4], [5]. Recently, multilevel converters have also been employed in medium-voltage and high-voltage applications, including flexible AC transmission systems, high-voltage direct current (HVDC), solar and wind power plants connected to the main grid, and medium-voltage drives[6], [7].

The MMC circuit topology is depicted in Figure 14.1. Power electronic converters that convert direct current (DC) power supplies to alternating current (AC) waveforms are known as inverters. The output of the inverter is determined by the type of inverter used. Inverters can be a sine wave, square wave, or quasi-square wave. Even when a sine-wave inverter is used, the output is not purely sinusoidal because it contains harmonics. These harmonics are periodic wave components that are multiples of the fundamental frequency and cause distortion in the output [8]. These harmonic distortions can be reduced by using the appropriate modulation technique or converter topology. MMCs have a large number of levels that are more efficient and have lower harmonic distortions [8], [9]. They are capable of handling high-voltage operations without the use of series-connecting switching devices. They offer lower

DOI: 10.1201/9781003367154-14

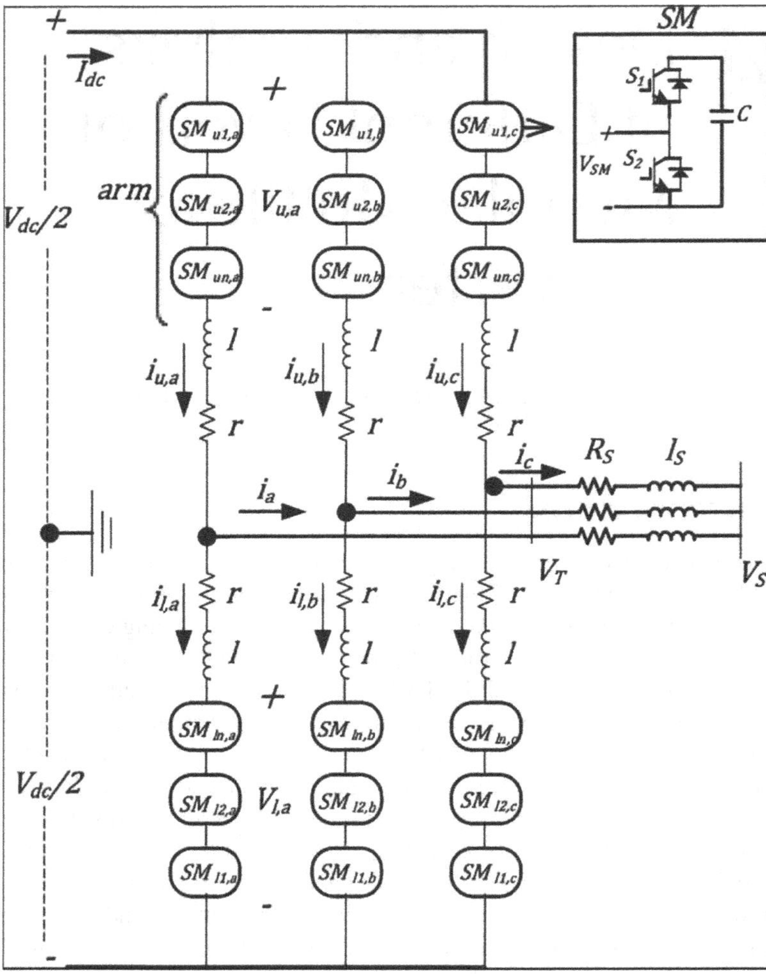

FIGURE 14.1 MMC circuit topology.

common-mode voltages and higher power quality. They have several advantages, including high modularity and scalability, transformer-free operation, lower switching losses, and lower filtering costs [10].

14.2 MODULAR MULTILEVEL CONVERTER

MMC includes legs that include both the upper and lower arm. Each arm contains submodules (SMs), which are the basic building blocks of MMC, shown in Figure 14.2. This fundamental building piece is where the modularity of the MMC derives from. Each SM features a well-known half-bridge construction made up of two power semiconductors, IGBT, and one SM capacitor.

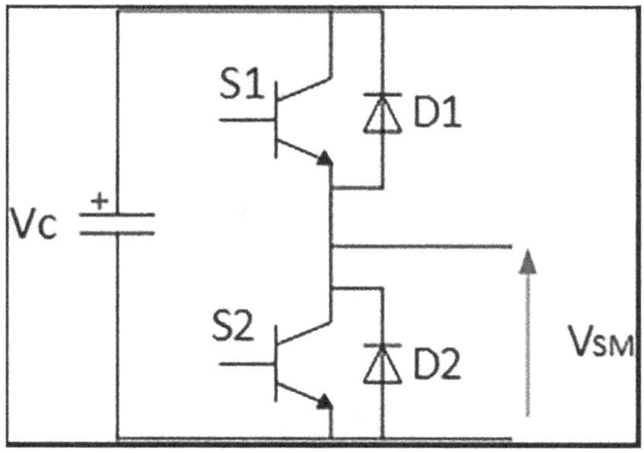

FIGURE 14.2 Half-bridge submodule structure.

The capacitor serves as the two-level converter's DC-link capacitor, although it is uniformly distributed throughout the MMC. It functions as an energy buffer and voltage source. According to PWM commands, the half-bridge IGBTs cut the capacitor voltage. The SM's transistors T1 and T2 operate in opposition to one another. To avoid shoot-through, T2 must be switched off while T1 is on. The capacitor voltage is seen at the SM terminals when T1 is turned on. The SM is referred to as being "inserted" in this situation. If not, it is considered to have been "bypassed."

14.3 MMC TOPOLOGIES

The number of SMs can be decreased or increased to get the required output voltage. Various SM configurations are proposed in the literature [11]. The topology of each switching SM contains a half-bridge (HB) converter, full-bridge (FB) converter, unidirectional SM [12], clamped-diode (CD) converter SM, three-level flying capacitor (FC), three-level neutral-point, and five-level cross-connected SM [13].

- **HB or chopper SM**: Figure 14.3a shows a HB structure with insulated-gate bipolar transistors (IGBTs), and every IGBT is connected in parallel with a reversed diode, Co denotes the paralleled capacitance [14].
- **FB converter**: As shown in Figure 14.3b, the full-bridge SM contains four IGBTs, four anti-parallel diodes, and a capacitor. This topology supports both positive and negative voltages.
- **Unidirectional SM converter**: As shown in Figure 14.3c, it is made up of one IGBT switch, which is connected in series with an anti-parallel diode. It reduces the number of semiconductors per module, but the switching states are limited due to the current direction.
- **CD converter SM** [15]: In Figure 14.3d, the normal operation of the CD converter represents two equivalent HB SMs. Both positive and negative terminals have been connected by the HBs. As a result, the insertion and bypass switch positions are switched.

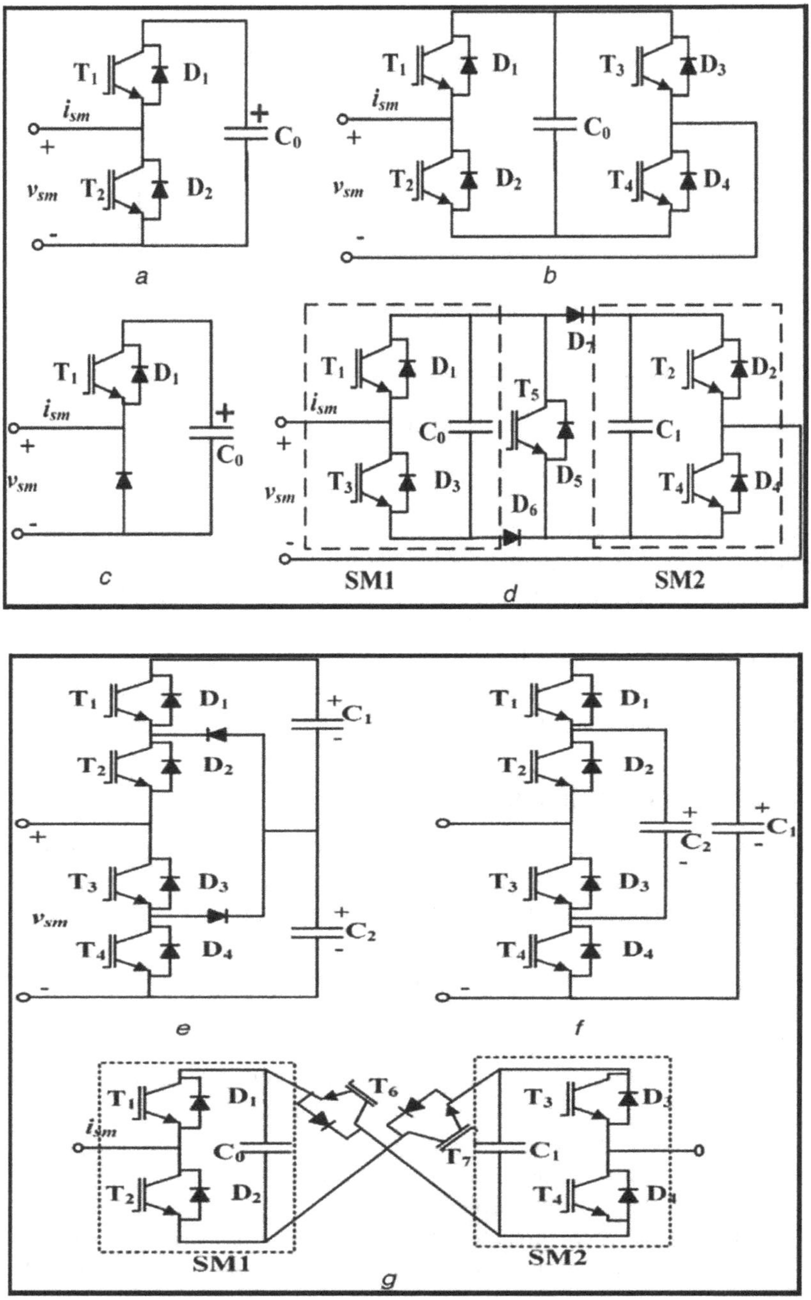

FIGURE 14.3 Switching module topologies: (a) HB or chopper converter, (b) FB converter, (c) unidirectional SM converter, (d) CD converter, (e) three-level FC, (f) three-level NPC, (g) five-level cross-connected SM [11].

- **Three-level FC and NPC** [11]: As shown in Figures 14.3e and 14.3f, it is generally made up of four IGBT switches, four anti-parallel diodes, and two DC capacitor storages, C1 and C2, and it has the same semiconductor losses as the HB module. On the other hand, the three-level NPC has higher semiconductor losses than the HB module but lower than the FB module.
- **Five-level cross-connected circuit** [11]: Figure 14.3g depicts a five-level cross-connected SM made up of two HB SMs connected back-to-back by two extra IGBTs with anti-parallel diodes. It has the same semiconductor losses as the clamp-double SM.

14.4 MODULATION TECHNIQUE

The nearest-level control (NLC), also known as the round method, uses the nearest voltage level that can be generated by converting to the desired output voltage reference [16]. The three phases are controlled separately using an independent comparison process. Figure 14.4 shows the block diagram and working of nearest-level modulation (NLM). The block diagram shows that at first the reference voltage is normalized with the capacitor voltage of the SM in the gain block, then the round function generates the closest integer number of the SMs to insert to approximate the reference voltage with the nearest voltage level [17].

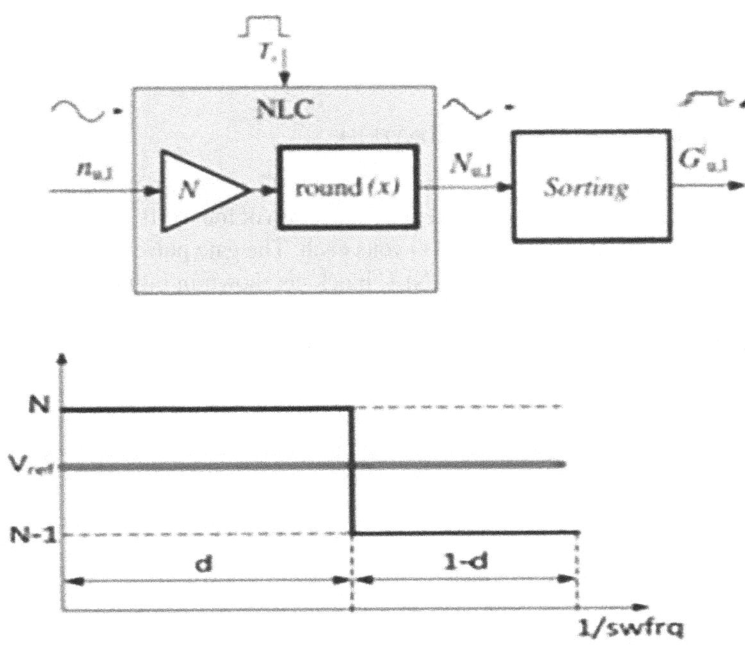

FIGURE 14.4 Block diagram and working principle of NLM.

The numbers of conducted SMs of upper- and lower-bridge arm n_p and n_n are calculated as shown in equation 14.1[18]:

$$\begin{cases} n_p = \dfrac{N}{2} - \left[\dfrac{u_m}{U_c}\right] \\ n_n = \dfrac{N}{2} + \left[\dfrac{u_m}{U_c}\right] \end{cases} \tag{14.1}$$

Here u_m is the instantaneous value of the modulation sine voltage. By comparing the modulation sinusoidal voltage um of the modulation module with the capacitance-voltage Uc of the SM, it is possible to determine the number of SMs np and nn that must be conducted by the upper and lower bridge arms in this modulation method. The capacitor voltage balance link is then input, which can be accomplished in two steps. The charge and discharge of the block molecular module, as well as the polarity of the bridge arm current, are determined in the first stage. The second step is to order the bridge arm SM's capacitance-voltage value for each sample period. To ensure that the capacitance-voltage values of each SM are similar and the effect of voltage sharing is realized, the SM with the lowest capacitance voltage is charged first during charging, and the SM with the highest capacitance voltage is charged first during discharging [18].

14.5 SIMULATION AND RESULTS

Five different levels of MMC have been simulated using MATLAB/SIMULINK R2021a to verify its performance with RL-load. In all five cases of MATLAB/SIMULINK models, the NLM is given to the gate pin of the IGBT. The simulation specification of parameters is given in Table 14.1.

14.5.1 Simulation of 5-Level Converter

In the Simulink model of five-level three-phase MMC, as shown in Figure 14.5, each arm of the converter leg consists of a series connection of four SMs, as shown in Figure 14.6. It has two voltage sources of 2,000 volts each. The gate pulses and the switching sequence is implemented through the NLC block, as shown in Figure 14.7. Now from the output voltage waveform, as shown in Figure 14.8, five different levels of voltages, 0 V, +1,000 V, +2,000 V, –1,000 V, and –2000 V, can be observed. Total harmonic

TABLE 14.1
Specification of Simulation Parameters

Simulation Step Size	Ts = $4e^{-2}$ s
Supply Voltage	V = 2,000 V
Fundamental Frequency	f = 50 Hz
Submodule Capacitor	C = 1 pF
R-L Load	R = 50Ω, L = 10^{e-3} H
Switching Element	IGBT

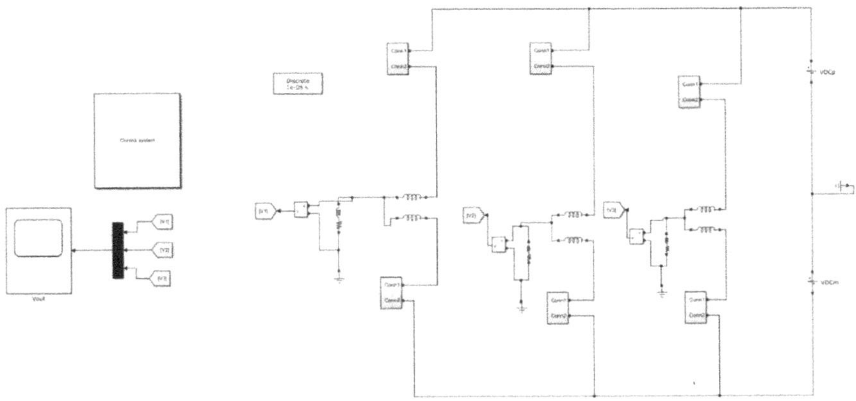

FIGURE 14.5 MATLAB/SIMULINK model of three-phase five-level MMC.

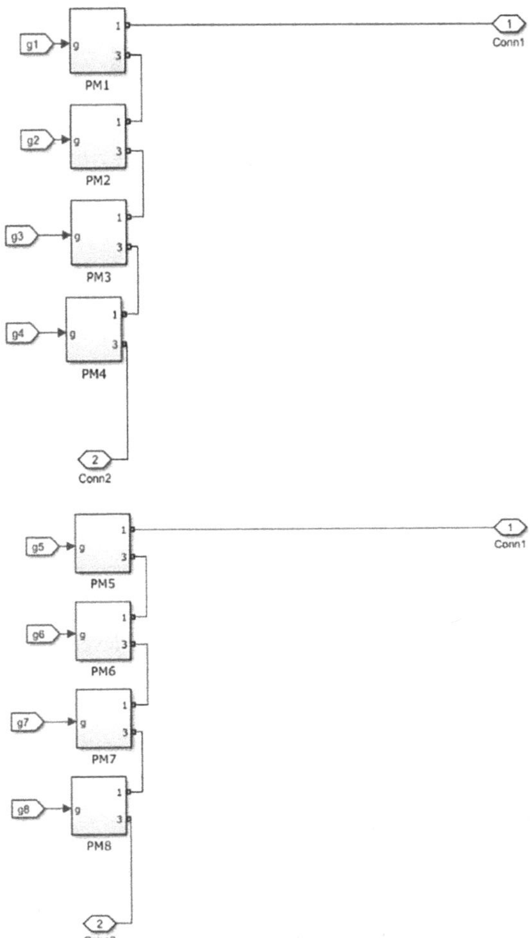

FIGURE 14.6 Upper- and lower-arm submodules of one phase of five-level MMC.

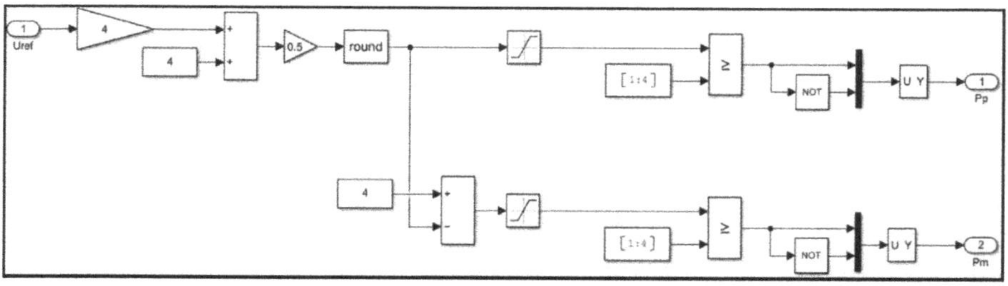

FIGURE 14.7 Nearest-level control for five-level MMC in Simulink.

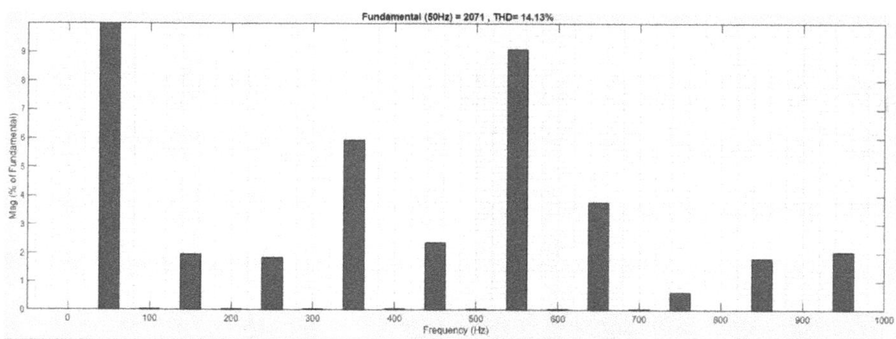

FIGURE 14.8 The output voltage waveform of five levels.

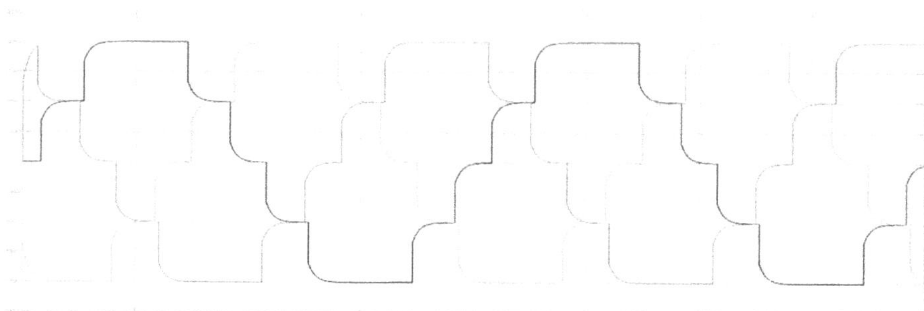

FIGURE 14.9 Harmonic spectrum for phase voltage of five-level MMC.

distortion (THD) analysis is performed from the fast Fourier transform (FFT) window, and the findings are displayed in the harmonic spectrum, as shown in Figure 14.9.

14.5.2 Simulation of 7-Level Converter

In the Simulink model of seven-level three-phase MMC, as shown in Figure 14.10, each arm of the converter leg consists of a series connection of six SMs, as shown in Figure 14.11. It has two voltage sources of 2,000 volts each. The gate pulses and the switching sequence is implemented through the NLC block, as shown in Figure 14.12. Now from the output voltage waveform, as shown in Figure 14.13, seven

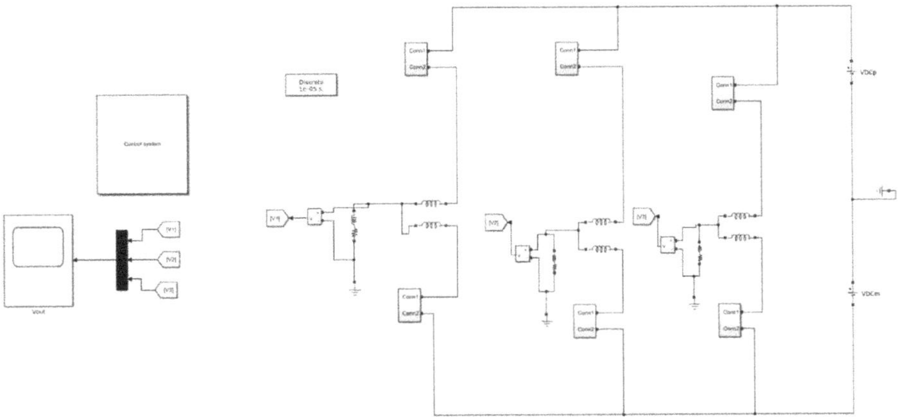

FIGURE 14.10 MATLAB/SIMULINK model of three-phase seven-level MMC.

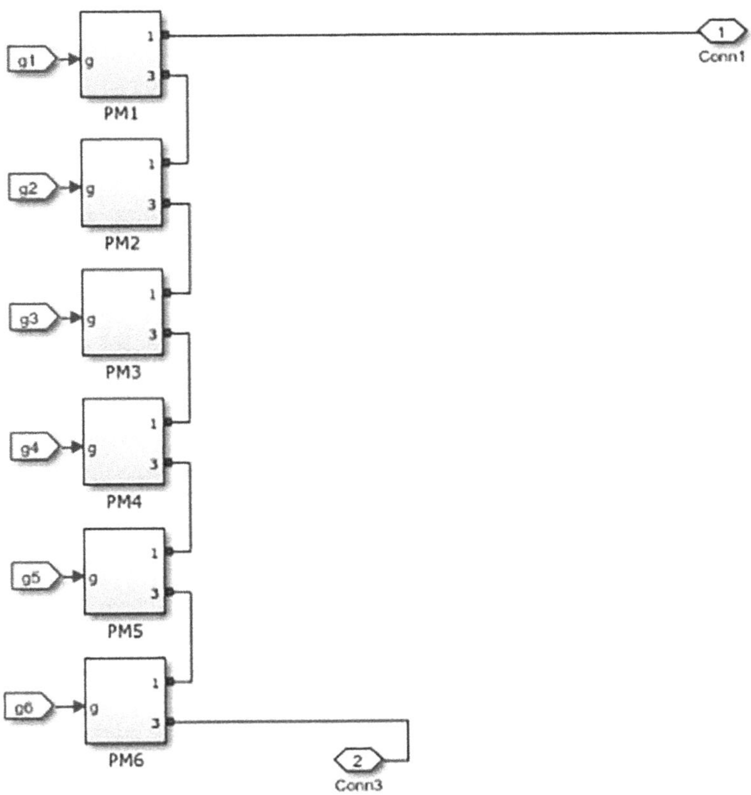

FIGURE 14.11 Upper- and lower-arm submodules of one phase of seven-level MMC.

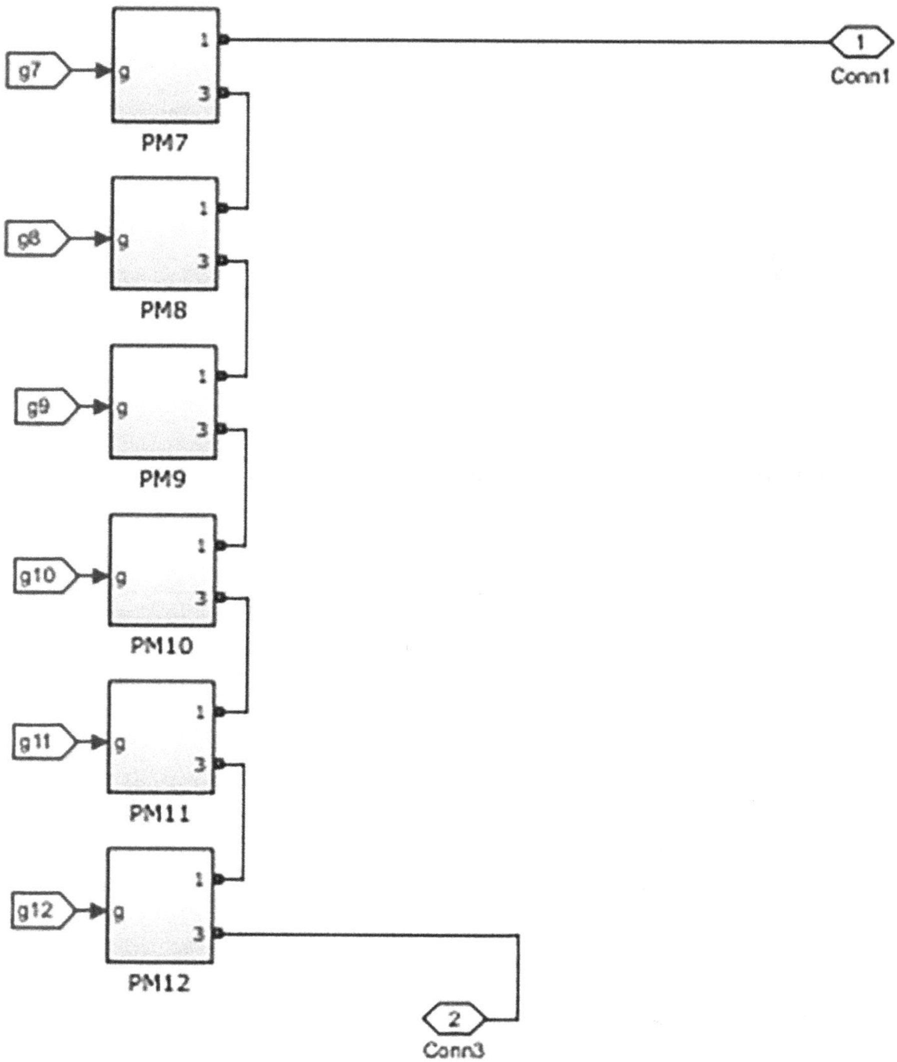

FIGURE 14.11 (Continued)

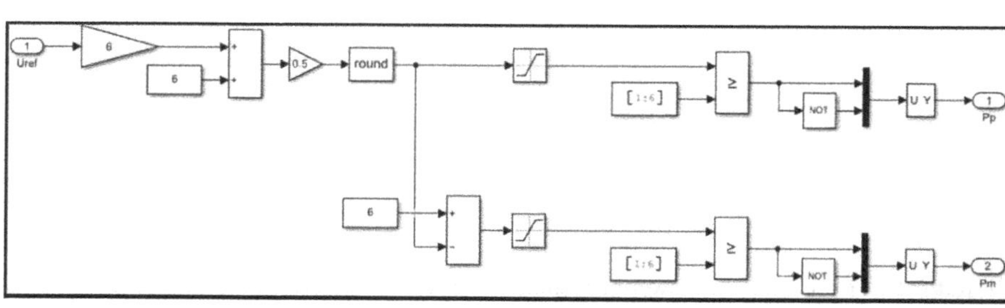

FIGURE 14.12 Nearest-level control of one phase for seven-level MMC in Simulink.

FIGURE 14.13 Three-phase output voltage for seven-level MMC.

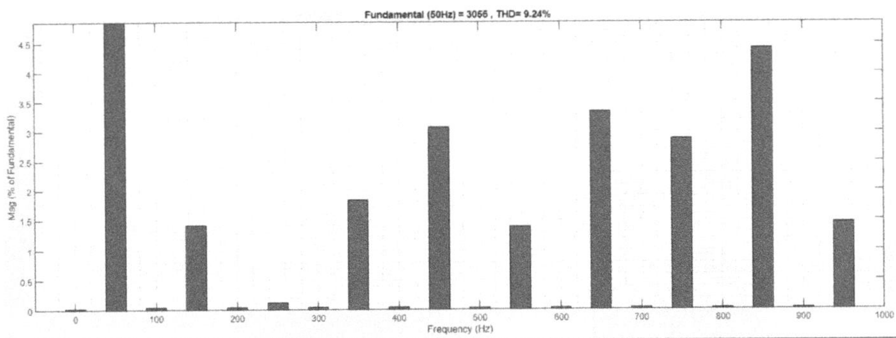

FIGURE 14.14 Harmonic spectrum for phase voltage of seven-level MMC.

different levels of voltages, 0 V, +1,000 V, +2,000 V, +3,000 V, −1,000 V, −2,000 V, and −3,000 V, can be observed THD analysis is performed from the FFT window, and the findings are displayed in the harmonic spectrum, as shown in Figure 14.14.

14.5.3 Simulation of 9-Level Converter

In the Simulink model of nine-level three-phase MMC, as shown in Figure 14.15, each arm of the converter leg consists of a series connection of eight SMs, as shown in Figure 14.16. It has two voltage sources of 2,000 volts each. The gate pulses and the switching sequence is implemented through the NLC block, as shown in Figure 14.17. Now from the output voltage waveform, as shown in Figure 14.18, nine different levels of voltages, 0 V, +1,000 V, +2,000 V, +3,000 V, +4,000 V, −1,000 V, −2,000 V, −3,000 V, and −4,000 V, can be observed. THD analysis is performed from the FFT window, and the findings are displayed in the harmonic spectrum, as shown in Figure 14.19.

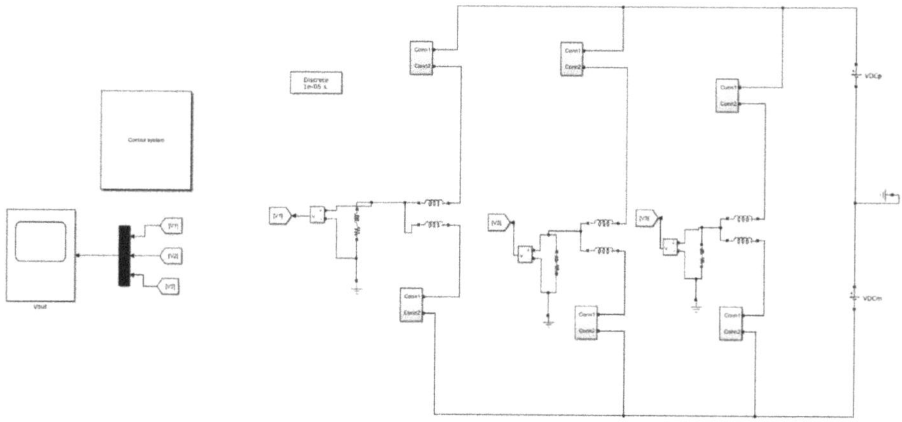

FIGURE 14.15 MATLAB/SIMULINK model of three-phase nine-level MMC.

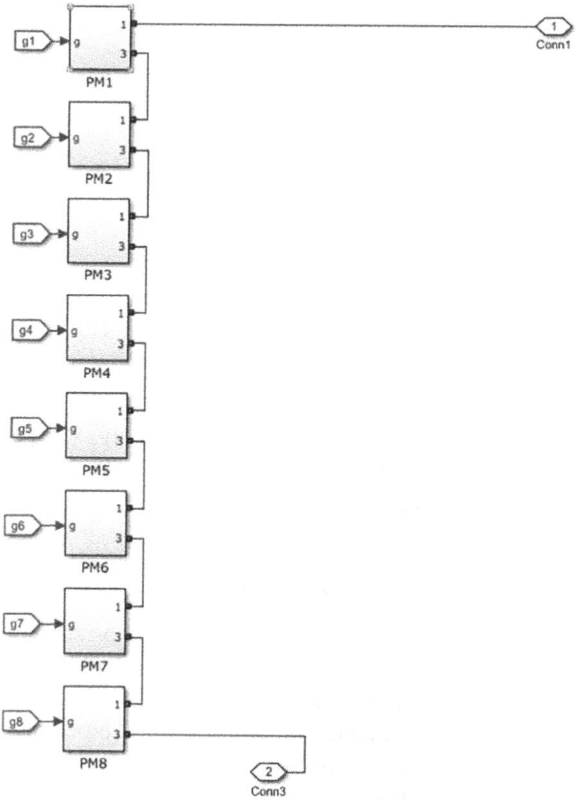

FIGURE 14.16 Upper- and lower-arm submodules of one phase of nine-level MMC.

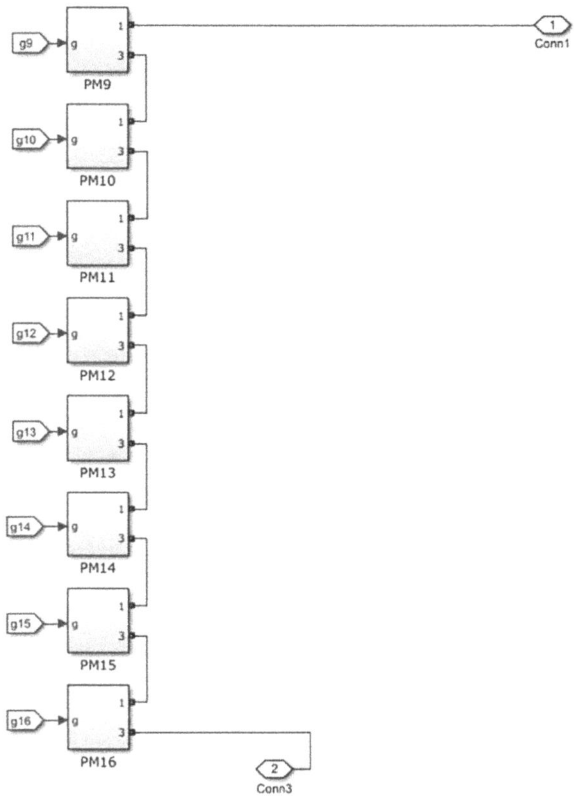

FIGURE 14.16 (Continued)

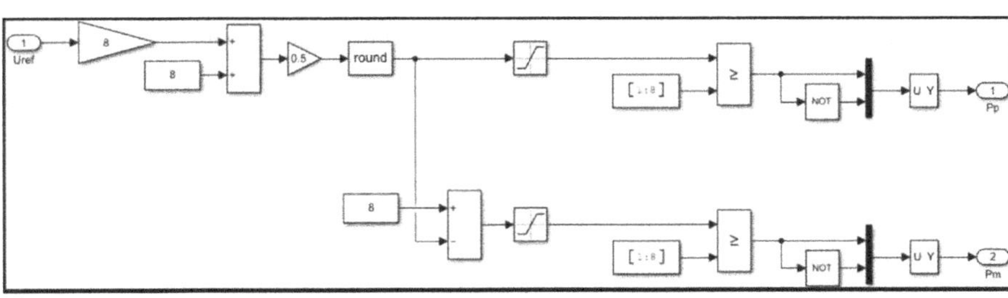

FIGURE 14.17 Nearest-level control of one phase for nine-level MMC in Simulink.

14.5.4 Simulation of 11-Level Converter

In the Simulink model of 11-level three-phase MMC, as shown in Figure 14.20, each arm of the converter leg consists of a series connection of ten SMs, as shown in Figure 14.21. It has two voltage sources of 2,000 volts each. The gate pulses and the switching sequence is implemented through the NLC block, as shown in Figure 14.22. Now

FIGURE 14.18 Three-phase output voltage for nine-level MMC.

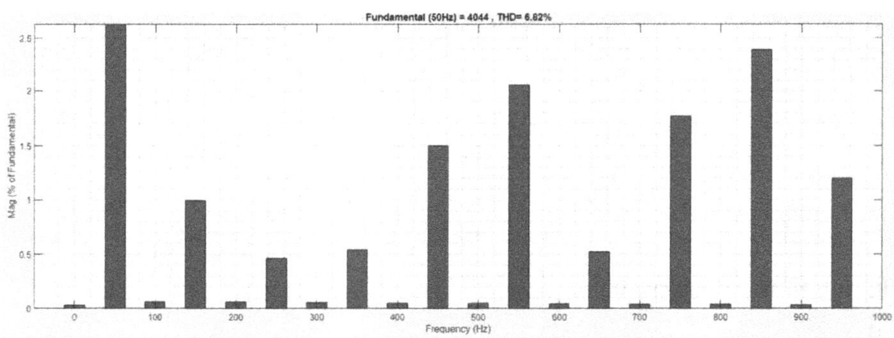

FIGURE 14.19 Harmonic spectrum for phase voltage of nine-level MMC.

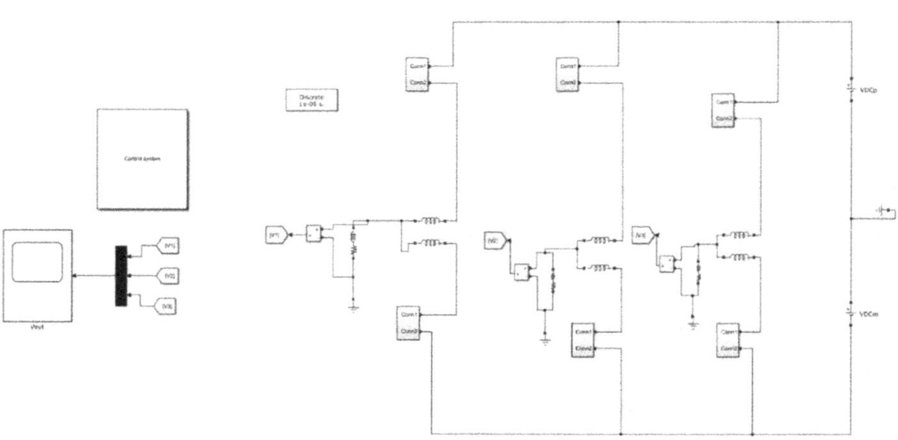

FIGURE 14.20 MATLAB/SIMULINK model of three-phase eleven-level MMC.

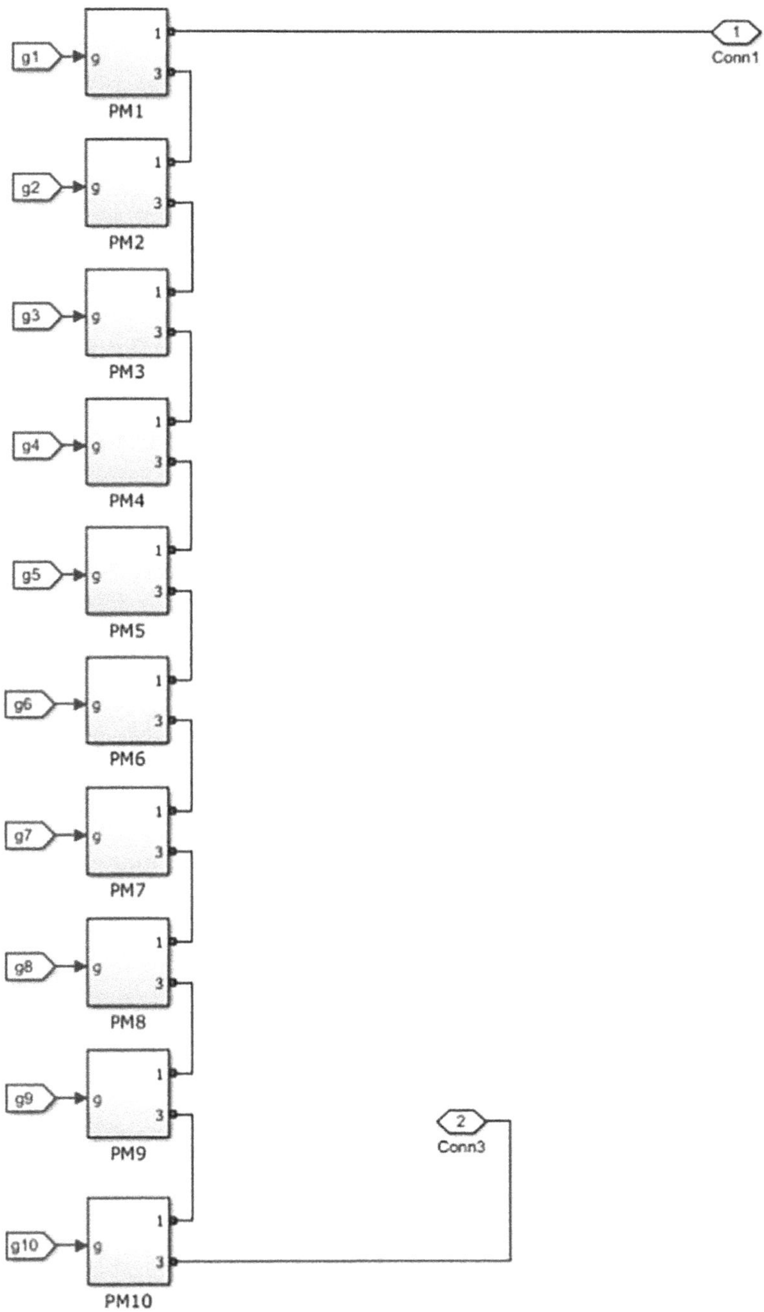

FIGURE 14.21 Upper- and lower-arm submodules of one phase of 11-level MMC.

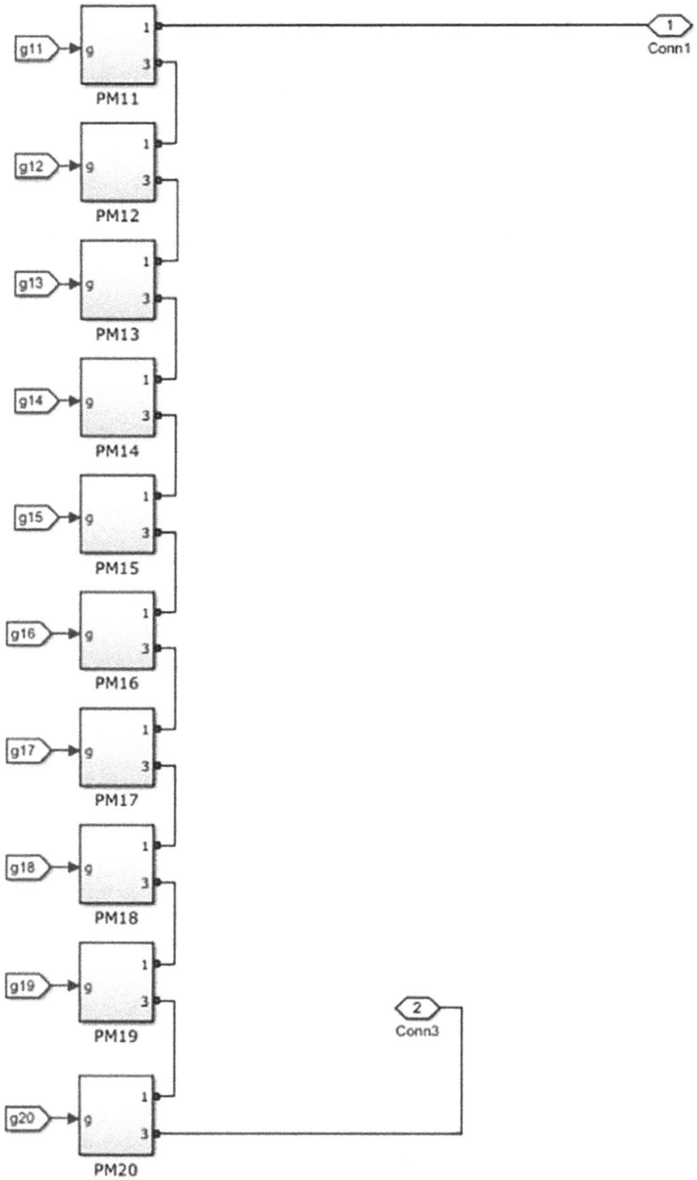

FIGURE 14.21 (Continued)

from the output voltage waveform, as shown in Figure 14.23, 11 different levels of voltages, 0 V, +1,000 V, +2,000 V, +3,000 V, +4,000 V, +5,000 V, –1,000 V, –2,000 V, –3,000 V, –4,000 V, and –5,000 V, can be observed. THD analysis is performed from the FFT window, and the findings are displayed in the harmonic spectrum, as shown in Figure 14.24.

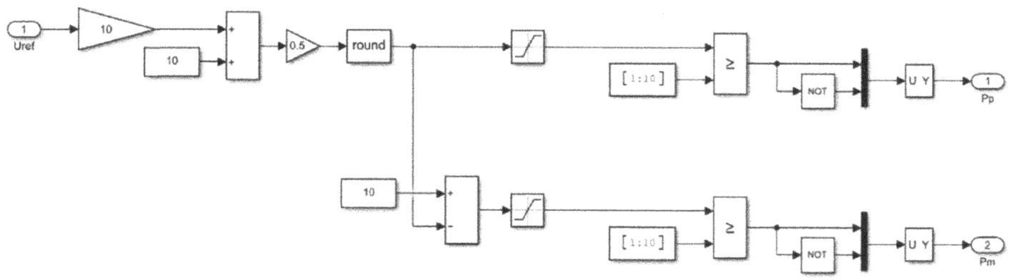

FIGURE 14.22 Nearest-level control of one phase for 11-level MMC in Simulink.

FIGURE 14.23 Three-phase output voltage for 11-level MMC.

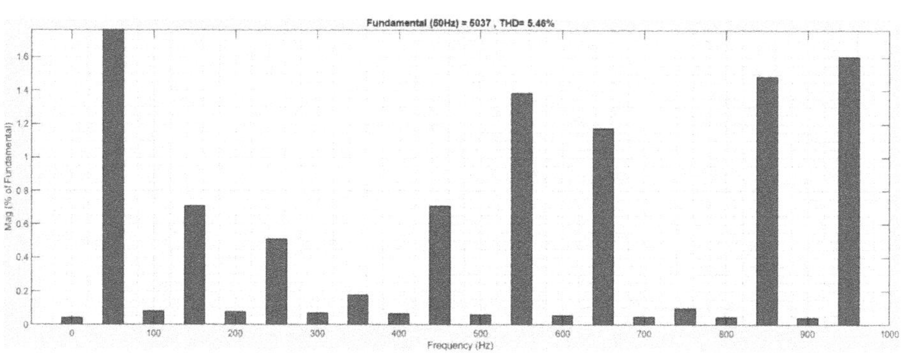

FIGURE 14.24 Harmonic spectrum for phase voltage of 11-level MMC.

14.5.5 Simulation of 13-Level Converter

In the Simulink model of 13-level three-phase MMC, as shown in Figure 14.25, each arm of the converter leg consists of a series connection of 12 SMs, as shown in Figure 14.26. It has two voltage sources of 2,000 volts each. The gate pulses and

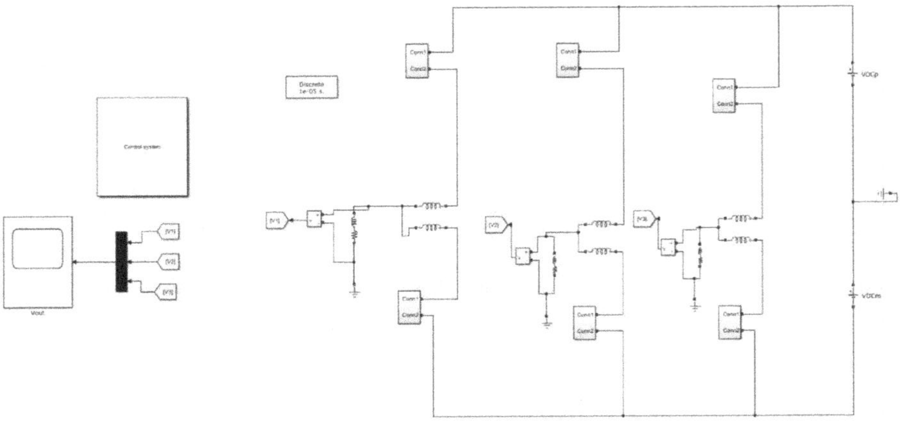

FIGURE 14.25 MATLAB/SIMULINK model of three-phase 13-level MMC.

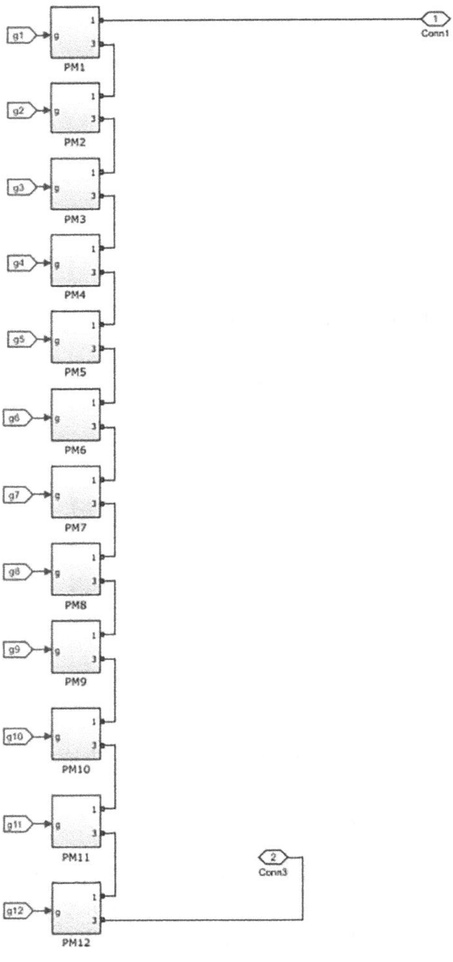

FIGURE 14.26 Upper- and lower-arm submodules of one phase of 13-level MMC.

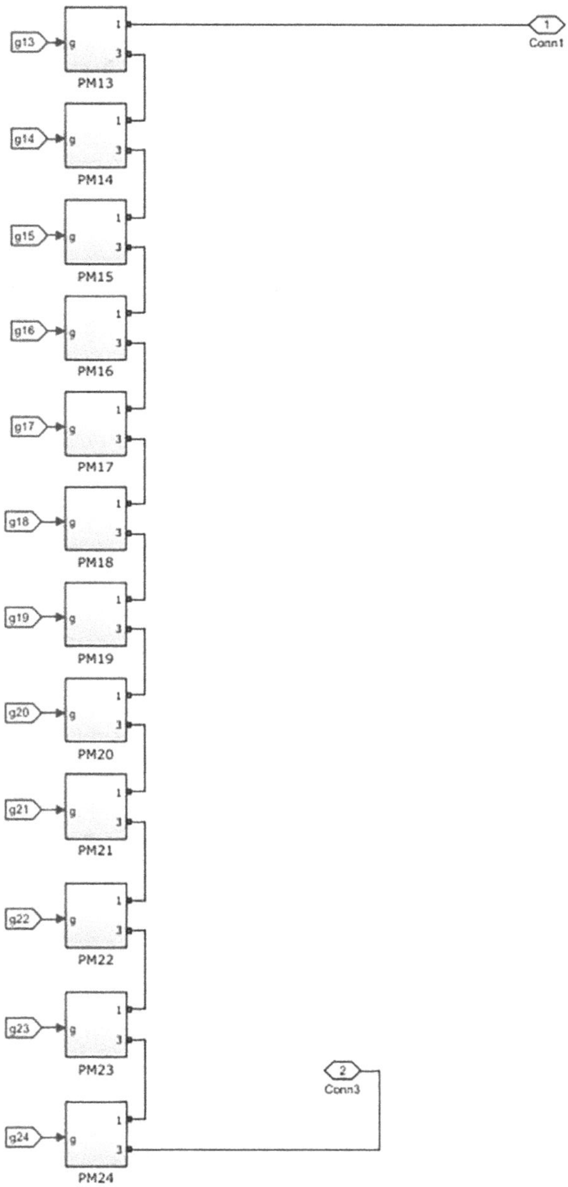

FIGURE 14.26 (Continued)

the switching sequence is implemented through the NLC block, as shown in Figure 14.27. Now from the output voltage waveform, as shown in Figure 14.28, 13 different levels of voltages, 0 V, +1,000 V, +2,000 V, +3,000 V, +4,000 V, +5,000 V, +6,000 V, −1,000 V, −2,000 V, −3,000 V, −4,000 V, −5,000 V, and −6,000 V, can be observed. THD analysis is performed from the FFT window, and the findings are displayed in the harmonic spectrum, as shown in Figure 14.29. The graph between actual THD

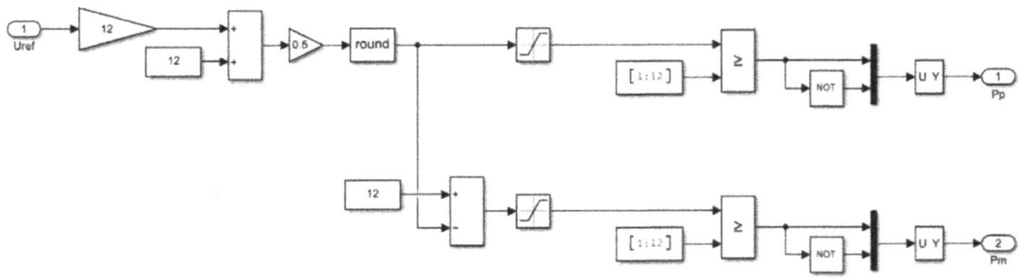

FIGURE 14.27 Nearest-level control of one phase for 13-level MMC in Simulink.

FIGURE 14.28 Three-phase output voltage for 13-level MMC.

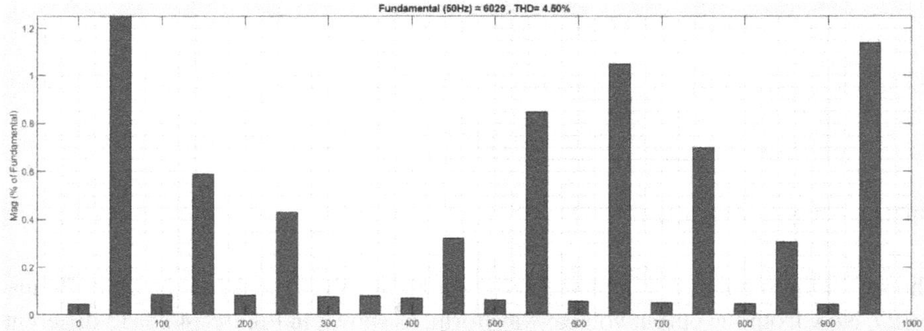

FIGURE 14.29 Harmonic spectrum for phase voltage of 13-level MMC

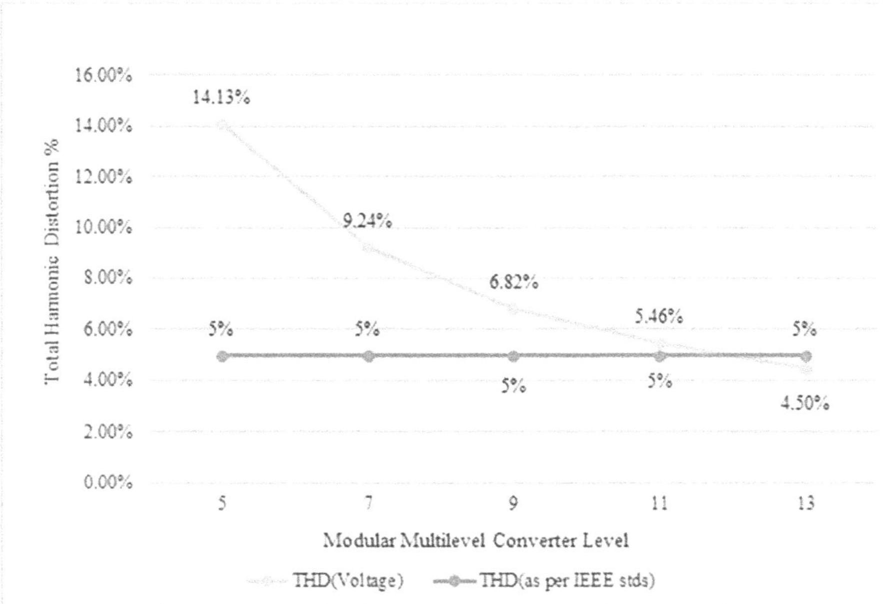

FIGURE 14.30 The graph between actual THD vs. obtained THD for different levels of MMC.

TABLE 14.2
Total Harmonic Distortion (THD) Produced by Different Levels of Modular Multilevel Converters

Level	THD (Voltage)	% Error (w.r.t. THD Value per IEEE Standards)
5	14.13%	−1.826
7	9.24%	−0.848
9	6.82%	−0.364
11	5.46%	−0.092
13	4.50%	+0.1

and obtained THD for different levels of MMC is shown in Figure 14.30. Moreover Table 14.2 lists the Total Harmonic Distortion (THD) Produced by Different Levels of Modular Multilevel Converters to show the percentage error in obtained values.

14.6 CONCLUSION

It is concluded from obtained simulation results that a MMC with H-bridge SM configuration requires at least 12 SMs per arm to produce THD within the IEEE standards (less than 5%) when it is used with NLM control technique. It is further analysed that MMC, through the NLM technique, is found better in achieving very minimum THD as compared to the conventional converters. However, in this work, the THD is reduced by 68% approximately with increased levels from 5 to 13, which results in mitigation of harmonics distortion and improved voltage waveform at the output side. This also fulfils IEEE 519 standard regarding harmonics Voltage limits. Also, it leads to applications where the proposed MMC-based HVDC structure can be used to replace conventional power networks, because of its capability to produce higher voltage levels.

REFERENCES

[1] S. Kouro *et al.*, "Recent Advances and Industrial Applications of Multilevel Converters," *IEEE Transactions on Industrial Electronics*, vol. 57, no. 8, pp. 2553–2580, Aug. 2010, doi:10.1109/TIE.2010.2049719.

[2] R. Marquardt, R. Marquardt, A. Lesnicar, and J. Hildinger, "Modulares Stromrichterkonzept für Netzkupplungsanwendung bei hohen Spannungen," *Proceedings of the ETG-Fachtagung*, Bad Nauheim, Germany, 2002.

[3] A. Lesnicar, R. Marquardt, A. Lesnicar, and R. Marquardt, "An Innovative Modular Multilevel Converter Topology Suitable for a Wide Power Range," *Power Tech Conference Proceedings, 2003 IEEE Bologna*, vol. 3, pp. 23–26, June 2003.

[4] S. Arazm and K. Al-Haddad, "ZPUC: A New Configuration of Single DC Source for Modular Multilevel Converter Applications," *IEEE Open Journal of the Industrial Electronics Society*, vol. 1, no. 1, pp. 97–113, 2020, doi:10.1109/OJIES.2020.2998694.

[5] J. Ebrahimi and H. R. Karshenas, "A New Reduced-Component Hybrid Flying Capacitor Multicell Converter," *IEEE Transactions on Industrial Electronics*, vol. 64, no. 2, pp. 912–921, Feb. 2017, doi:10.1109/TIE.2016.2618876.

[6] J. Qin and M. Saeedifard, "Reduced switching-frequency voltage-balancing strategies for modular multilevel hvdc converters," *IEEE Transactions on Power Delivery*, vol. 28, no. 4, pp. 2403–2410, 2013, doi:10.1109/TPWRD.2013.2271615.

[7] M. Sleiman, K. Al-Haddad, H. F. Blanchette, and H. Y. Kanaan, "Insertion Index Generation Method Using Available Leg-Average Voltage to Control Modular Multilevel Converters," *IEEE Transactions on Industrial Electronics*, vol. 65, no. 8, pp. 6206–6216, Aug. 2018, doi:10.1109/TIE.2017.2784408.

[8] M. Kumar, Z. A. Memon, M. A. Uqaili, and M. H. Baloch, "An Overview of Uninterruptible Power Supply System with Total Harmonic Analysis & Mitigation: An Experimental Investigation for Renewable Energy Applications," *IJCSNS International Journal of Computer Science and Network Security*, vol. 18, no. No. 6, Jun. 2018.

[9] D. Wu and L. Peng, "Characteristics of Nearest Level Modulation Method with Circulating Current Control for Modular Multilevel Converter," *IET Power Electronics*, vol. 9, no. 2, pp. 155–164, Feb. 2016, doi:10.1049/iet-pel.2015.0504.

[10] D. M. Soomro, S. K. Alswed, M. N. Abdullah, N. H. M. Radzi, and M. H. Baloch, "Optimal Design of a Single-Phase APF Based on PQ Theory," *International Journal of Power Electronics and Drive Systems (IJPEDS)*, vol. 11, no. 3, p. 1360, Sep. 2020, doi:10.11591/ijpeds.v11.i3.pp1360-1367.

[11] M. Priya, P. Ponnambalam, and K. Muralikumar, "Modular-Multilevel Converter Topologies and Applications—A Review," *IET Power Electronics*, vol. 12, no. 2, pp. 170–183, Feb. 2019, doi:10.1049/iet-pel.2018.5301.

[12] M. A. Perez, S. Bernet, J. Rodriguez, S. Kouro, and R. Lizana, "Circuit Topologies, Modeling, Control Schemes, and Applications of Modular Multilevel Converters," *IEEE Transactions on Power Electronics*, vol. 30, no. 1, pp. 4–17, Jan. 2015, doi:10.1109/TPEL.2014.2310127.

[13] S. Debnath, J. Qin, B. Bahrani, M. Saeedifard, and P. Barbosa, "Operation, Control, and Applications of the Modular Multilevel Converter: A Review," *IEEE Transactions on Power Electronics*, vol. 30, no. 1, pp. 37–53, Jan. 2015, doi:10.1109/TPEL.2014.2309937.

[14] Y. Wang, Z. Yuan, and J. Fu, "A Novel Strategy on Smooth Connection of an Offline MMC Station Into MTDC Systems," *IEEE Transactions on Power Delivery*, vol. 31, no. 2, pp. 568–574, Apr. 2016, doi:10.1109/TPWRD.2015.2437393.

[15] S. Debnath, J. Qin, B. Bahrani, M. Saeedifard, and P. Barbosa, "Operation, Control, and Applications of the Modular Multilevel Converter: A Review," *IEEE Transactions on Power Electronics*, vol. 30, no. 1, pp. 37–53, Jan. 2015, doi:10.1109/TPEL.2014.2309937.

[16] Gum Tae Son *et al.*, "Design and Control of a Modular Multilevel HVDC Converter with Redundant Power Modules for Noninterruptible Energy Transfer," *IEEE Transactions on Power Delivery*, vol. 27, no. 3, pp. 1611–1619, Jul. 2012, doi:10.1109/TPWRD.2012.2190530.

[17] S. Ali, Z. Ling, K. Tian, and Z. Huang, "Recent Advancements in Submodule Topologies and Applications of MMC," *IEEE Journal of Emerging and Selected Topics in Power Electronics*, vol. 9, no. 3, pp. 3407–3435, Jun. 2021, doi:10.1109/JESTPE.2020.2990689.

[18] J. Lu, Z. He, W. Xu, S. Liu, H. Pang, and X. Zhou, "Comparison Analysis of the Flexible Sub-module voltage modulation and NLM for MMC-DC grid," in *2019 4th IEEE Workshop on the Electronic Grid (eGRID)*, Nov. 2019, pp. 1–5, doi:10.1109/eGRID48402.2019.9092632.

15 Effect of Locational Variation in Mechanical Properties of a Bone in Fabricating Anatomical Locational Bio-Implants

Sachin Kalsi, Jagjit Singh, N. K. Sharma

15.1 INTRODUCTION

Bone is a composite material with mechanical, chemical, composition, and biological properties varying per anatomical location and orientation of lamellae. There also found a variation in the properties due to geographical conditions, species, genetic factors, and so on. In addition, the bone structure is found to be heterogeneous and hierarchical with variation of properties from nano scale to macro scale. The mechanical properties of a bone, including stiffness, toughness, elasticity, plasticity, modulus of resilience, and so on, vary, which is required for scientific purposes. For designing and fabricating the bone bio-implants, planning treatment strategies to heal bone fractures requires evaluating the change in mechanical properties of a bone. From these aspects, the assessment of the fracture and mechanical properties of bone has become significant in many biological and bioengineering studies for better understanding the degradation of bone quality due to ageing, disease, and therapeutic treatment. These properties provide elaborate how to prevent bone fracture and prepare strategies for bone treatment. Because of the complex structure of bone, it is difficult to duplicate material properties of bone, to build a fine prosthetic bone implant [1]. Significant technical advances have been made in recent years in experimental techniques to investigate the mechanical properties of complex biological materials. In the experimental studies, the mechanical properties [2] [3] [4] [5] of a cortical bone have been evaluated and correlated with the composition [6] [7] [8] [9], microstructure [10] [11], and so on. Some of the studies were performed using specimens from different ages [12] [1], genders [13], species [14], and so on to measure the variation in the mechanical properties of a bone using different techniques and methods. The elastic and plastic properties [15] [16] [17] [18] [19] of a bone have also been evaluated using experimental and numerical studies. Some studies involve fracture mechanics [20] [21] [22], viscoelasticity [23] [24] [25], screw implant interface [26] [27] [28] [29] [30], and implants [31] [32] [33] [34]

DOI: 10.1201/9781003367154-15

using experimental methods. Few of the studies involve developing and proposing FEM models [35] for compressive, tensile, shear, and other properties of a bone. The present study is centred around evaluating the variation in mechanical behaviour of a cortical bone with implant screws in it and the external compressive load is applied on it. The study involves FEM analysis of a cortical bone along with implant screw for three locations of a diaphysis of a bone—upper, middle, and lower. For finding an appropriate bone bio-implant per anatomical location, it is essential to have a thorough understanding of variation in the mechanical behaviour of bone.

15.2 METHODOLOGY

In the present study, the effect of compressive loading on three different locational specimens of a cortical bone along with screw implanted in them were considered to check the variation in mechanical behaviour of mentioned three types of specimens. The three types of specimens include upper, middle, and lower diaphysis. The size of three specimen is 10 × 10 × 4 mm and the implanted screw has a screw shaft length of 8 mm, screw shaft diameter of 2 mm, screw head diameter of 3 mm, and screw head height of 4 mm. The CAD model of specimen with implanted screws at the centre of each specimen was modelled in SOLIDWORKS 2016, as shown in Figure 15.1(a), and further, structural analysis was performed in FEM software, such as ANSYS 18.1. The material properties of a bone for mentioned locations, upper, middle, and lower diaphysis, were considered in the experimental study, as shown in Table 15.1. The material of a screw implant is pure titanium alloy (CP Ti) since it has high corrosion resistivity and high impact toughness with material properties [36] shown in Table 15.1.

The load of maximum 8,000 N was applied with fixed support, as shown in Figure 15.1 (c) and Figure 15.1(d), respectively. Automatic mesh with the number of elements as 1,716 and the number of nodes as 6,904 were applied on the assembly model. The type of element chosen is 3D—tetrahedral and hexahedral—with element size of 0.01 mm, as shown in Figure 15.1 (b). All meshing measuring values—Jacobian ratio, aspect ratio, and skewness—come within the limits as defined in the literature.

TABLE 15.1
Material Properties of a Cortical Bone and Implanted Screw [36]

Name of Part/Subject	Name of Location	Elastic Modulus (GPa)	Poisson Ratio	Density (g/cm^3)
Cortical Bone	Upper diaphysis	20.6	0.4	1.87
	Middle diaphysis	24.3	0.4	1.98
	Lower diaphysis	16.1	0.4	1.74
Screw Implant (Pure Titanium Alloy, CP Ti)	NA	113	0.37	4.4

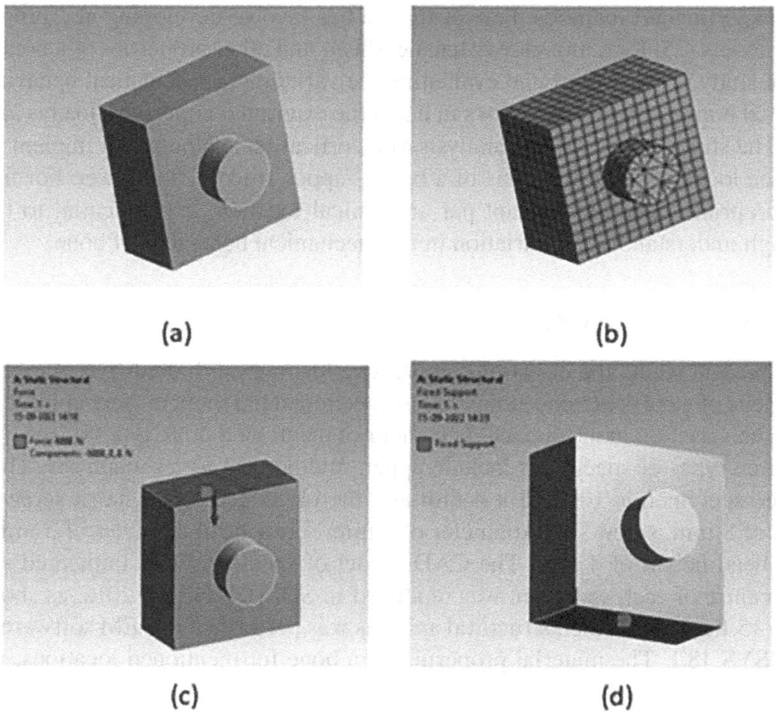

FIGURE 15.1 (a) CAD model; (b) mesh model; (c) compressive load of 8,000 N; (d) fixed support.

The different properties of a cortical bone, including von Mises stress, total deformation, directional deformation, factor of safety, normal stress, and strain energy, were measured using FEM structural analysis.

15.3 RESULTS AND DISCUSSION

The structural analysis of different specimens obtained from three different locations of diaphysis—upper, middle, and lower—are shown in Figure 15.2, Figure 15.3, and Figure 15.4, respectively. The details of each specimen are discussed further here.

15.3.1 Upper Diaphysis

The von Mises stress for the specimen from the upper diaphysis was 458.79 MPa (maximum) and 10.322 MPa (minimum). The maximum stress was found at the interface of cortical bone and screw head, as shown in Figure 15.2(a). The maximum deformation at the upper part of a specimen, as shown in Figure 2(b), is 0.079 mm; the factor of safety, per Figure 15.2(c), is 0.35 (minimum); the strain energy value is 0.639 MJ, as shown in Figure 15.2(d); the value of normal stress in the direction of load applied is 61.36 MPa (Figure 15.2(e)) at the screw head and the value of directional deformation in the direction of load applied is 0.075 mm (Figure 15.2(f)).

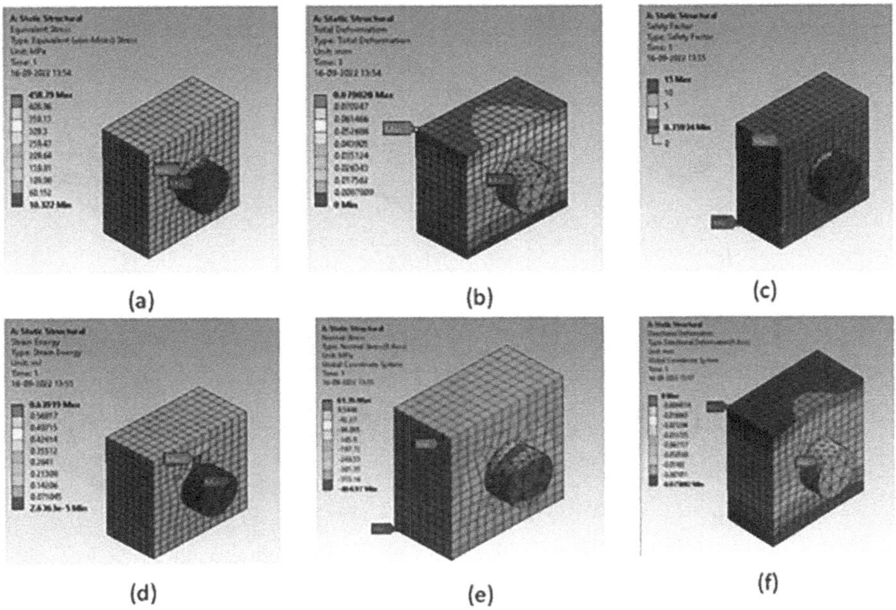

FIGURE 15.2 Results of the upper diaphysis specimen: (a) von Mises stress; (b) total deformation; (c) factor of safety; (d) strain energy; (e) normal stress; (f) directional deformation.

15.3.2 Middle Diaphysis

As shown in Figure 15.3, the von Mises stress was 432.32 MPa (maximum) and 8.64 MPa (minimum) at the interface of screw and cortical bone. The total deformation is 0.06 mm, factor safety is 0.48, strain energy value is 0.52 MJ, normal stress is 59.6 MPa, and the value of deformation in the direction of applied load is 0.06 mm.

15.3.3 Lower Diaphysis

As shown in Figure 15.4; the maximum stress found at the interface of screw and cortical bone with a value of 496.06 MPa and minimum value to be 10.64 MPa. The value of total deformation is 0.10 mm; factor of safety as 0.39; strain energy value is 0.84 MJ; the value of normal stress in the direction of loading is 68.98 MPa and deformation in the direction of loading is 0.09 mm.

As shown in Table 15.2, it was observed from the structural analysis that more value of von Mises stress was found in the specimen extracted from the lower diaphysis location, and that from the middle diaphysis shows a least value. While comparing the total deformation and directional deformation, the value was found to be more for the lower diaphysis and less for the middle diaphysis. Normal stress relatively found more in the specimen extracted from lower diaphysis. The factor of safety was found to be more for the middle diaphysis specimen and less for the upper diaphysis specimen. The strain energy is stored more in the specimen from the lower diaphysis and less in the specimen from the middle diaphysis.

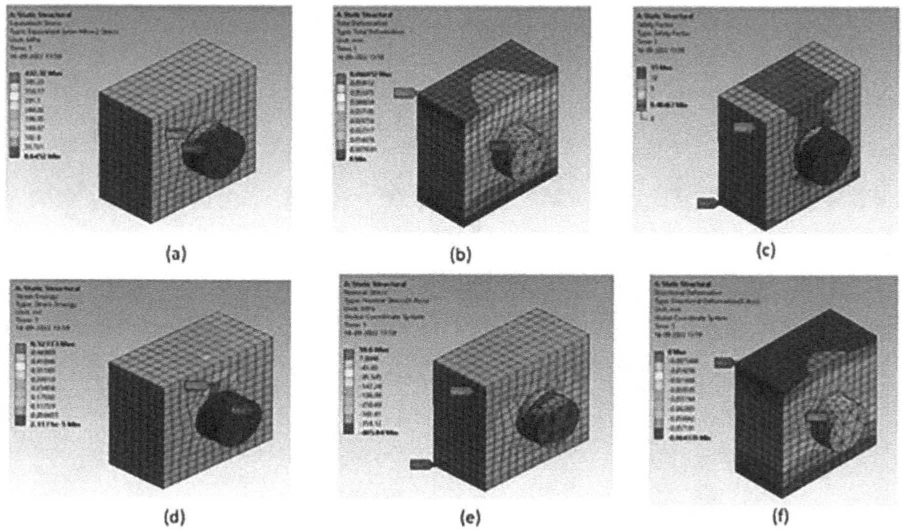

FIGURE 15.3 Results of the middle diaphysis specimen: (a) von Mises stress; (b) total deformation; (c) factor of safety; (d) strain energy; (e) normal stress; (f) directional deformation.

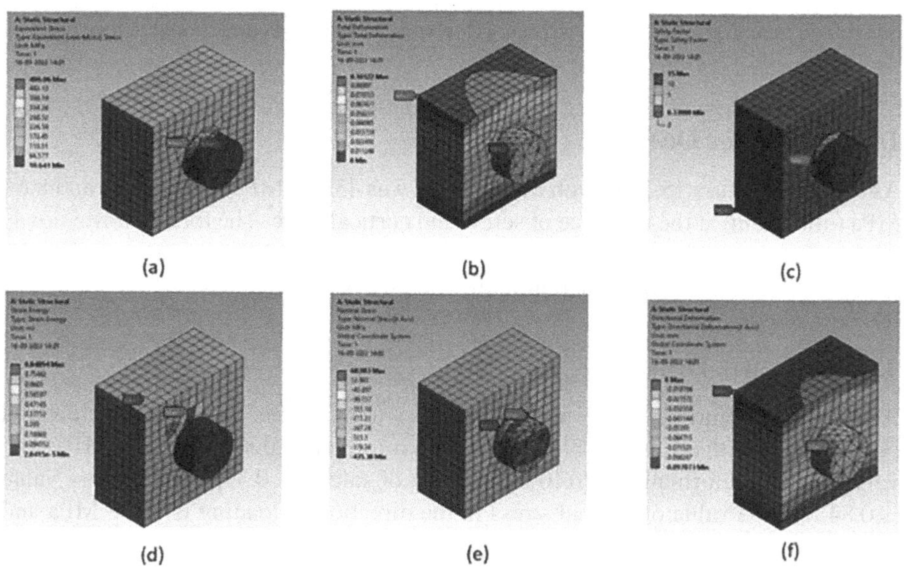

FIGURE 15.4 Results of the lower diaphysis specimen: (a) von Mises stress; (b) total deformation; (c) factor of safety; (d) strain energy; (e) normal stress; (f) directional deformation.

It can be observed from the results while designing bio-implants, the post-effect shall also be evaluated per the anatomical location of a bone. The current study shall be useful in predicting the effect of compressive loading on the bone with implant on the mechanical behaviour of bone since most of the time, a bone comes into contact with compressive load in daily activities.

TABLE 15.2
Results of Different Specimens

Specimen Name	Stress (Max.) MPa	Stress (Min.) MPa	Total Deformation (mm)	Factor of Safety (FOS)	Strain Energy (MJ)	Normal Stress (MPa)	Directional Deformation (mm)
UD	458.79	10.322	0.079	0.35	0.63	61.36	0.075
MD	432.32	8.64	0.06	0.48	0.52	59.6	0.06
LD	496.06	10.64	0.1	0.39	0.84	68.98	0.09

15.4 CONCLUSIONS

In the current study, specimens were obtained from three different locations of a bone—lower, upper, and middle diaphysis. The compressive load was applied for all specimens having screwed bio-implants. The modelling of screwed implant was done by using Solidworks, whereby structural analysis FEM software (ANSYS) was used. It was observed from the structural analysis that the maximum von Mises stress was found in the specimen extracted from the lower diaphysis and the minimum value was found in the middle diaphysis. When comparing the deformation, the value was found to be more for the lower diaphysis and less for the middle diaphysis. Normal stress relatively was found more in the specimen extracted from the lower diaphysis. The factor of safety was found to be more for the middle diaphysis specimen and less for the upper diaphysis specimen. The strain energy is stored more in the specimen from the lower diaphysis and less for the middle diaphysis. It was observed from the results that when designing implanted screws, the effect of variation in mechanical properties should also be considered.

REFERENCES

[1] S. Jaramillo Isaza, P.-E. Mazeran, K. El Kirat, and M.-C. Ho Ba Tho, "Time-dependent mechanical properties of rat femoral cortical bone by nanoindentation: An age-related study," *J Mater Res*, vol. 29, no. 10, pp. 1135–1143, 2014, doi:10.1557/jmr.2014.104.

[2] N. Chennimalai Kumar, J. A. Dantzig, I. M. Jasiuk, A. G. Robling, and C. H. Turner, "Numerical modeling of long bone adaptation due to mechanical loading: Correlation with experiments," *Ann Biomed Eng*, vol. 38, no. 3, pp. 594–604, 2010, doi:10.1007/s10439-009-9861-4.

[3] H. Gustafson, G. Siegmund, and P. Cripton, "Comparison of strain rosettes and digital image correlation for measuring vertebral body strain," *J Biomech Eng*, vol. 138, no. 5, 2016, doi:10.1115/1.4032799.

[4] Y. Chen *et al.*, "Experimental investigation on the mechanical behavior of bovine bone using digital image correlation technique," *Appl Bionics Biomech*, vol. 2015, 2015, doi:10.1155/2015/609132.

[5] N. J. Wachter *et al.*, "Correlation of bone mineral density with strength and micro-structural parameters of cortical bone in vitro," *Bone*, vol. 31, no. 1, pp. 90–95, 2002, doi:10.1016/S8756-3282(02)00779-2.

[6] H. B. Hunt and E. Donnelly, "Bone quality assessment techniques: Geometric, compositional, and mechanical characterization from macroscale to nanoscale," *Clin Rev*

Bone Miner Metab, vol. 14, no. 3, pp. 133–149, 2016. Humana Press Inc., doi:10.1007/s12018-016-9222-4.

[7] J. T. Rexwinkle, N. C. Werner, A. M. Stoker, M. Salim, and F. M. Pfeiffer, "Investigating the relationship between proteomic, compositional, and histologic biomarkers and cartilage biomechanics using artificial neural networks," *J Biomech*, vol. 80, pp. 136–143, 2018, doi:10.1016/j.jbiomech.2018.08.032.

[8] A. C. Abraham, A. Agarwalla, A. Yadavalli, J. Y. Liu, and S. Y. Tang, "Microstructural and compositional contributions towards the mechanical behavior of aging human bone measured by cyclic and impact reference point indentation," *Bone*, vol. 87, pp. 37–43, 2016, doi:10.1016/j.bone.2016.03.013.

[9] I. Dickson, "The composition and antigenicity of sheep cortical bone matrix proteins," *Calcif Tissue Res*, vol. 16, no. 1, pp. 321–333, 1974, doi:10.1007/BF02008240.

[10] O. L. Katsamenis, H. M. H. Chong, O. G. Andriotis, and P. J. Thurner, "Load-bearing in cortical bone microstructure: Selective stiffening and heterogeneous strain distribution at the lamellar level," *J Mech Behav Biomed Mater*, vol. 17, pp. 152–165, 2013, doi:10.1016/j.jmbbm.2012.08.016.

[11] M. D. Clark, N. P. Davis, A. F. Walsh, B. Savilonis, C. Kirker-Head, and C. Les, "The effect of microstructure on the mechanical properties of equine bone," *Proceedings of the IEEE Annual Northeast Bioengineering Conference*, NEBEC, vol. 30, pp. 152–153, 2004.

[12] T. Diab, S. Sit, D. Kim, J. Rho, and D. Vashishth, "Age-dependent fatigue behaviour of human cortical bone," *Eur J Morphol*, vol. 42, no. 1–2, pp. 53–59, 2005, doi:10.1080/09243860500095539.

[13] P. J. Thurner, "Commentary on: Mechanical properties of cortical bone and their relationships with age, gender, composition and microindentation properties in the elderly," *Bone*, vol. 87, pp. 159–160, 2016, doi:10.1016/j.bone.2016.04.009.

[14] R. A. Ayers, M. R. Miller, S. J. Simske, and R. W. Norrdin, "Correlation of flexural structural properties with bone physical properties: A four species survey," *Biomed Sci Instrum*, vol. 32, pp. 251–260, 1996.

[15] J. Singh, N. K. Sharma, M. D. Sarker, S. Naghieh, S. S. Sehgal, and D. X. B. Chen, "Assessment of elastic-plastic fracture behavior of cortical bone using a small punch testing technique," *J Biomech Eng*, vol. 142, no. 1, pp. 1–18, 2020, doi:10.1115/1.4043870.

[16] D. Garcia, P. K. Zysset, M. Charlebois, and A. Curnier, "A 1D elastic plastic damage constitutive law for bone tissue," *Arch Appl Mech*, vol. 80, no. 5, pp. 543–555, 2010, doi:10.1007/s00419-009-0382-2.

[17] J. Singh, N. K. Sharma, and S. S. Sehgal, "Application of small punch testing to evaluate elastic and plastic parts of fracture energy of cortical bone," *Mater Today Proc*, vol. 5, no. 9, pp. 18442–18450, 2018, doi:10.1016/j.matpr.2018.06.185.

[18] J. Singh, N. K. Sharma, and S. S. Sehgal, "Application of small punch testing to evaluate elastic and plastic parts of fracture energy of cortical bone," *Mater Today Proc*, vol. 5, no. 9, pp. 18442–18450, 2018, doi:10.1016/j.matpr.2018.06.185.

[19] S. Xie, K. Manda, and P. Pankaj, "Time-dependent behaviour of bone accentuates loosening in the fixation of fractures using bone-screw systems," *Bone Joint Res*, vol. 7, no. 10, pp. 580–586, 2018, doi:10.1302/2046-3758.710.BJR-2018-0085.R1.

[20] B. Yu *et al.*, "Biomechanical comparison of conventional and anatomical calcaneal plates for the treatment of intraarticular calcaneal fractures—a finite element study," *Comput Methods Biomech Biomed Engin*, vol. 19, no. 13, pp. 1363–1370, 2016, doi:10.1080/10255842.2016.1142534.

[21] T. Tang, V. Ebacher, P. Cripton, P. Guy, H. McKay, and R. Wang, "Shear deformation and fracture of human cortical bone," *Bone*, vol. 71, pp. 25–35, 2015, doi:10.1016/j.bone.2014.10.001.

[22] J. Yamashita, X. Li, B. R. Furman, H. Ralph Rawls, X. Wang, and C. Mauli Agrawal, "Collagen and bone viscoelasticity: A dynamic mechanical analysis," *J Biomed Mater Res*, vol. 63, no. 1, pp. 31–36, 2002, doi:10.1002/jbm.10086.
[23] T. P. M. Johnson, S. Socrate, and M. C. Boyce, "A viscoelastic, viscoplastic model of cortical bone valid at low and high strain rates," *Acta Biomater*, vol. 6, no. 10, pp. 4073–4080, 2010, doi:10.1016/j.actbio.2010.04.017.
[24] S. Bernard, J. Schneider, P. Varga, P. Laugier, K. Raum, and Q. Grimal, "Elasticity–density and viscoelasticity–density relationships at the tibia mid-diaphysis assessed from resonant ultrasound spectroscopy measurements," *Biomech Model Mechanobiol*, vol. 15, no. 1, pp. 97–109, 2016, doi:10.1007/s10237-015-0689-6.
[25] W. Pomwenger, K. Entacher, H. Resch, and P. Schuller-Götzburg, "Influence of glenoid implant depth on the bone–polymethylmethacrylate interface | Einfluss der Implantatstiefe auf die Knochen-Polymethylmethacrylat Fixierung am Glenoid," *Obere Extrem*, vol. 14, no. 4, pp. 284–291, 2019, doi:10.1007/s11678-019-0512-6.
[26] C. M. Bellini *et al.*, "Comparison of tilted versus nontilted implant-supported prosthetic designs for the restoration of the edentuous mandible: A biomechanical study," *Int J Oral Maxillofac Surg*, vol. 24, no. 3, pp. 511–517, 2009.
[27] L. Capek, A. Simunek, P. Henys, and L. Dzan, "The role of implant's surface treatment to its preload," *Comput Methods Biomech Biomed Engin*, vol. 17, no. Supp 1, pp. 8–9, 2014, doi:10.1080/10255842.2014.931051.
[28] M. Dalstra and R. Huiskes, "Prestresses around the acetabulum generated by screwed cups," *Clin Mater*, vol. 16, no. 3, pp. 145–154, 1994, doi:10.1016/0267-6605(94)90110-4.
[29] M. Migliorati *et al.*, "On the stability efficiency of anchorage self-tapping screws: Ex vivo experiments on miniscrew implants used in orthodontics," *J Mech Behav Biomed Mater*, vol. 81, no. February, pp. 46–51, 2018, doi:10.1016/j.jmbbm.2018.02.019.
[30] C. Nelson Elias, D. Jogaib Fernandes, D. Souza Zanivan, and Y. Resende Fonseca, "Extensiometric analysis of strain in craniofacial bones during implant-supported palatal expansion," *J Mech Behav Biomed Mater*, vol. 76, pp. 104–109, 2017, doi:10.1016/j.jmbbm.2017.05.028.
[31] A. Vemuganti, S. Siegler, A. Abusafieh, and S. Kalidindi, "Development of self-anchoring bone implants. II. Bone-implant interface characteristics in vitro," *J Biomed Mater Res*, vol. 38, no. 4, pp. 328–336, 1997, doi:10.1002/(SICI)1097-4636(199724)38:4<328::AID-JBM4>3.0.CO;2-Q.
[32] Y. Çiftçi and Ş. Canay, "The effect of veneering materials on stress distribution in implant-supported fixed prosthetic restorations," *Int J Oral Maxillofac Implants*, vol. 15, no. 4, pp. 571–582, 2000.
[33] A. M. Tsitsiashvili, A. S. Silant'ev, A. M. Panin, and S. D. Arutyunov, "Short dental implant biomechanics in the mandible bone tissue | Biomekhanika korotkogo dental'nogo implantata v kostnoi tkani nizhnei chelyusti," *Stomatologiia (Mosk)*, vol. 98, no. 6 Vyp 2, pp. 33–36, 2019, doi:10.17116/stomat20199806233.
[34] A. A. de Sousa and B. S. C. Mattos, "Finite element analysis of stability and functional stress with implant-supported maxillary obturator prostheses," *J Prosthet Dent*, vol. 112, no. 6, pp. 1578–1584, 2014, doi:10.1016/j.prosdent.2014.06.020.
[35] A. R. Manral, N. Gariya, and K. C. Nithin Kumar, "Material optimization for femur bone implants based on vibration analysis," *Mater Today Proc*, vol. 28, pp. 2393–2399, 2020, doi:10.1016/j.matpr.2020.04.714.
[36] A. Kumar, A. Rathi, J. Singh, and N. K. Sharma, "Studies on titanium hip joint implants using finite element simulation," *Lect Notes Eng Comput Sci*, vol. 2224, pp. 986–990, 2016.

16 A Review on Mixed-Reality Technology in Medical Anatomy Structure

Anjali, Gurjinder Singh, Jasminder Kaur Sandhu, Shinnu

16.1 INTRODUCTION

Mixed reality (MR) is one of the most advanced technologies and shows an immersive environment with the combination of real and digital world. It interacts and manipulates with both computer-generated information and physical items by applying 2D or 3D imaging technologies [1]. The main functionality of the MR technology is the computer-related harmonization system which displays virtual scenes and real-world images in front a person eyes [2]. This technology arises from the virtual reality (VR) and augmented reality (AR) continuum technologies which concept is shown in Figure 16.1.

The use of AR, MR, and VR technologies enhances the world, along with more advanced features which is applied in the healthcare field, the movie and television industry, virtual tours, and the gaming and sports sector [3]. There are several devices and tools available nowadays which help to create an imaginary world environment, such as Google Cardboard, Oculus Rift, and Quest devices, with the enhanced, fully immersive MR, such as in HoloLens, Metaverse 2, Google Glass, AR glasses, and holograms [4]. For different head-mounted displays need different system specifications, and it recommends the graphical processing unit (GPU) to render the high-pixel-resolution immersive images and videos such as for Oculus devices, HTC Vive, and Microsoft virtual tools.

An environmental study of real-world interaction includes AR, VR, and MR. MR involves the concept of both virtual and augmented surroundings in which humans interact with the imaginary and real objects [7]. So this encompasses the concept in two parts—video and optical see-through (VOST). The VOST glasses or headset has translucent glasses and uses a camera to capture real and imaginary video content with the mobile-based application [8]. MR is an emerging technology that integrates augmented reality and virtual reality. The Microsoft HoloLens and other MR devices are expanding the potential for mixed reality experiences in the healthcare field [9, 5]. MR has the ability to map everything that exists in the physical

DOI: 10.1201/9781003367154-16

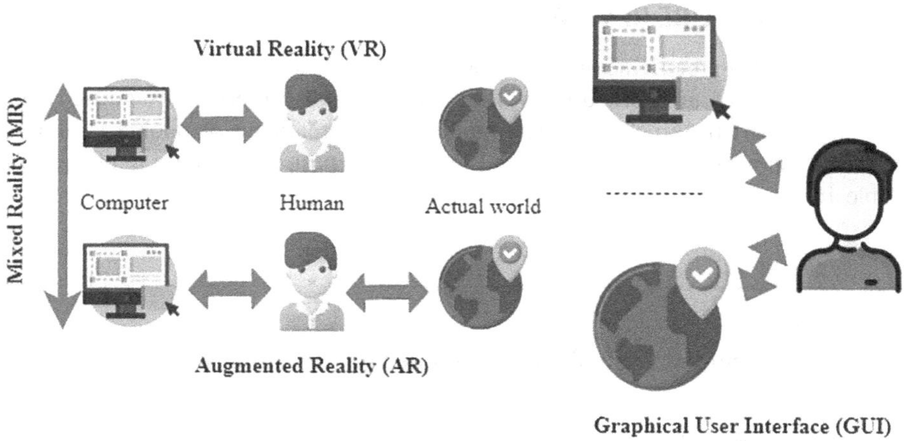

FIGURE 16.1 Mixed reality: human-computer interaction via interface [5].

environment, and human abilities include technology-based camera features and sensors that are used to capture voice and body motions [10]. This technology uses spatial music as well as GPS locations and positions to demonstrate the concepts of 360° images in an immersive setting.

16.2 TRANSFORMING THE VIRTUAL WORLD IN HEALTHCARE

Expert doctors and medical students learn several techniques to make the procedures and treatments simple for the patients. According to the recent research, training was also provided to medical students and doctors regarding the virtual treatment [5, 11]. For medical training, education, and dentistry, AR and VR are becoming easier and more accessible as they combine digital information with the real world.

1. The digital world helps visually impaired people with detecting objects by using three-dimensional recognition applications. It can have a better impact on blind persons by helping them recognize faces, easily navigate their surroundings, and find lost objects.
2. Imaginary interaction through augmentation is used in telemedicine. Sometimes surgeons can do train people remotely as procedures in the operation theatre can be projected into MR displays [12].
3. CT scan and MRI can be visualized with the use of stereoscopic projection at the time of surgical procedure. This projection system helps the neuro-surgeons collaborate in every kind of surgery.
4. Therapists also use AR to tackle the mental health problems of those with phobias, such as acrophobia, agoraphobia, and arachnophobia [13]. AR technology, which provides information from the digital world, diverts patients' attention from their discomfort and entirely motivates them to engage in environments set in an imaginary world.

16.3 LITERATURE REVIEW

The use of MR in medicine makes all procedures simple and safe for both patients and doctors. This area broadly involves simulation of surgeries, providing robot training, expert training of medical students, and treatment of phobia [14]. The MR developments MR in the medical field in the past decades are explained in Table 16.1.

TABLE 16.1
Study of Augmented, Virtual, and Mixed Reality Technology in the Medical Field

S. No,	Author(s) and References	Year of Publication	Technologies	Summary
1	Christopoulos, Athanasios, et al. [13]	2022	AR and VR	Nurses, as well as doctors and surgeons, have been trained for medical therapy via the use of augmented and virtual reality.
2	Tang, Kevin S., et al. [2]	2020	AR applications (Blender, Google Glass, AR glasses, Microsoft HoloLens)	The use of AR technologies in the medical field is growing rapidly. Through the use of AR, virtual information and structures are able to be projected onto actual things, improving or changing the environmental surroundings.
3	Ma, D., Gausemeier et al. [7]	2011	MR, 3D graphic models	To measure the behavioural responses of the clinical patients, VR tools were used to diagnose the patients by studying the anatomy of the brain. VR applications were used for radiology, such as X-ray images, CT scanning, MRI, and ultrasound.
4	Sahija, D. et al. [15]	2022	MR and AR	MR was integrated into medical instruments and machines for the care of patients, improving the healthcare services received by patients.
5	Dhar, P., Rocks et al. [14]	2021	AR technologies (HoloHuman, HoloPatient, OculAR SIM)	The use of AR in the field of medical training and education enhances the learning experience of students. With AR techniques, every person can improve their knowledge and skills in any field, such engineering, medicine, banking, teaching, learning, and marketing.

TABLE 16.1 *(Continued)*
Study of Augmented, Virtual, and Mixed Reality Technology in the Medical Field

S. No,	Author(s) and References	Year of Publication	Technologies	Summary
6	Herron, J. et al. [4]	2016	AR technologies (Google Glass, Microsoft HoloLens)	With technological enhancements, people grow more in their work with the use of mobile applications, e-books, and 3D technologies, such as Google Glass and HoloLens.
7	Moro, C., Štromberga et al. [9]	2017	VR (Oculus Rift), MR, 2D and 3D learning models (Unity V5)	MR enhances the education and classroom experience of medical students by using a 3D environment, increasing the level of interaction and engagement in the lecture hall.
8	Heinrich, F. et al. [11]	2019	AR and MR (head-mounted displays)	Augmentation visualization is used to simulate medical procedures involving syringes. To use the AR concept to visualize the medical instrument guidance for patients directly, in order to aid in the treatment of medical issues.
9	Jaris Gerup et al. [16]	2020	AR, MR, and VR (2D or 3D visual perception)	There are several applications used such as computer-based and marker-based applications for the virtual surgery. Applying these 2D and 3D techniques helps with preventing complications and managing the treatment of damaged tissues.
10	Tadatsugu Morimoto et al. [17]	2022	Extended reality (AR, VR, and MR)	The use of Extended Reality (XR) technology for the treatment of neuropathic pain which is the part of spinal cord injury. For the education of medical students, extended reality techniques were applied in medicinal examination of and the surgical and rehabilitative procedures involving the spinal cord.

16.4 MIXED REALITY SYSTEM IN MEDICAL FIELD

In the medical sector, human-computer interaction enables providers to analyse data, control the structure and movements of the patients, and coordinate the processing system in a more effective and convenient manner. MR technology is used for therapeutic purposes, such as tracking the eyesight of patients with low-vision problems [15, 18]. MR is the extended version of AR, which enables humans to interact with the digital system in real time to help them interact with and control the real world with machines and medical instrumentations [19]. In this, AR and VR revolutionize future medicine and provide advanced applications and software to cover the several fields, including the visualization of three-dimensional information from diagnosis, training medical students and physicians, computerized surgeries with the use of 3D diagnoses, plastic surgeries, telemedicine, physical therapy and rehabilitation, and psychiatric treatments, such as the treatment of phobias [20].

An immersive experience is done by the use of graphical interface in the real world and is accessible through mobile and computer screens. To track and recognize

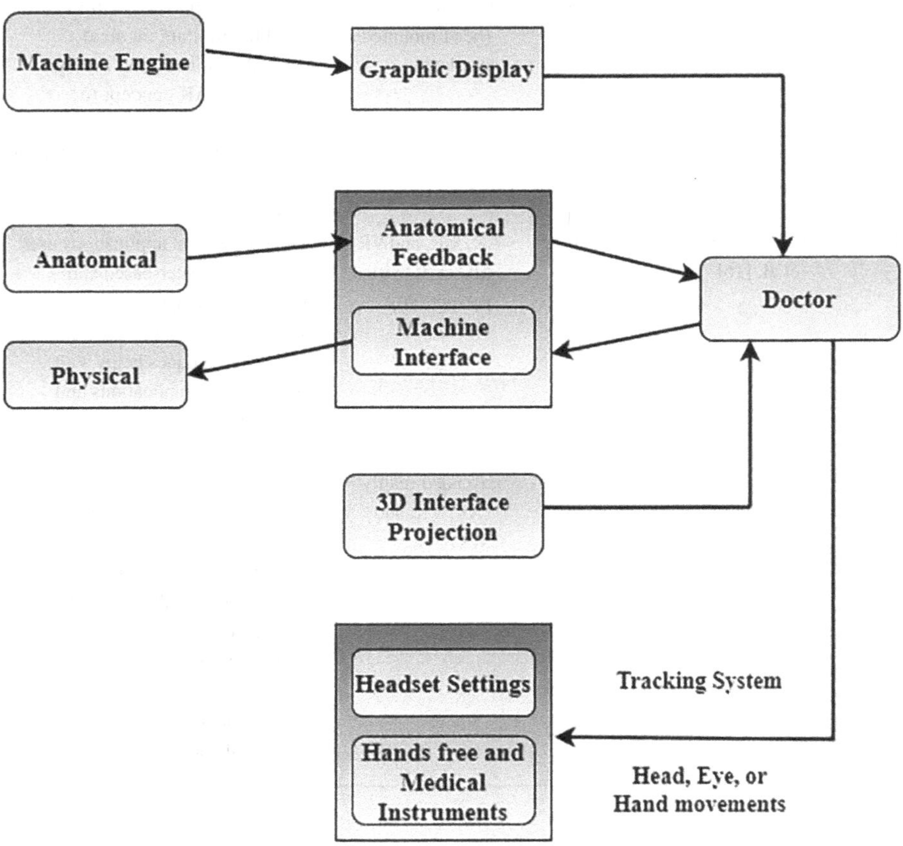

FIGURE 16.2 Medical treatment using the mixed reality system.

TABLE 16.2
Features of Virtual, Augmented, and Mixed Reality

Feature	Virtual Reality	Augmented Reality	Mixed Reality
Tools	Immersive technology	Holographic images	Both holographic and immersive
Display	Opaque	See-through	Blend of virtual and actual images together
Interaction	More interaction with imaginary world	High interaction with the real world and display like actual informational concept	Both real and imaginary worlds interact simultaneously
Multimedia Type	3D world, UIs, text, audios, animations, and videos	3D world, UIs, texts, audios, animations, and videos	Substantial user interaction with the environment, animations, videos, texts, lighting, and spatial awareness
Uses	Virtual training, data visualization, and technological education	Remote meetings, job site virtual inspection, and technical calls support	Medical tools and techniques in the training of doctors and students, architecture visualization, and machine manufacturing training

the movements of the hand, head, and eyes with the use of graphical machine interface [17]. The flow diagram of the medical treatment is shown in Figure 16.2. In the first phase, tracking technology would be able to sense the movements of the person in real time by using sensors and capturing images through cameras. In the second step, the poses of the human are detected during any disease treatment [16]. The third step sets the medical instruments on their specified positions to detect hand postures and equipment movements through the 3D interface projection.

16.4.1 Feature Comparison of Extended Reality Techniques

There are three technologies under extended reality (XR), and the features of these enhance the experience of the users, which has several purposes, such as entertainment and educational and business trainings [21]. The basic features of VR, AR, and MR are given in Table 16.2.

16.5 CONCLUSION

The development of MR enhances the visualization of techniques in the medical field to make patient care better. This technology, 3D visualization interface, is used in several mobile and computer applications, providing humans with enhanced experiences, such as in Microsoft HoloLens, holographic images, and 3D modelling software like Vuforia and Unity 3D. These software help the visually impaired and hearing-impaired people, allow the observing and performing of surgeries remotely, and help with physical and rehabilitative therapies. But till now challenges faced in the procedures of patients' surgeries. The study of AR/VR, which builds an imaginary environment or sets the surroundings through headsets and displays 3D visualization for the treatment of patient problems, has been pursued by doctors. Due to the existence of an imaginary world, immersive technology is now widely used to monitor problems precisely and easily. In the future, AR/VR can enhance the experience of medical trainers, allowing them to easily and quickly deliver much information to the specialists and medical students, increase the communication level, and integrate virtual and real spaces.

REFERENCES

[1] Cox, K., Privitera, M. B., Alden, T., Silva, J. R., & Silva, J. N. A. (2019). Augmented reality in medical devices. In Mary Beth Privitera, (Ed.), *Applied Human Factors in Medical Device Design* (pp. 327–337). Cambridge, Massachusetts, US: Academic Press.

[2] Tang, K. S., Cheng, D. L., Mi, E., & Greenberg, P. B. (2020). Augmented reality in medical education: a systematic review. *Canadian Medical Education Journal*, *11*(1), e81.

[3] Bichlmeier, C., Heining, S. M., Feuerstein, M., & Navab, N. (2009). The virtual mirror: a new interaction paradigm for augmented reality environments. *IEEE Transactions on Medical Imaging*, *28*(9), 1498–1510.

[4] Herron, J. (2016). Augmented reality in medical education and training. *Journal of Electronic Resources in Medical Libraries*, *13*(2), 51–55.

[5] Tang, Y. M., & Ho, H. L. (2020). 3D modeling and computer graphics in virtual reality. In Branislav Sobota and Dragan Cvetković, (Eds.), *Mixed Reality and Three-Dimensional Computer Graphics* (p. 143). London: IntechOpen Limited.

[6] Moro, C., Phelps, C., Redmond, P., & Stromberga, Z. (2021). HoloLens and mobile augmented reality in medical and health science education: a randomised controlled trial. *British Journal of Educational Technology*, *52*(2), 680–694.

[7] Ma, D., Gausemeier, J., Fan, X., & Grafe, M. (Eds.). (2011). *Virtual Reality & Augmented Reality in Industry*. Berlin, Heidelberg: Springer.

[8] Hu, H. Z., Feng, X. B., Shao, Z. W., Xie, M., Xu, S., Wu, X. H., & Ye, Z. W. (2019). Application and prospect of mixed reality technology in medical field. *Current Medical Science*, *39*(1), 1–6.

[9] Moro, C., Štromberga, Z., Raikos, A., & Stirling, A. (2017). The effectiveness of virtual and augmented reality in health sciences and medical anatomy. *Anatomical Sciences Education*, *10*(6), 549–559.

[10] Sveinsson, B., Koonjoo, N., & Rosen, M. S. (2021). ARmedViewer, an augmented-reality-based fast 3D reslicer for medical image data on mobile devices: a feasibility study. *Computer Methods and Programs in Biomedicine*, *200*, 105836.

[11] Heinrich, F., Joeres, F., Lawonn, K., & Hansen, C. (2019). Comparison of projective augmented reality concepts to support medical needle insertion. *IEEE Transactions on Visualization and Computer Graphics*, *25*(6), 2157–2167.

[12] von Jan, U., Noll, C., Behrends, M., & Albrecht, U. V. (2012). mARble–augmented reality in medical education. *Biomedical Engineering/Biomedizinische Technik*, *57*(SI-1-Track-A).

[13] Christopoulos, A., Pellas, N., Kurczaba, J., & Macredie, R. (2022). The effects of augmented reality-supported instruction in tertiary-level medical education. *British Journal of Educational Technology*, *53*(2), 307–325.

[14] Dhar, P., Rocks, T., Samarasinghe, R. M., Stephenson, G., & Smith, C. (2021). Augmented reality in medical education: students' experiences and learning outcomes. *Medical Education Online*, *26*(1), 1953953.

[15] Sahija, D. (2022). Critical review of mixed reality integration with medical devices for patientcare. *International Journal for Innovative Research in Multidisciplinary Field*, 10.

[16] Gerup, J., Soerensen, C. B., & Dieckmann, P. (2020). Augmented reality and mixed reality for healthcare education beyond surgery: an integrative review. *International Journal of Medical Education*, *11*, 1.

[17] Morimoto, T., Kobayashi, T., Hirata, H., Otani, K., Sugimoto, M., Tsukamoto, M., . . . & Mawatari, M. (2022). XR (extended reality: virtual reality, augmented reality, mixed reality) technology in spine medicine: status quo and quo vadis. *Journal of Clinical Medicine*, *11*(2), 470.

[18] Kuehn, B. M. (2018). Virtual and augmented reality put a twist on medical education. *Jama*, *319*(8), 756–758.

[19] Sielhorst, T., Feuerstein, M., & Navab, N. (2008). Advanced medical displays: a literature review of augmented reality. *Journal of Display Technology*, *4*(4), 451–467.

[20] Sutherland, J., Belec, J., Sheikh, A., Chepelev, L., Althobaity, W., Chow, B. J., . . . & La Russa, D. J. (2019). Applying modern virtual and augmented reality technologies to medical images and models. *Journal of Digital Imaging*, *32*(1), 38–53.

[21] Chien, C. H., Chen, C. H., & Jeng, T. S. (2010, March). An interactive augmented reality system for learning anatomy structure. In *Proceedings of the International Multiconference of Engineers and Computer Scientists* (Vol. 1, pp. 17–19). Hong Kong, China: International Association of Engineers.

17 Evaluation of the Strength Characteristics of Geopolymer Concrete Produced with Fly Ash and Slag

Nerswn Basumatary, Paramveer Singh, Kanish Kapoor, S. P. Singh
Department of Civil Engineering, Dr B. R. Ambedkar National Institute of Technology, Jalandhar, India

17.1 INTRODUCTION

Geopolymers have been widely appreciated because of their ability to change the use of ordinary Portland cement (OPC) in the concrete manufacturing. The OPC use has been heavily contested around the world due to its high energy consumption and contribution to global warming through the production of carbon dioxide (CO_2), which accounts for roughly 5–7% of worldwide CO_2 emissions [1]. The main cause of CO_2 emission due to the manufacture of OPC has been attributed to limestone calcination and high energy consumption during manufacturing [2]. The geopolymer binders are generally alumina- and silica-rich minerals which are present in industrial by products such as fly ash (FA) and ground granulated blast furnace slag (GGBS), metakaolin (MK), and so on [3]. The utilization of these by products as binders in geopolymers helps to reduce the landfills, which helps lead to sustainable environment [4]. Generally, FA was used extensively as a primary binder in geopolymer concrete as it is rich in alumina and silica and its easy availability [5]. Considering all these factors, a sustainable scientific approach to replace OPC as a binder by using industrial wastes such as FA has been widely appreciated for the past few decades. In addition to this, the alkali activators, such as sodium hydroxide (NaOH) and sodium silicate (Na_2SiO_3), are generally used for geopolymer reaction to form polymeric chain of Si–O–Al–O bonds as a final product [6]. Geopolymers' polymerization process and pace of strength growth are influenced by a number of variables, including the chemical composition of the source materials, alkaline activators, and curing conditions. [7]. The one of the concerned factors of strength enhancement of geopolymer concrete is ambient curing condition. For early strength enhancement, generally heat is provided to enhance the geopolymer synthesis, but

 DOI: 10.1201/9781003367154-17

for in situ condition, it is not feasible to provide the heat curing [8]. To overcome that, calcium-rich minerals are combined with FA to enhance the geopolymer reaction. The GGBS is extensively used as a calcium-rich mineral along with FA to enhance the geopolymer reaction [9]. The use of GGBS improves the geopolymer concrete's strength characteristics. Rao and Rao [10] achieved the optimum compressive strength at a 50% replacement of FA with GGBS. Nath and Sarker [11] used ambient curing temperature and attained 55 MPa strength with 30% GGBS incorporation with FA. Nagajothi and Elavelin [12] revealed that compressive strength increased up to 40% at ambient curing. The addition of GGBS during geopolymer concrete preparation produces calcium silicate gel along with polymeric chain of alumina and silica, which consequently enhances the compressive strength of the concrete. With this in mind, the addition of GGBS during geopolymer concrete preparation at ambient curing is beneficial in the case of a high strength requirement and should give promising results [9]. The key objective of the study to evaluate the strength properties of geopolymer concrete at ambient curing with the addition of GGBS. The compressive strength and split tensile strength have been evaluated at 7 and 28 days of curing along with non-destructive testing—that is, ultrasonic pulse velocity (UPV) test.

17.2 EXPERIMENTAL PROGRAMME

17.2.1 Source Materials

The FA was used as primary binder and GGBS was partially replaced with FA at different replacement levels. The FA were collected from Rajpura thermal plant, Punjab, and GGBS was obtained from Aastra Chemicals, India. The particle distribution curve of FA and GGBS is given in Figure 17.1. The physical and chemical properties of FA and GGBS are shown in Table 17.1.

17.2.2 Alkaline Activator

For alkali-activated solutions, the combination of sodium hydroxide (NaOH) and sodium silicate (Na_2SiO_3) with 16.3% sodium oxide (Na_2O), 29.6% silicon oxide (SiO_2), and 54.1% water (H_2O) was used in the experimental work. For preparing NaOH solution, NaOH pellets were dissolved in tap water and stirred until they dissolved completely. To make a solution of 10 M concentration, the solid percentage in NaOH solution was taken as 31.6% of the total weight of NaOH pellets in alkali solutions. The solution was kept for 24 hours, and the sodium silicate was mixed with sodium hydroxide solution before casting of specimens.

17.2.3 Aggregates

The coarse aggregates of nominal maximum size 12.5 mm and natural fine aggregates were obtained from the local market and were used in this investigation. The specific gravity and water absorption of coarse aggregates are 2.63 and 0.55%.

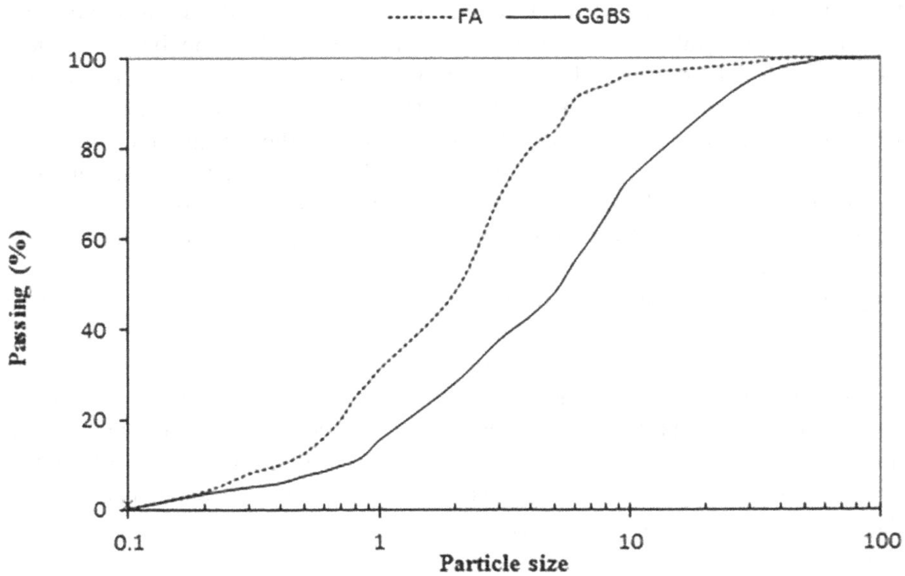

FIGURE 17.1 Particle size distribution curve of FA and GGBS used in the experimental work.

TABLE 17.1
Physical Properties and Chemical Composition of FA and GGBS

Properties	FA	GGBS
Specific gravity	2.2	2.85
Fineness, (cm^2/gm)	4100	3900
Chemical composition	**FA (%)**	**GGBS (%)**
SiO_2	56.50	33.1
Al_2O_3	17.70	18.2
Fe_2O_3	11	0.31
MgO	2.30	7.6
CaO	3.20	35.3

Similarly, the specific gravity and water absorption of fine aggregates were 2.71 and 1.5%. The particle distribution curve of the coarse and fine aggregates is given in Figure 17.2.

17.3 METHODOLOGY

17.3.1 Mixing Proportions and Method

The mix design method of geopolymer concrete used in the experimental work was attained from literature [13]. The total binder content was taken 420 kg/m^3 out of which GGBS was substituted with FA at different replacement levels—0%,

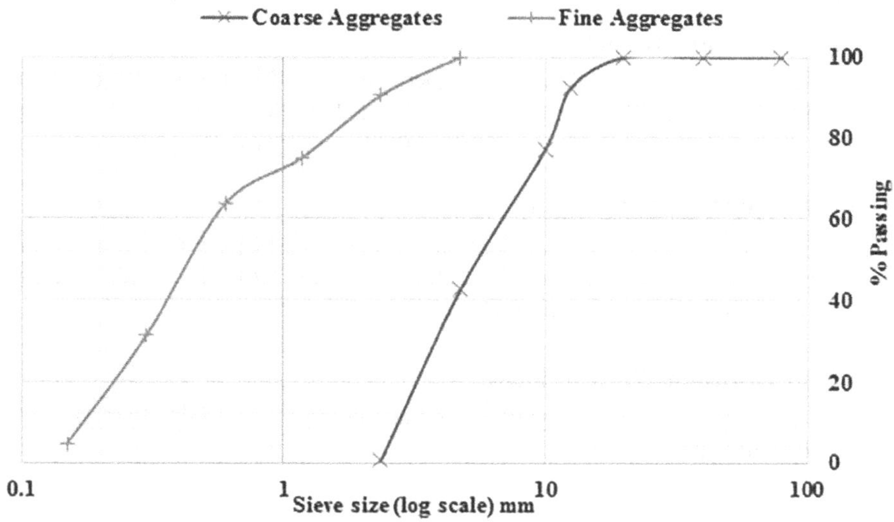

FIGURE 17.2 Gradation curves of the coarse and fine aggregates.

TABLE 17.2
Mix Proportioning of Geopolymer Mixes in kg/m³

Mix ID	FA	GGBS	NH	NS	Additional water	Coarse Aggregates	Fine Aggregates
F100G0	420	0	54	135	10.56	935	742
F90G10	378	54.4	54	135	10.56	935	742
F80G20	420	108.8	54	135	10.56	935	742

10%, and 20%. The alkali content was taken as 45% of total binder content. The NaOH was kept constant at 10 M, and Na_2SiO_3 was taken 2.5 times of NaOH quantity. The required volume of coarse aggregate and fine aggregate was calculated based on the maximum nominal size of aggregate of 12.5 mm reference to IS 10262: 2019 [14]. The proportion of geopolymer mixes used in the experimental work are shown in Table 17.2. For mixing, firstly, coarse and fine aggregates were mixed with blended binders in drum mixer for 5 minutes. Secondly, the alkali activators were added in the dry mix and further mixed for 10 minutes for the consistent mix. The additional water was also fixed base on required workability based on mixed design method. The prepared fresh geopolymer concrete were poured in the moulds for further curing. The moulds were sealed with poly wrap to avoid moisture loss and cured at ambient conditions at room temperature up to their testing period.

17.3.2 Test Methods

17.3.2.1 Compressive Strength

The compressive strength of all mixes was performed on 100 mm cube specimen per IS 516 [15]. The 7-and-28-day testing was performed on specimens, and the averages of the three specimens were taken as final result of compressive strength of each mix.

17.3.2.2 Split Tensile Strength

The split tensile strength was evaluated on cylinder specimens with 100 mm diameter and 200 mm height in accordance with IS 5816:1959 [16]. The testing was done on specimens at 7 days and 28 days of curing.

17.3.2.3 Ultrasonic Pulse Velocity

The ultrasonic pulse velocity was measured according to IS: 13311 (Part 1)—1992 [17] on 100 mm cube specimens.

17.4 RESULTS AND DISCUSSION

17.4.1 Compressive Strength

The compressive strength of geopolymer mixes were show in Figure 17.3. The compressive strength of control mix, F100G0, is 9.5 MPa and 16.8 MPa at 7 and 28 days of curing, respectively. Further, with addition of GGBS at 10%, for mix F90G10, the compressive enhanced by 94.73% at 7 days of curing and 46.42% at 28 days of curing. Likewise, for mix F80G20 with 20% GGBS, at 7 and 28 days of curing, the compressive strength increased by 184.21% and 82.85%, respectively. Faried et al. [18] also concluded that the substitution of GGBS aids in C-S-H gel polymerization to achieve the higher strength. The results demonstrate that the addition of GGBS in geopolymer concrete greatly increases compressive strength at early curing stages. The GGBS provided additional nucleation sites along with geopolymer matrix, which helps to improve the compressive strength of geopolymer concrete [11].

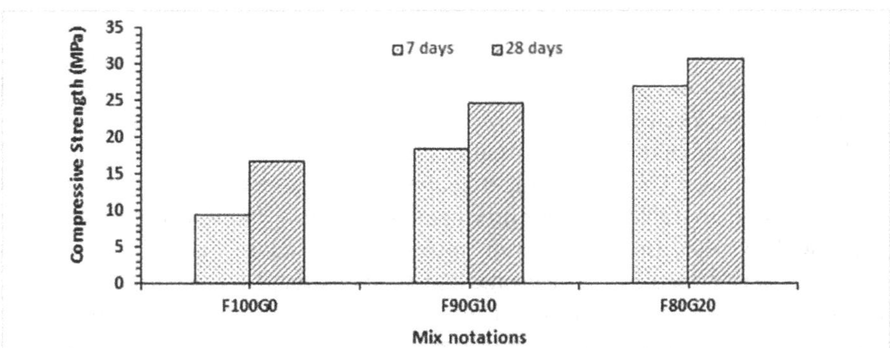

FIGURE 17.3 Development of compressive strength in geopolymer mixes at 7 and 28 days of curing.

17.4.2 Split Tensile Strength

The split tensile strength results of geopolymer mixes are shown in Figure 17.4. The split tensile strength also increases in the same trend of compressive strength. For mix F90G10 with 10% GGBS content, the split tensile strength increased by 96.66% at 7 days and 56% at 28 days of curing. Similarly, the substitution of FA by 20% GGBS, the compressive strength increased by 143% and 82.53% at 7 days and 28 days of curing, respectively. The inclusion of GGBS provides Ca content in addition to the geopolymer matrix, increasing the strength of geopolymer concrete.

17.4.3 Ultrasonic Pulse Velocity

The ultrasonic pulse velocity results of geopolymer mixes with varying content of GGBS are given in Figure 17.5. The observed results show that presence of GGBS enhance the UPV values of geopolymer mixes up to 6% at 28 days of curing. For

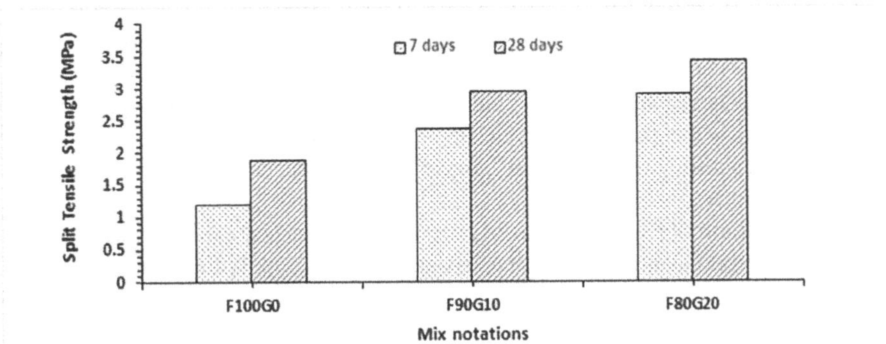

FIGURE 17.4 Development of split tensile strength in geopolymer mixes at 7 and 28 days of curing.

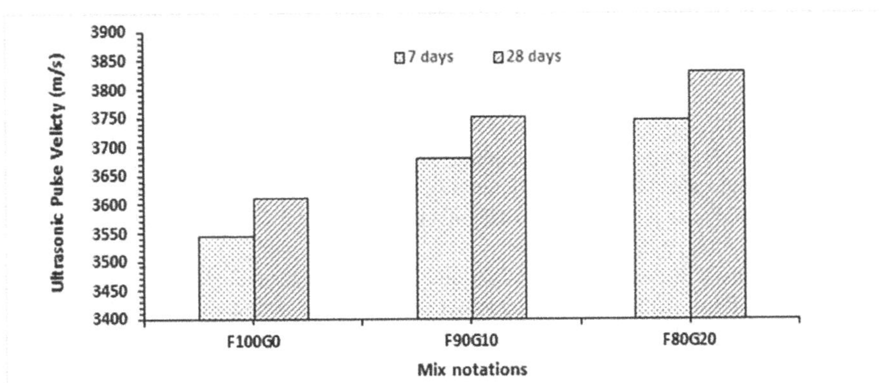

FIGURE 17.5 UPV values for various geopolymer mixes at 7 and 28 days of curing.

mix F90G10, the UPV values increased by 3.8% and 5.6% at 7 and 28 days of curing, respectively. For mix F80G20, the UPV values increased by 3.9% and 6% at 7 and 28 days of curing, respectively. The UPV values lies in the range of good category for all geopolymer mixes per IS: 13311 (Part 1)—1992.

17.5 CONCLUSION

The overall results conclude that addition of GGBS in FA-based geopolymer concrete the overall performance enhanced significantly. The conclusions of experimental work are given here:

1. The maximum compressive strength was achieved with addition of 20% GGBS—for mix F90G10 in geopolymer concrete. The addition of GGBS provide additional calcium/sodium aluminate silicate gel along with geopolymer matrix. The further addition of GGBS may improve more compressive strength.
2. The similar trend was observed for split tensile strength as that of compressive strength of the geopolymer concrete. The maximum strength of 3.92 MPa was obtained at 28 days of curing with addition of 20% GGBS.
3. All geopolymer mixes UPV values fall within the range of the good category with the inclusion of GGBS. The addition of GGBS densify the microstructure of geopolymer concrete; therefore, UPV values increase significantly.
4. The addition of GGBS also overcomes the issue of heat curing of geopolymer concrete as it helps to gain early strength of geopolymer concrete.

REFERENCES

[1] M. B. Ali, R. Saidur, and M. S. Hossain, "A review on emission analysis in cement industries," *Renewable Sustainable Energy Rev.*, vol. 15, no. 5, pp. 2252–2261, 2011.

[2] M. Schneider, M. Romer, M. Tschudin, and H. Bolio, "Sustainable cement production-present and future," *Cem. Concr. Res.*, vol. 41, no. 7, pp. 642–650, 2011.

[3] A. Karthik, K. Sudalaimani, C. T. Vijayakumar, and S. S. Saravanakumar, "Effect of bio-additives on physico-chemical properties of fly ash-ground granulated blast furnace slag based self cured geopolymer mortars," *J. Hazard. Mater.*, vol. 361, pp. 56–63, 2019.

[4] L. N. Assi, K. Carter, E. Deaver, and P. Ziehl, "Review of availability of source materials for geopolymer/sustainable concrete," *J. Clean. Prod.*, vol. 263, p. 121477, 2020.

[5] A. Narayanan and P. Shanmugasundaram, "An experimental investigation on flyash-based geopolymer mortar under different curing regime for thermal analysis," *Energy Build.*, vol. 138, pp. 539–545, 2017.

[6] H. Y. Zhang, V. Kodur, B. Wu, L. Cao, and F. Wang, "Thermal behavior and mechanical properties of geopolymer mortar after exposure to elevated temperatures," *Constr. Build. Mater.*, vol. 109, pp. 17–24, 2016.

[7] S. K. John, Y. Nadir, and K. Girija, "Effect of source materials, additives on the mechanical properties and durability of fly ash and fly ash-slag geopolymer mortar: A review," *Constr. Build. Mater.*, vol. 280, p. 122443, 2021.

[8] A. Hassan and M. Arif, "Effect of curing condition on the mechanical properties of fly ash-based geopolymer concrete," *SN Appl. Sci.*, vol. 1, no. 12, pp. 1–9, 2019.
[9] A. Mehta and R. Siddique, "Properties of low-calcium fly ash based geopolymer concrete incorporating OPC as partial replacement of fly ash," *Constr. Build. Mater.*, vol. 150, pp. 792–807, 2017.
[10] G. Mallikarjuna Rao and T. D. Gunneswara Rao, "A quantitative method of approach in designing the mix proportions of fly ash and GGBS-based geopolymer concrete," *Aust. J. Civ. Eng.*, vol. 16, no. 1, pp. 53–63, 2018.
[11] P. Nath and P. K. Sarker, "Effect of GGBFS on setting, workability and early strength properties of fly ash geopolymer concrete cured in ambient condition," *Constr. Build. Mater.*, vol. 66, pp. 163–171, 2014.
[12] S. Nagajothi and S. Elavenil, "Effect of GGBS addition on reactivity and microstructure properties of ambient cured fly ash based geopolymer concrete," *Silicon*, vol. 13, pp. 507–516, 2021.
[13] K. K. Paramveer Singh, "Development of mix design method based on statistical analysis of different factors for geopolymer concrete," *Front. Struct. Civ. Eng.*, vol. 16, pp. 1315–1335, 2022.
[14] IS: 10262-2019, "Concrete mix proportioning—guidelines," *Bur. Indian Stand. New Delhi*, New Delhi, India, 2019, https://www.services.bis.gov.in/php/BIS_2.0/bisconnect/knowyourstandards/Indian_standards/isdetails/MTA4Mw==
[15] IS : 516 (1959), "Method of tests for strength of concrete," *Bur. Indian Stand. New Delhi*, p. New Delhi, India, 2004.
[16] IS:5816-1999, "Indian standard splitting tensile strength of concrete-method of test," *Bur. Indian Stand*, New Delhi, pp. 1–14, 1999. https://www.services.bis.gov.in/php/BIS_2.0/bisconnect/knowyourstandards/Indian_standards/isdetails/
[17] IS: 13311(Part I), "Non-destructive testing of concrete methods of test (Ultrasonic pulse velocity)," *Bur. Indian Stand.*, New Delhi, 1992. https://www.services.bis.gov.in/php/BIS_2.0/bisconnect/knowyourstandards/Indian_standards/isdetails/
[18] A. Serag Faried, W. H. Sofi, A. Z. Taha, M. A. El-Yamani, and T. A. Tawfik, "Mix design proposed for geopolymer concrete mixtures based on ground granulated blast furnace slag," *Aust. J. Civ. Eng.*, vol. 18, no. 2, pp. 205–218, 2020.

18 Plane Wave Propagation in Thermoelastic Diffusion Medium Using TPLT and TPLD Models

K. D. Sharma, Puneet Bansal, Vandana Gupta

18.1 INTRODUCTION

The non-Fourier models have grown in prominence as a result of their efforts to resolve the issue of heat conduction's limitless speed. In the event of high-speed energy transit and frictional heat transfer, these can partially justify the thermal interactions [1]. Additionally, the transition from macro to micro and nano scale is necessary because of its growing demand for micro and nano engineering. Consequently, changes have been made to the heat transfer equation to include the atomic level properties.

The thermoelasticity theory was first thoroughly formulated on the base of Fourier's law by Biot [2]. Though, infinite speed was a problem for the theory, which is frequently referred to as the classical thermoelastic theory. This is because this theory incorporates Fourier's law. Thus, the non-Fourier heat conduction models–based generalized thermoelastic theories replaced the traditional thermoelastic theory as the preferred theory. Numerous changes have been made to the thermoelastic theory, including the inclusion of phase delays, new consecutive variables, modified consecutive equations, and the inclusion of non-local phenomena. The researchers were drawn to these modified models to examine their many mathematical characteristics.

Tzou [3] initially introduced the idea of phase lag with heat flow, which broadened scope of the concept in the Cattaneo-Vernotte (CV) model. The generalized thermoelasticity theory developed from the Tzou [4, 5] heat conduction model was then provided by Chandrasekharaiah [6]. In the future, Roychoudhuri [7] expanded on Tzou's theory and created a new generalized thermoelastic theory called three-phase-lag thermoelasticity theory.

Using a linked thermoelastic issue, Nowacki [8]–[11] established thermoelastic diffusion theory. This suggests that thermoelastic waves can propagate at unlimited speeds. Sherief et al. [12] permits the limited wave propagation speeds in thermoelastic diffusion medium. Uniqueness, reciprocity theorems are further shown in this study under constrained assumptions on the elastic coefficients.

DOI: 10.1201/9781003367154-18

The fractional order derivative effect on thermoelastic diffusive waves was given by Kumar and Gupta [13]. Chiriţă [14] for an anisotropic and inhomogeneous material, developed uniqueness and continuous dependency results for thermoelastic dual-phase-lag model. Youssef and Alghamdi [15] introduced a mathematical model of thermoelastic skin tissue using dual-phase-lag thermoelasticity. Bazarra [16] took into account the interaction between dual-phase-lag models and the impacts of microtemperatures. Marwan and Zenkour [17] studied the Green-Naghdi models for thermoelastic diffusion having dual-phase lags. Tian, Peng, and He [18] analysed the dual-phase-lag thermoelastic diffusion for a size-dependent microplate using fractional-order heat conduction model.

In the current work, homogeneous isotropic thermoelastic diffusion medium is considered. The field equations, this medium having three-phase-lag models (TPLT and TPLD), are delivered, and plane wave propagation and its characteristics are investigated.

18.2 THREE-PHASE-LAG DIFFUSION MODEL

As Roy Choudhuri [7] derived the three-phase-lag thermal model (TPLT), following the same way, three-phase-lag diffusion (TPLD) model can be derived as follows:

A modified form of Fick's law which includes chemical potential gradient $P_{,i}$ and chemical potential displacement gradient $\kappa_{,i}$ among the constitutive variables is

$$\eta_i = -DP_{,i} - D^* \kappa_{,i} \tag{18.1}$$

where $\dot{\kappa} = P$

The mass concentration law is

$$-\eta_{i,i} = \dot{C} \tag{18.2}$$

where D is the diffusivity, P is the chemical potential per unit mass, C is the concentration and η_i is the flow of diffusing mass vector.

Introducing phase lag of diffusing mass vector η, chemical potential P, and chemical potential displacement κ in equation (18.1), we arrive at

$$\eta_i\left(R,t+\tau_\eta\right) = -DP_{,i}\left(R,t+\tau_P\right) - D^*\kappa_{,i}\left(R,t+\tau_\kappa\right) \tag{18.3}$$

The potential gradient and potential displacement gradient at a point R(r) at time $t+\tau_P$ and at time $t+\tau_\kappa$ results in a mass flux at the same point at time $t+\tau_\eta$. The phase lag of diffusing mass flux vector τ_η represents the delayed time required for the diffusion of the mass flux, the phase lag of chemical potential τ_P represents the delayed time required for the establishment of the potential gradient, and the phase lag of potential displacement gradient τ_κ represents the delayed time required for the establishment of the potential displacement gradient. For the special case of $\tau_\eta = \tau_P = \tau_\kappa = \tau$ (though not necessarily equal to zero), equation (18.3) simplifies to (18.1).

The Taylor series expansion of (3) up to the first-order terms in $\tau_\eta, \tau_P, \tau_\kappa$ leads to the following generalized:

Fick's law valid at point R and time t as:

$$\eta_i\left(1+\tau_\eta\left(\partial / \partial t\right)\right)=-D\mathrm{P}_{,i}\left(1+\tau_P\left(\partial / \partial t\right)\right)-D^{*}\kappa_{,i}\left(1+\tau_\kappa\left(\partial / \partial t\right)\right) \tag{18.4}$$

By taking the gradient of both sides of (4) and using equation (18.2), we arrive at

$$\left(1+\tau_\eta\left(\partial / \partial t\right)\right)\dot{C}=D\left(1+\tau_P\left(\partial / \partial t\right)\right)\mathrm{P}_{,ii}+D^{*}\left(1+\tau_\kappa\left(\partial / \partial t\right)\right)\kappa_{,ii} \tag{18.5}$$

18.3 HYPERBOLIC THREE-PHASE-LAG DIFFUSION MODEL

Retaining terms of the order τ_η^2 in the Taylor's expansion of the generalized diffusion law (5), we have

$$\begin{aligned}&\left(1+\tau_P\left(\partial / \partial t\right)\right)\mathrm{DP}_{,ii}+\left(1+\tau_\kappa\left(\partial / \partial t\right)\right)\mathrm{D}^{*}\kappa_{,ii}-\left(\partial / \partial t\right)\\&\left(1+\tau_\eta\left(\partial / \partial t\right)+\left(\tau_\eta^2 / 2\right)\left(\partial / \partial t\right)^2\right)C=0\end{aligned} \tag{18.6}$$

The relation between chemical potential P and mass concentration C is

$$P=-\beta_2\,\mathrm{e}_{kk}-\mathrm{aT}+\mathrm{bC} \tag{18.7}$$

Using equation (18.7) in (18.6), we arrive at the mass diffusion equation in this case, namely,

$$\begin{aligned}&\left(\left(\partial / \partial t\right)+\tau_P\left(\partial^2 / \partial t^2\right)\right)\left(D\beta_2 e_{kk,ii}+DaT_{,ii}-DbC_{,ii}\right)\\&+\left(1+\tau_\kappa\left(\partial / \partial t\right)\right)\left(D\beta_2 e_{kk,ii}+DaT_{,ii}-DbC_{,ii}\right)\\&+\left(\partial^2 / \partial t^2\right)\left(1+\tau_\eta\left(\partial / \partial t\right)+\tau_\eta^2\left(\partial^2 / \partial t^2\right)\right)C=0\end{aligned} \tag{18.8}$$

18.4 GOVERNING EQUATIONS

The basic equations for homogeneous isotropic thermoelastic diffusion with TPLT and TPLD models are together with the equations (18.2) and (18.7).

18.4.1 The Constitutive Relations

$$\sigma_{ij}=2\mu e_{ij}+\delta_{ij}\left[\lambda e_{kk}-\beta_1 T-\beta_2 C\right] \tag{18.9}$$

$$-q_{i,i}=\rho T_0\dot{S} \tag{18.10}$$

$$\rho T_0 S=\left(1+\tau_q\left(\partial / \partial t\right)+\left(\tau_q^2 / 2\right)\left(\partial / \partial t\right)^2\right)\left(\rho C_E T+\beta_1 T_0 e_{kk}+aT_0 C\right) \tag{18.11}$$

18.4.2 Equations of Motion

$$\sigma_{ij,j} + \rho F_i = \rho \ddot{u}_i \tag{18.12}$$

18.4.3 Equation of Heat Conduction

$$q_{i,i} = -\left[\left(1+\tau_t\left(\partial/\partial t\right)\right)KT_{,ii} + \left(1+\tau_v\left(\partial/\partial t\right)\right)K^* v_{,ii}\right] \tag{18.13}$$

where $\dot{v} = T$

Equations (18.6) and (18.9)–(18.13) constitute the basic equations for homogeneous isotropic thermoelastic diffusion with thermal and diffusion phase lags, where λ, μ are the Lame's constants; ρ is the density assumed to be independent of time; u_i are the components of displacement vector u; K is the coefficient of thermal conductivity; C_E is the specific heat at constant strain; τ_t, τ_q, τ_v are the phase lags of the temperature gradient, heat flux, and thermal displacement gradient, respectively; $T = \Theta - T_0$ is small temperature increment; Θ is the absolute temperature of the medium; T_0 is the reference temperature of the body chose such that $\left|\left(T/T_0\right)\right| << 1$, a and b are, respectively, the coefficients describing the measure of thermodiffusion and mass diffusion effects, respectively; σ_{ij}, e_{ij} are the components of the stress and strain, respectively; e_{kk} is the dilatation; S is the entropy per unit mass; $\beta_1 = (3\lambda + 2\mu)v_t$ and $\beta_2 = (3\lambda + 2\mu)v_c, v_t$ is the coefficient of thermal linear expansion; v_c is the coefficient of linear diffusion expansion. In these equations, a comma followed by a suffix denotes spatial derivative and a superposed dot denotes the derivative with respect to time.

After using the chemical potential as a state variable instead of the concentration, the equations (18.7), (18.9) and (18.11) can be rewritten as:

$$C = \gamma_2 \, \mathrm{e}_{kk} + d\,\mathrm{T} + nP \tag{18.14}$$

$$\sigma_{ij} = 2\mu e_{ij} + \delta_{ij}\left[\lambda_0 e_{kk} - \gamma_1 T - \gamma_2 P\right] \tag{18.15}$$

$$\rho S = \left(1 + \tau_q\left(\partial/\partial t\right) + \left(\tau_q^2/2\right)\left(\partial^2/\partial t^2\right)\right)\left(\varepsilon T + \gamma_1 e_{kk} + dP\right) \tag{18.16}$$

where

$$\gamma_1 = \beta_1 + (a/b)\beta_2, \gamma_2 = \left(\beta_2/b\right), \lambda_0 = \lambda - \left(\beta_2^2/b\right)$$
$$\varepsilon = \left(\rho C_E/T_0\right) - \left(a^2/b\right), \mathrm{d} = (a/b), n = (1/b)$$

18.5 PLANE WAVE PROPAGATION

From equations (18.9)–(18.13), we have

$$(\lambda + \mu)u_{j,ij} + \mu u_{i,jj} - \beta_1 T_{,i} - \beta_2 C_{,i} = \rho \ddot{u}_i \tag{18.17}$$

$$\left(K^* \tau_v^* + K\tau_t^*\left(\partial/\partial t\right)\right)T_{,ii} = \tau_q^*\left(\rho C_E \ddot{T} + \beta_1 T_0 \ddot{e}_{kk} + aT_0 \ddot{C}\right) \tag{18.18}$$

$$\left(D^* \tau_K^* + D\tau_P^*\left(\partial/\partial t\right)\right)\left(\beta_2 e_{kk,ii} + aT_{,ii} - bC_{,ii}\right) + \tau_\eta^* \ddot{C} = 0 \tag{18.19}$$

where

$$\tau_t^* = 1+\tau_t(\partial/\partial t), \tau_q^* = 1+\tau_q(\partial/\partial t)+(\tau_q^2/2)(\partial^2/\partial t^2), \tau_v^* = 1+\tau_v(\partial/\partial t)$$
$$\tau_P^* = 1+\tau_P(\partial/\partial t), \tau_\eta^* = 1+\tau_\eta(\partial/\partial t)+(\tau_\eta^2/2)(\partial^2/\partial t^2), \tau_\kappa^* = 1+\tau_\kappa(\partial/\partial t)$$

For 2D problem, we take

$$u = (\mathrm{u}_1, 0, \mathrm{u}_3), T(x_1, x_3, t) \tag{18.20}$$

Dimensionless quantities are taken as

$$\begin{aligned} &x_1' = (\omega_1^* x_1 / c_1), x_3' = (\omega_1^* x_3 / c_1), u_1' = (\omega_1^* u_1 / c_1), u_3' = (\omega_1^* u_3 / c_1),\\ &t' = \omega_1^* t, T' = (\beta_1 T / \rho c_1^2), C' = (\beta_2 C / \rho c_1^2), \tau_q' = \omega_1^* \tau_q, \tau_t' = \omega_1^* \tau_t,\\ &\tau_v' = \omega_1^* \tau_v, \tau_P' = \omega_1^* \tau_P, \tau_\eta' = \omega_1^* \tau_\eta, \tau_\kappa' = \omega_1^* \tau_\kappa, \sigma_{ij}' = (\sigma_{ij} / \rho c_1^2) and\\ &\omega_1^* = (\rho C_E c_1^2 / K), c_1^2 = ((\lambda + 2\mu)/\rho) \end{aligned} \tag{18.21}$$

Relation between u_1, u_3 and potential functions ϕ, ψ is

$$u_1 = (\partial\phi/\partial x_1) - (\partial\psi/\partial x_3), u_3 = (\partial\phi/\partial x_3) + (\partial\psi/\partial x_1) \tag{18.22}$$

Using (21) in (17)–(19) and with the help of (20) and (22), we obtain

$$\nabla^2\phi - T - C = \ddot{\phi} \tag{18.23}$$

$$\nabla^2\psi - (\ddot{\psi}/\delta^2) = 0 \tag{18.24}$$

$$(\zeta_1\tau_v^* + \tau_t^*(\partial/\partial t))\nabla^2 T = \tau_q^*(\zeta_2\ddot{T} + \zeta_3\nabla^2\ddot{\phi} + \zeta_4\ddot{C}) \tag{18.25}$$

$$\begin{aligned} &(\tau_\kappa^* + \tau_P^*\zeta_5(\partial/\partial t))\nabla^4\phi - (\tau_\kappa^*\zeta_7 + \tau_P^*\zeta_6(\partial/\partial t))\nabla^2 C\\ &+(\tau_\kappa^*\zeta_9 + \tau_P^*\zeta_8(\partial/\partial t))\nabla^2 T + \tau_\eta^*\zeta_{10}\ddot{C} = 0 \end{aligned} \tag{18.26}$$

Where,

$$\begin{aligned} &c_2 = \sqrt{(\mu/\rho)}, \delta^2 = (c_2^2/c_1^2), \zeta_1 = (K^*/K\omega_1^*), \zeta_2 = 1, \zeta_3 = (\beta_1^2 \mathrm{T}_0 / \rho K \omega_1^*)\\ &\zeta_4 = (a\beta_1 \mathrm{T}_0 c_1^2 / K\beta_2\omega_1^*), \zeta_5 = (D\omega_1^*/D^*), \zeta_6 = (Db\rho c_1^2\omega_1^* / D^*\beta_2^2), \zeta_7 = (b\rho c_1^2/\beta_2^2)\\ &\zeta_8 = (aD\rho c_1^2 / D^*\beta_1\beta_2), \zeta_9 = (a\rho c_1^2/\beta_1\beta_2), \zeta_{10} = (\rho c_1^4 / D^*\beta_2^2)\\ &e_{kk} = (\partial u_1/\partial x_1) + (\partial u_3/\partial x_3), \nabla^2 \equiv (\partial^2/\partial x_1^2) + (\partial^2/\partial x_3^2) \end{aligned}$$

18.6 PROBLEM SOLUTION

For plane harmonic wave propagation ($x_1 x_3$ – plane)

$$\{\phi, \psi, T, C\}(x_1, x_3, t) = \{\bar{\phi}, \bar{\psi}, \bar{T}, \bar{C}\} \exp\left[i\left(\xi x_m n_m - \omega t\right)\right] \tag{18.27}$$

Here ω and ξ are the angular frequency and complex wave number. $\bar{\phi}, \bar{\psi}, \bar{T}, \bar{C}$ are undetermined amplitude vectors that are independent of time t and coordinates x_i, n_m is the unit normal vector.

Upon using (27) in (23)-(26) and for non-trivial solution, we have

$$A_1 \xi^6 + A_2 \xi^4 + A_3 \xi^2 + A_4 = 0 \tag{18.28}$$

and the equation

$$\left(-\xi^2 + \left(\omega^2 / \delta^2\right)\right)\bar{\psi} = 0 \tag{18.29}$$

where

$$q_1^* = \omega^2 \tau_q^{10} \zeta_3, q_2^* = \tau_\nu^{10} \zeta_1 - i\omega \tau_t^{10}, q_3^* = \omega^2 \tau_q^{10} \zeta_2, q_4^* = \omega^2 \tau_q^{10} \zeta_4$$
$$q_5^* = -i\omega \tau_P^{10} \zeta_5 + \tau_\kappa^{10}, q_6^* = -i\omega \tau_P^{10} \zeta_8 + \zeta_9 \tau_\kappa^{10}, q_7^* = -\omega^2 \tau_\eta^{10} \zeta_{10}$$
$$q_8^* = i\omega \tau_P^{10} \zeta_6 - \zeta_7 \tau_\kappa^{10}, \tau_t^{10} = 1 - i\omega \tau_t, \tau_q^{10} = 1 - i\omega \tau_q - \omega^2 \tau_q^2$$
$$\tau_\nu^{10} = 1 - i\omega \tau_\nu, \tau_P^{10} = 1 - i\omega \tau_P, \tau_\eta^{10} = 1 - i\omega \tau_\eta - \omega^2 \tau_\eta^2$$
$$\tau_\kappa^{10} = 1 - i\omega \tau_\kappa$$

$$A_1 = q_2^*\left(q_8^* + q_4^* q_5^*\right), A_2 = q_2^* q_7^* + q_3^* q_8^* - q_4^* q_6^* + \omega^2 q_2^* q_8^* + q_1^* q_8^* - q_4^* q_5^* + q_4^*\left(q_3^* q_5^* - q_1^* q_6^*\right)$$
$$A_3 = q_3^* q_7^* + \omega^2\left(q_2^* q_7^* + q_3^* q_8^* - q_4^* q_6^*\right) + q_1^* q_8^*, A_4 = \omega^2 q_3^* q_7^*$$

From equations (18.28) and (18.29), we obtain three quasi waves explicitly, longitudinal qP, thermal qT and mass-diffusion qMD, and one transverse wave.

18.6.1 Phase Velocity

$$V_i = \frac{\omega}{\left|\mathrm{Re}\left(\xi_i\right)\right|}, i = 1, 2, 3 \tag{18.30}$$

Here $V_i\ (i = 1, 2, 3)$ are the velocities of qP, qT, and qMD waves.

18.6.2 Attenuation Coefficient

$$Q_i = \mathrm{Im}\left(\xi_i\right), \quad i = 1, 2, 3 \tag{18.31}$$

Here, $Q_i\ (i = 1, 2, 3)$ are the attenuation coefficients of qP, qT, and qMD waves.

18.7 NUMERICAL RESULTS

For copper material (Sherief et al. [12]):

$$\lambda = 7.76\times10^{10}\,Kgm^{-1}s^{-2},\ \mu = 3.86\times10^{10}\,Kgm^{-1}s^{-2},\ T_0 = 0.293\times10^{3}\,K,$$
$$C_E = .3831\times10^{3}\,JKg^{-1}K^{-1}$$
$$\alpha_t = 1.78\times10^{-5}\,K^{-1},\ \rho = 8.954\times10^{3}\,Kgm^{-3},\ K = 0.383\times10^{3}\,Wm^{-1}K^{-1}$$
$$K^* = \left(\left(\lambda+2\mu\right)C_E/4\right)Wm^{-1}s^{-1}K^{-1}$$

We take diffusion parameters:

$$\alpha_c = 1.98\times10^{-4}\,Kg^{-1}m^{3},\ a = 1.2\times10^{4}\,m^{2}s^{-2}K^{-1},\ b = 9\times10^{5}\,Kg^{-1}m^{5}s^{-2}$$
$$D = 0.85\times10^{-8}\,Kgsm^{-3},\ D^* = 0.65\times10^{3}\,Kgm^{-3}$$
$$\tau_v = 0.9s,\ \tau_t = 1.0s,\ \tau_q = 1.1s,\ \tau_\kappa = 1.2s,\ \tau_P = 1.3s,\ \tau_\eta = 1.4s$$

Phase velocity and attenuation coefficient for altered values of frequency (ω) ranging from 0 to 5 Hz for different values of chemical potential displacement gradient: $D^* = \left(0.65, 0.065\right)\times10^{3}$. In all the figures circles and dash lines corresponds to $D^* = \left(0.65, 0.065\right)\times10^{3}$, respectively.

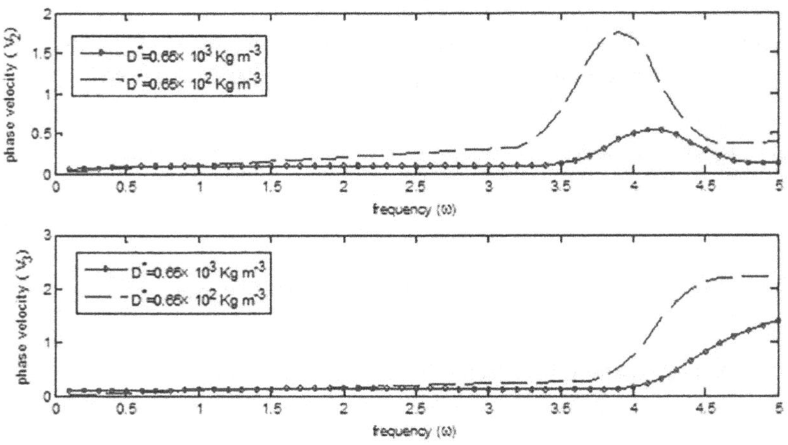

FIGURE 18.1 (a, b, c) Variation of phase velocities (V_1, V_2, V_3) with frequency (w).

18.7.1 Phase Velocity

Figure 18.1(a) illustrates that phase velocity initially increases, decreases for $\omega \geq 4$. Figure 18.1(b) shows that phase velocity increases, becomes maximum at ω = 3, and thereafter decreases and becomes stationary. Figure 18.1(c) shows that phase velocity increases, attains maxima, and becomes stationary having this value of maxima. In Figures 18.1 (a, b, c), increase in the value of D^* decreases phase velocity. For small frequency, the phase velocity for qP wave is much larger than for qT and qMD waves.

18.7.2 Attenuation Coefficient

Figure 18.2(a) depicts that with increase in frequency, Q_1initially increases, attains maximum for ω=3, and then decreases till it becomes constant. Figure 18.2(b) shows some oscillation in values of Q_3, in the range $0 < \omega < 1$, then decreases and attains minimum value. It is clear from Figure 18.2(c) that Q_3 initially increases, attains maxima, and thereafter behaves stationary. In Figures 18.1 (a, b, c), increase in D^* decreases phase velocity. In Figures 18.2 (a, b, c), increase in D^* increases attenuation coefficient. For $\omega \leq 2$, the value of attenuation coefficients is small in comparison to its values for $\omega > 2$.

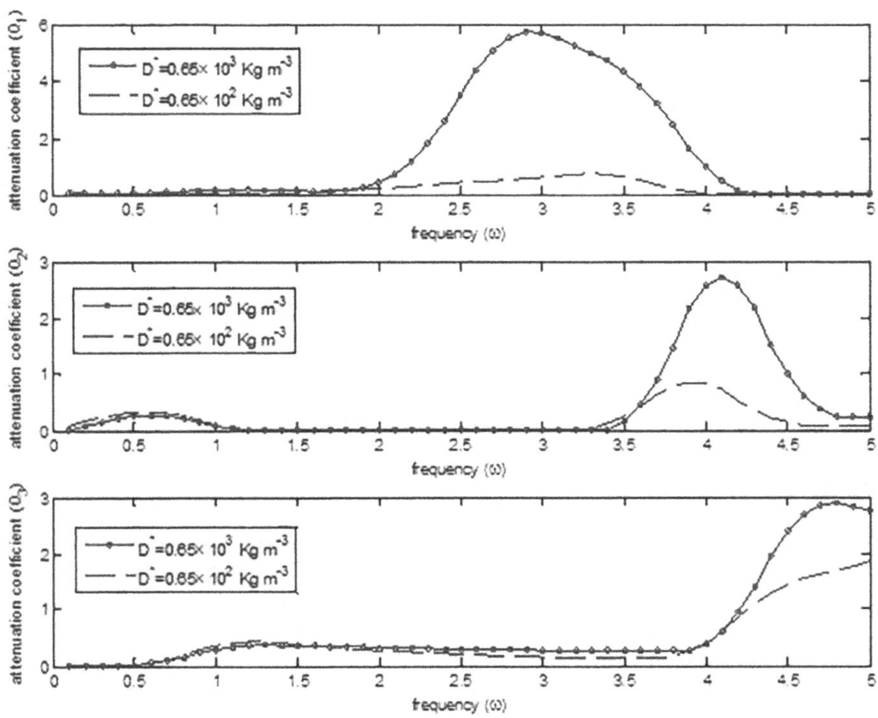

FIGURE 18.2 (a, b, c) Variation of attenuation coefficients (Q_1, Q_2, Q_3) with frequency (w).

18.8 CONCLUSION

1. For the two-dimensional problem, plane wave propagation in thermoelastic diffusion with TPLT and TPLD models was studied, and it was found that quasi-longitudinal wave (qP), quasi-thermal waves (qT), a quasi-mass diffusion wave (qMD), and one transverse wave propagate in this medium.
2. On computing the characteristics of waves, we found that phase velocity is inversely proportional to frequency, whereas attenuation coefficient is directly proportional to frequency.
3. The obtained coefficients are sensitive to the value of diffusion parameter D^*, and this analysis is significant in continuum mechanics.
4. The TPLT and TPLD models are useful in the further study of thermoelastic diffusion mediums.

REFERENCES

[1] Kumar R and Gupta V, Dual-phase-lag model of wave propagation at the interface between elastic and thermoelastic diffusion media. *Journal of Engineering Physics and Thermophysics* 88: 247–259, 2015.

[2] Biot MA, Thermoelasticity and irreversible thermodynamics. *Journal of Applied Physics* 27: 240–253, 1956.

[3] Tzou DY, Thermal shock phenomena under high rate response in solids. *Annual Review of Heat Transfer* 4(4), 1992.

[4] Tzou DY, A unified field approach for heat conduction from macro-to micro-scales. *Journal of Heat Transfer* 117(1): 8–16, 1995. http://doi.org/10.1115/1.2822329.

[5] Tzou DY, The generalized lagging response in small-scale and high-rate heating. *International Journal of Heat and Mass Transfer* 38(17): 3231–3240, 1995. http://doi.org/10.1016/0017-9310(95)00052-B.

[6] Chandrasekharaiah DS, Hyperbolic thermoelasticity: A review of recent literature. *Applied Mechanics Reviews* 51(12): 705–729, 1998. http://doi.org/10.1115/1.3098984.

[7] Choudhuri SKR, On a thermoelastic three-phase-lag model. *Journal of Thermal Stresses* 30(3): 231–238, 2007. http://doi.org/10.1080/01495730601130919.

[8] Nowacki W, Dynamical problems of thermodiffusion in elastic solids. *Proceedings of Vibration Problems* 15: 105–128, 1974a.

[9] Nowacki W, Dynamical problems of thermodiffusion in solids I. *Bulletin of the Polish Academy of Sciences, Series IV(Technical Sciences)* 22: 55–64, 1974b.

[10] Nowacki W, Dynamical problems of thermodiffusion in solids II. *Bulletin of the Polish Academy of Sciences, Series IV(Technical Sciences)* 22: 205–211, 1974c.

[11] Nowacki W, Dynamical problems of thermodiffusion in solids III. *Bulletin of the Polish Academy of Sciences, Series IV(Technical Sciences)* 22: 257–266, 1974d.

[12] Sherief HH, Hamza FA and Saleh HA, The theory of generalized thermoelastic diffusion. *International Journal of Engineering Science* 42: 591–608, 2004.

[13] Kumar R and Gupta V, Uniqueness, reciprocity theorem, and plane waves in thermoelastic diffusion with a fractional order derivative. *Chinese Physics B* 22, 2013. https://doi.org/10.1088/1674-1056/22/7/074601.

[14] Chiriţă S, On the time differential dual-phase-lag thermoelastic model. *Meccanica* 52: 349–361, 2016. https://doi.org/10.1007/s11012-016-0414-2.

[15] Youssef HM and Alghamdi NA, Modeling of one-dimensional thermoelastic dual-phase-lag skin tissue subjected to different types of thermal loading. *Scientific Reports* 10: 3399, 2020. https://doi.org/10.1038/s41598-020-603426.

[16] Bazarra N, Copetti MIM, Fernández JR, Quintanilla R, Numerical analysis of a dual-phase-lag model with microtemperatures. *Applied Numerical Mathematics* 166: 1–25, 2021.

[17] Kutbi MA and Zenkour AM, Refined dual-phase-lag Green–Naghdi models for thermoelastic diffusion in an infinite medium. *Waves in Random and Complex Media* 32: 1–21, 2020.

[18] Tian L, Peng W and He T, Dual-phase-lag thermoelastic diffusion analysis of a size-dependent microplate based on modified fractional-order heat conduction model. *Journal of Applied Mathematics and Mechanics* 102, 2022. https://doi.org/10.1002/zamm.202200124.

19 Electro-Discharge Coating of the Surface Using the WC-Cu P/M Electrode Tool

Harvinder Singh, Santosh Kumar, Satish Kumar, Rakesh Kumar

19.1 INTRODUCTION

Electro-discharge machining (EDM), called non-conventional metal removal process, utilized to machined extremely hard metals that are highly complicated to machine. EDM is a heating process, which comprises vaporization and melting of a pair of electrode and workpiece. Several studies have been carried out by several researchers employing the electrode materials Cu + $ZrBr_2$, WC + Fe, and W + Cu. Numerous authors have reported on the bonding power, corrosion resistance, and tribological characteristics of EDM alloyed layers [1–6]. Moro et al. [7] investigated the use of electrical discharge coating (EDC) in place of PVD or CVD process to extend the life of cutting tools (CVD). Tool electrodes are created by semi-sintering TiC powder on an S45C (JIS) substrate for 1 hour at 900°C. The experiment was conducted for 16 minutes at an 8 amp discharge current, a tonality time of 8 s, and a duty factor of 5.9%. The correlation between an electrode's maximum thickness and rate of wear has been studied. Peak current use ranged from 1 to 18 amps, and on-time was between 6 and 18 seconds, according to Simao et al.'s investigation on hardened AISI D2 Sendzimir rolls' surface modification/alloying and coupled electrical discharge texturing, which is frequently used to produce SS strip. Using green compact and heated electrodes mainly made of WC/Co and TiC/WC/Co in powder metallurgy (PM) has been demonstrated to significantly improve roll life and performance. The topographical property of an EDT sample treated with PM electrodes were investigated to be equivalent to those obtained when using conventional copper or graphite tools, utilizing the identical pulse generator settings. The Ti, W, and C contents of the PM electrodes, combined with the C breakdown of the dielectric medium, resulted in a variety of compounds that were delivered to the workpiece surface after sparking, according to GDOES [8] analysis. The roll white layer's microhardness increased when sintered tool electrodes (TiC/WC/Co) were applied up to 950 HK (0.025). When using common tool electrodes, this result measured was

 DOI: 10.1201/9781003367154-19

much higher than the previously reported average "roll white layer hardness was 600 HK (0.025) or the heated AISI D2 roll matrix hardness was 800 HK (0.025)" [9–17]. Various researchers [18–22] reported that the resistance against corrosion can be enhanced by adding chromium particles or by chromium coatings. Simao et al. [23] used design of experiment (Taguchi approach) for the determination of most effective variables and their associated levels on specific output measures. However, the current investigation analyses the electro-discharge coating of the surface using the WC-Cu P/M electrode tool.

19.2 EXPERIMENTAL DETAILS AND PROCEDURE

19.2.1 Green Compact of Heated P/M Tool Preparation

Since the compact does not sinter in green compacted instruments, the technique basically corresponds to the moniker "green compact." Additionally, the particles were only weakly compressed utilizing a die and press, which allowed the tool metal to readily separate from the tool electrode and deposit across the work surface during the deposition process. During the procedure, a power press with a 15-ton weight capacity was employed. Machining is used to make the tool extension, and brazing is used to attach tools made in presses.

1. Tool fabricated using pure Cu
2. Tool manufactured of (W + Cu): by blending of 70% W + 30% Cu with 50% W + 50% Cu by wt%

The details regarding powder companion, proportion, and press capacity are summarized in Table 19.1.

The characteristics of substrate and tool material is given in Table 19.2.

A tool electrode with dimensions of 15 mm in diameter and 10 mm in height is present on every sample electrode. It is not at all necessary to create heighted tool electrodes and retain them in the EDM machine since green compacted tool electrodes are very sensitive when they are packed loosely. For the aforementioned reasons, tool electrode extensions have been created, and powder green compressed electrodes are brazed onto the tips of the extensions.

TABLE 19.1
Press Capacity, Ratio, and Powder Compaction

Capacity of Press	15-25 Tons
Load used	2.7 and 3.6 tons (depends on dimensions of compact)
Holding duration	2 min
Powders ratio (W:Cu)	50:50 wt% and 70:30 wt%
Compaction pressures	150 and 200 MPa

TABLE 19.2
Characteristics of Work Material of the Tools

Material	Density (gm/cm^3)	Melting Temp. (K)	Specific Heat	Thermal Conductivity	Coefficient of Thermal Expansion (/K)	Particle size (Microns)	Mesh Size
Cu powder	8.97	1,355	385	393	16.5	44	325
W powder	19.29	3,683	138	166	4.5	44	325
MS substrate	6.92	1,644	490	20	12	–	–

TABLE 19.3
Some Constant Parameters of EDM

Voltage (V)	40 V
Duty factor	50%
Ton	100 μs
Duration of exp.	20 min

TABLE 19.4
The Experiment Analysis of EDM

Expt. No.	Powder Ratio (W:Cu) wt%	Pressure (MPa)	Current (A)
1	50:50	150	2
2	50:50	150	4
3	50:50	150	6
4	50:50	150	8
5	50:50	200	2
6	50:50	200	4
7	50:50	200	6
8	50:50	200	8
9	70:30	150	2
10	70:30	150	4
11	70:30	150	6
12	70:30	150	8
13	70:30	200	2
14	70:30	200	4
15	70:30	200	6
16	70:30	200	8

19.2.2 Experimental Procedures

The major parameters that apply to all experimental settings are shown in Table 19.3 and Table 19.4. Electronic weight machine was used to measure the weight of the tools and workpieces, and the results are accurate to three decimal places. The workpieces' and tools' weights have been measured before and after the coating, and the quantity of deposition has been estimated. Figure 19.1 depicts the coated surface at various currents and the surface coated by a tool compacted with W and Cu mixed powder at currents of 8 A and compaction pressures of 150 and 200 MPa, respectively.

19.3 RESULTS AND DISCUSSION

The experimental details are mentioned in Table 19.5.

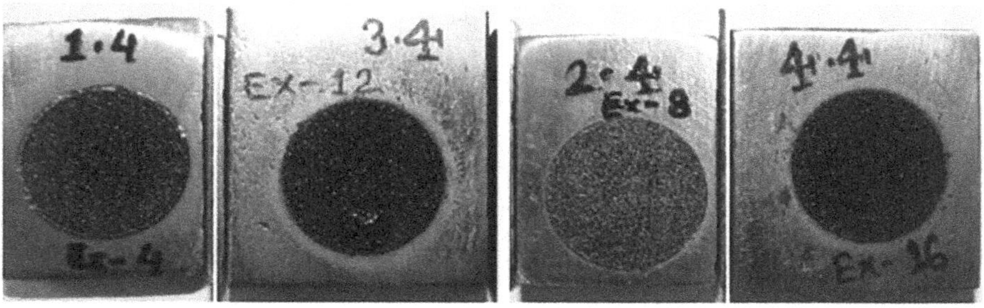

FIGURE 19.1 WC-Cu-coated MS surface prepared by utilizing the tool electrode processed with W:Cu = (a) 50:50, (b) 70:30 wt%, and 150 MPa, respectively, compaction pressures and peak current 8 amp (EDM parameters) deals upon outcomes obtained (Figures 19.2 to 19.5).

TABLE 19.5
Experimental Details

	Weight of the Workpiece			Weight of the Tool		
Expt. No.	Before (A)	After (B)	Deposition (B-A)	Before (C)	After (D)	Tool Wear (C-D)
1	26.118	26.124	0.006	115.213	115.193	0.020
2	23.793	23.807	0.014	115.183	115.113	0.070
3	23.328	23.358	0.030	115.103	114.843	0.260
4	17.718	17.835	0.117	114.843	112.071	2.772
5	19.405	19.408	0.003	114.307	114.298	0.009
6	20.801	20.809	0.008	114.298	114.258	0.040
7	22.148	22.160	0.012	114.241	114.111	0.130
8	18.323	18.353	0.030	114.090	113.102	0.988
9	24.683	24.693	0.010	118.303	118.246	0.057
10	27.705	27.750	0.045	122.797	122.540	0.257
11	28.221	28.367	0.146	122.540	121.335	1.205
12	27.741	28.123	0.382	121.335	118.308	3.027
13	25.949	25.996	0.047	137.996	137.841	0.155
14	27.585	27.705	0.120	137.841	136.891	0.950
15	27.112	27.341	0.229	136.891	134.868	2.023
16	26.757	27.190	0.433	134.868	131.330	3.538

19.3.1 Major Consequences of Composition on Coating Deposition

When the compaction pressure (150 MPa or 200 MPa) remains constant, the deposition rate rises as the current does. The rate of deposition rises gradually at lower currents (up to 6 amp) but rapidly increases at higher currents (6 to 8 amp). It is clear from the graphs that the deposit rate is greater when the compact includes more

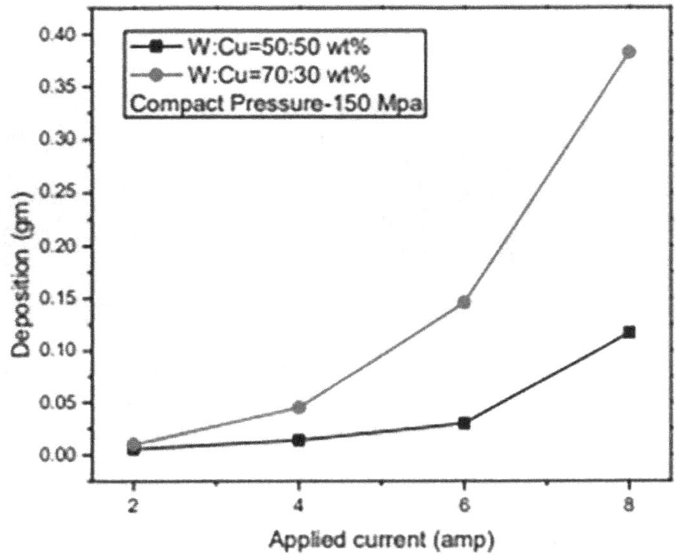

FIGURE 19.2 Consequence of composition for 150 MPa fixed pressure on deposition.

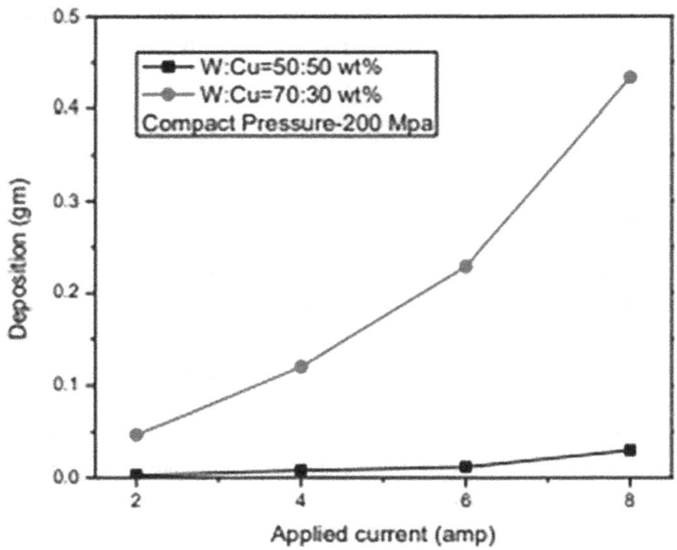

FIGURE 19.3 Consequence of composition for 200 MPa compact pressure on deposition.

tungsten by weight than copper. The deposition curve is sharp for both the compositions even at higher currents. In contrast to greater copper amounts, the curve is steeper with larger tungsten amounts.

19.3.2 Effect of Composition on Tool Wear

The impact of "current corresponding to tool wear rate" is shown in Figures 4.3.1 and 4.3.2. It is obvious that a rise in current results in a high rate of tool wear. As

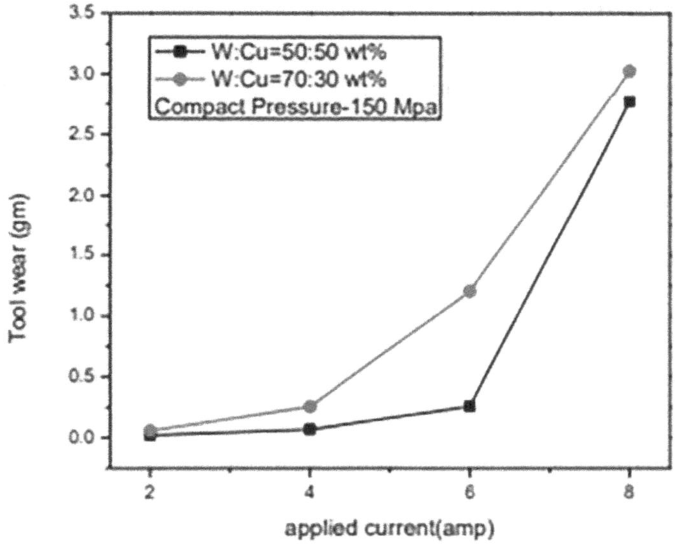

FIGURE 19.4 Consequence of composition for 150 MPa fixed pressure on tool wear.

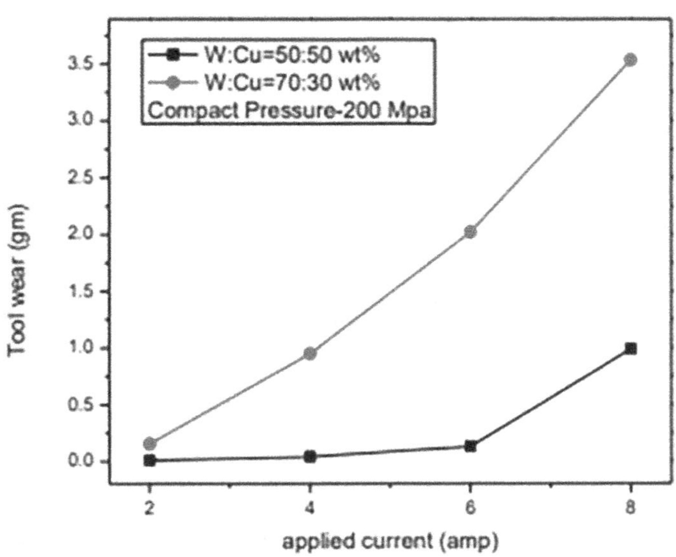

FIGURE 19.5 Consequence of composition for 200 MPa fixed pressure on tool wear.

current climbs from 2 to 6 amps, the rate at which tools wear out increases gradually. As current continues to grow, however, the rate at which tools wear out increases as well, leading to more deposition and increased wear on the tools themselves. Tool wear is considerably greater in compositions with higher tungsten content. At higher compaction pressures compared to lower pressures, deposition rate and tool wear are both more gradual. Based on these results, it can be said that the compaction pressure should be kept within limits that do not affect the deposition's quality.

19.3.3 XRD Analysis

The X-ray diffraction (XRD) image is shown in Figure 19.6.

The substrate surface is scanned for compositional components using XRD. In this study project, the coating's XRD analysis produces the graph, which displays the diffraction apex of WC, C, and W. In tabular form, the decompositional elements corresponding to the angle are shown.

19.3.4 SEM

A SEM may generate SE, BSE, light, distinctive X-rays, transmitted electrons, and sample current, among other forms of signals. All SEMs come equipped with secondary electron detectors; however, it is uncommon for a single instrument to include detectors for every potential signal. The signals are the consequence of interactions

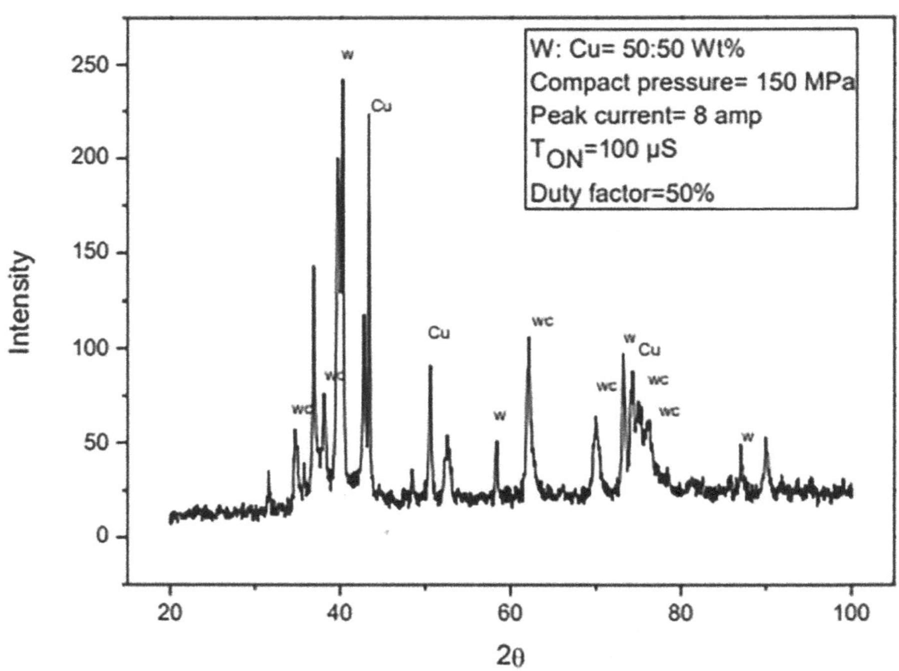

FIGURE 19.6 XRD image at 50:50 W:Cu wt%, 150 MPa (coated specimen) originated with peak current of 8 amp, and compact pressure.

between the atoms and the electron beam on or near the sample surface. SEM may provide high-resolution pictures of a substrate surface in the most typical or standard detection mode, exposing features as small as 1 nm (Figure 19.7 to Figure 19.10).

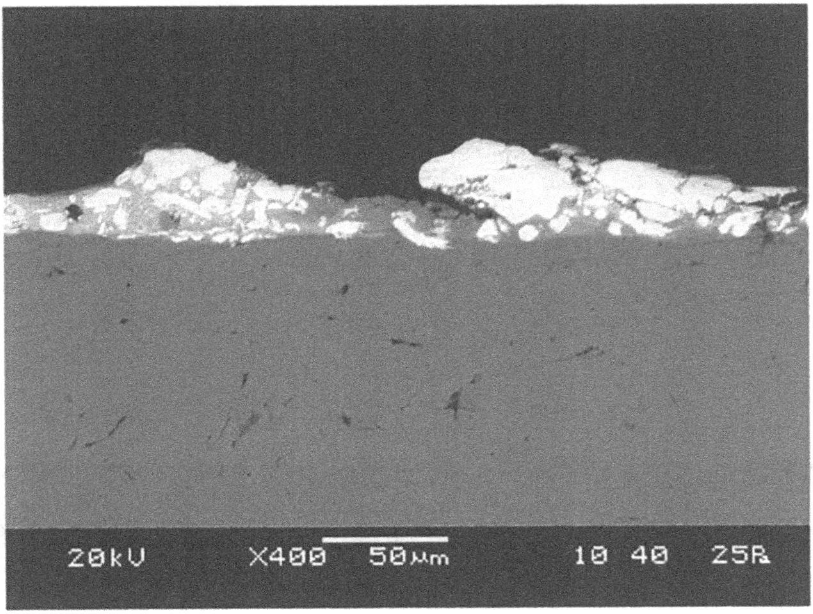

FIGURE 19.7 SEM image: at 4 amp and 200 MPa (50:50).

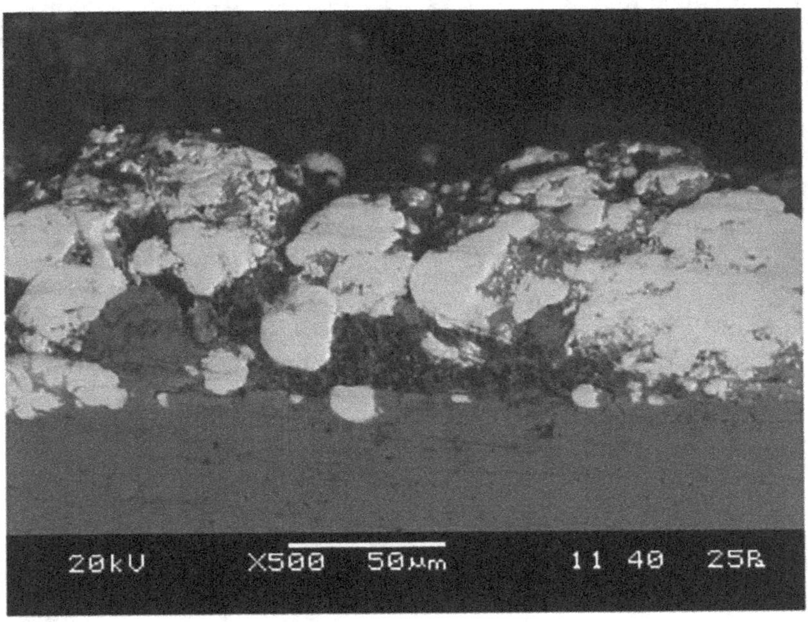

FIGURE 19.8 SEM image: at 4 amp at 200 MPa (70:30).

FIGURE 19.9 SEM image: at 6 amp and 200 MPa (50:50).

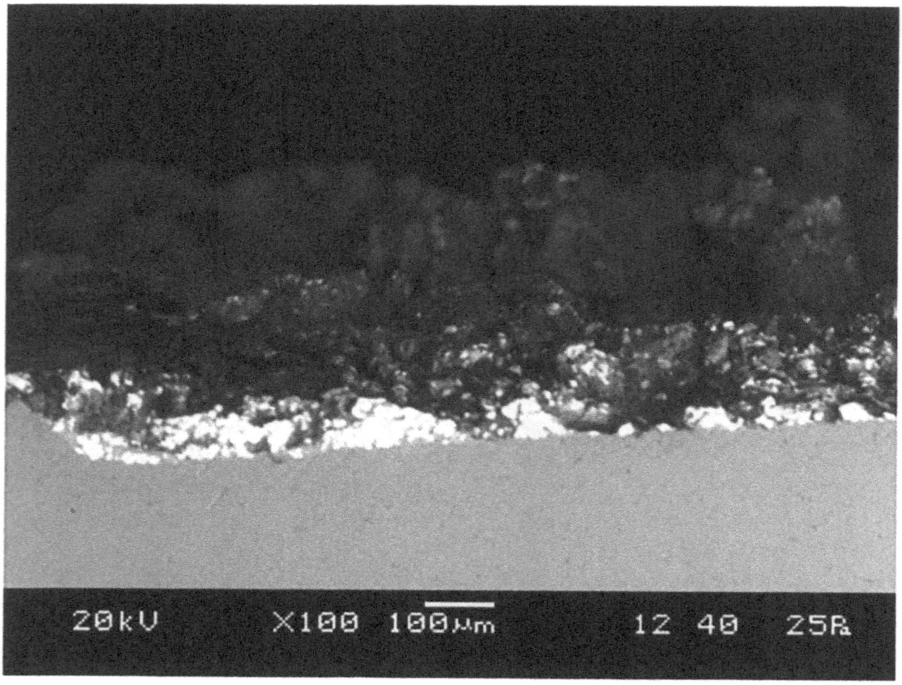

FIGURE 19.10 SEM image: at 6 amp and 200 MPa (70:30).

These pictures were captured using a SEM, and analysis led to the results listed here:

1. When there is a significant percentage of tungsten in a powder, the rate of deposition is substantially greater.
2. The composition with more tungsten powder has a larger average layer thickness over the substrate surface.
3. The sole variable that influences the mean/average layer thickness of deposition with fixed pressure is composition; the other variables are current and compaction pressure.

19.4 CONCLUSIONS

1. Lower compaction pressures and a green compact tool electrode result in a thicker layer of coating covering the surface.
3. SEM study of the coating reveals that it was uniform, and XRD study reveals the existence of WC, W, and Cu in the coating.
4. When the current was raised while the compacting pressure and composition remained constant, the coating's average layer thickness rose.

REFERENCES

1. Mahajan, A., Kumar, S., Singh, H., Kumar, S., et al., 2021, "Mechanical properties assessment of TIG welded SS 304 joints," *Materials Today: Proceedings*, 56, pp. 3037–3046. https://doi.org/10.1016/j.matpr.2021.12.133
2. Singh, H., Kumar, S. and Kumar, R., 2022, "Friction stir welding in al alloys: A study," *IUP Journal of Mechanical Engineering*, 15(1), pp. 49–61.
3. Khan, A.A., Ndaliman, M.B., and Ali, M.Y., 2011. "An introduction to electrical discharge machining." In Ali, M.Y., NurulAmin, A.K.M., and Adesta, E.Y.T., (Eds.), *Advanced machining process*. 1st ed. Kuala Lumpur, Malaysia: IIUM Press, 65–69.
4. Kumar, R., Singh, H., Kumar, S. and Chohan, J.S., 2022, "Effects of tool pin profile on the formation of friction stir processing zone in AA1100 aluminium alloy," *Material Today Proceeding*, 48, pp. 1594–1603.
5. Kumar, S., Mahajan, A., Kumar, S. and Singh, H., 2021, "Friction stir welding: Types, merits & demerits, applications, process variables & effect of tool pin profile," *Material Today Proceeding*, 56(1–2), pp.1–13. http://doi.org/10.1016/j.matpr.2021.12.097
6. Singh, S., Kumar, H., Kumar, S. and Chaitanya, S., 2022, "A systematic review on recent advancements in Abrasive Flow Machining (AFM)," *Material Today Proceeding*, 56(5), pp. 3108–3116.
7. Patowari, P.K., Saha, P. and Mishra, P.K., 2010, "Artificial neural network model in surface modification by EDM using tungsten–copper powder metallurgy sintered electrodes," *The International Journal of Advanced Manufacturing Technology*, 51, pp. 627–638.
8. Aspinwall, D.K., Dews, R.C., Lee, H.G. and Simao, J., 2003, "Electrical discharge surface alloying of TI and Fe workpiece materials using refractory powder compact electrodes and Cu wire," *CIRP Annals*, 52(1), pp. 151–156.
9. Lee, H.G., Simao, J., Aspinwall, D.K., Dewesa, R.C. and Voice, W., 2004, "Electrical discharge surface alloying," *Journal of Materials Processing Technology*, 149, pp. 334–340.

10. Zaw, H.M., Fuh, J.Y.H., Neeb, A.Y.C. and Lub, L., 1999, "Formation of a new EDM electrode material using sintering techniques," *The International Journal of Advanced Manufacturing Technology*, 89–90, pp. 182–186.
11. Shunmugam, M.S., Philip, P.K. and Gangadhar, A., 1994, "Improvement of wear resistance by EDM with tungsten carbide P/M electrode," *Wear*, 171, pp. 1–5.
12. Samueli, M.P. and Philip, P.K., 1997, "Power metallurgy tool electrodes for electrical discharge machining," *International Journal of Machine Tools and Manufacture*, 37(11), pp. 1625–1633.
13. Moroa, T., Mohri, N., Otsubo, H., Goto, A. and Saito, N., 2004, "Study on the surface modification system with electrical discharge machine in the practical usage," *Journal of Materials Processing Technology*, 149, pp. 65–70.
14. Simao, J., Aspinwalla, D., El-Menshawyb, F. and Meadows, K., 2002, "Surface alloying using PM composite electrode materials when electrical discharge texturing hardened AISI D2," *Journal of Materials Processing Technology*, 127, pp. 211–216.
15. Kumar, S., Singh, R., Singh, T.P. and Sethi, B.L., 2009, "Surface modification by electrical discharge machining," *Journal of Materials Processing Technology*, 209, pp. 3675–3687.
16. Gangadhar, A., Shunmugam, M.S. and Philip, P.K., 1991, "Surface modification in electro discharge processing with a powder compact tool electrode," *Wear*, 143, pp. 45–55.
17. Kumara, S. and Batra, U., 2012, "Surface modification of die steel materials by EDM method using tungsten powder-mixed dielectric," *Journal of Manufacturing Processes*, 14, pp. 35–40.
18. Kumar, S., 2022, "Influence of processing conditions on the mechanical, tribological and fatigue performance of cold spray coating: A review," *Surface Engineering*, 38(4), pp. 324–365. https://doi.org/10.1080/02670844.2022.2073424
19. Kumar, S. and Kumar, R., 2021, "Influence of processing conditions on the properties of thermal sprayed coating: A review," *Surface Engineering*, 37(11), pp. 1339–1372. http://doi.org/10.1080/02670844.2021.1967024
20. Kumar, S., Handa, A., Chawla, V., Grover, N.K. and Kumar, R., 2021, "Performance of thermal-sprayed coatings to combat hot corrosion of coal-fired boiler tube and effects of process parameters and post coating heat treatment on coating performance: A review," *Surface Engineering*, 37(7), pp. 833–860. http://doi.org/10.1080/02670844.2021.1924506
21. Kumar, S., Kumar, M. and Handa, A., 2020, "Erosion corrosion behavior and mechanical property of wire arc sprayed Ni-Cr and Ni-Al coating on boiler steels in actual boiler environment," *Material at High Temperature*, 37(6), pp. 370–384. https://doi.org/10.1080/09603409.2020.1810922
22. Kumar, S., Kumar, M. and Handa, A., 2019, "Comparative study of high temperature oxidation behavior of wire arc sprayed Ni-Cr and Ni-Al coatings," *Engineering Failure Analysis*, 106, pp. 104173–104189.
23. Simao, J., Lee, H.G., Aspinwall, D.K., Dewes, R.C. and Aspinwall, E.M., 2003, "Workpiece surface modification using electrical discharge machining," *International Journal of Machine Tools and Manufacture*, 43(2), pp. 121–128.

20 Structural and Magnetic Studies on Vanadium-Doped Manganese Nanoferrite Synthesized by Co-precipitation Method

R. Suruthy, B. J. Kalaiselvi,
B. Uthayakumar, S. Sukandhiya

20.1 INTRODUCTION

In recent years, the study of spinel ferrites got the attention of many researchers to enhance the spinel nanoferrite due to its potential medicinal and technological applications, such as MRI technology, magnetically guided drug delivery systems, magnetic recording, ferrofluids, radar-absorbing coatings, sensors, catalysts, and pigments [1, 2]. The preparation of new materials with enhanced properties with the help of new dopants with controlled size is one of the emerging works. Nowadays, the cubic spinel structure is one of the most promising and interesting structure in magnetic materials. Spinel ferrite has three types: (1) normal spinel, (2) inverse spinel, and (3) mixed inverse spinel.

Unique research works were done to better the structural, magnetic properties and electrical properties of nanoferrites by different research groups. Some of them were quoted here related to the present study. The crystallite size and shape play a major role in biological applications. Neda Akhlaghi *et al.* prepared $MnFe_2O_4$ nanoferrites by various methods with controllable size and desired morphology and the structure obtained from co-precipitation method is cubic spinel structure [3]. $MnFe_2O_4$ is unstable in atmospheric air and Mn^{2+} ions in the surface oxidize to form Mn^{3+} ions resulting in the dissociation of the formed $MnFe_2O_4$. A.V. [4] Gopalan *et al.* reported that bulk material manganese ferrite exists with 20% cations residing on the octahedral site and 80% residing on tetrahedral sites [5]. Sukandhiya *et al.* synthesized cobalt-doped nickel–chromium nanoferrite by co-precipitation method. For $Co_{0.6}Ni_{0.4}Cr_{0.5}Fe_{1.5}O_4$ the magnetic saturation value as 80.38 (emu/g) and the magnetic crystalline anisotropy as 17219.49 (erg/cm^3) [6]. The aim of the present

DOI: 10.1201/9781003367154-20

work is to synthesize vanadium-doped manganese ferrite material with x = 0.0, 0.2, 0.4, 1.0, prepared at 1,123 K. To confirm, its magnetic and electrical property of the prepared sample, the following characterization, like XRD, SEM with EDX, FTIR, XPS, and VSM, was done.

20.2 MATERIALS AND METHODS

20.2.1 Materials for Synthesis

$V_xMn_{1-x}Fe_2O_4$ nanoferrites, at various concentrations, are synthesized from precursors $MnCl_{2.}4H_2O$, $FeCl_3$. $6H_2O$, VCl_3, and NaOH. The analytical grade of these precursors, purchased from SIGMA ALDRICH, Germany, with 98% purity, was used in the synthesis without further purification.

20.2.2 Experimental Procedure

An efficient synthesizing involves by choosing precursors and their chemical composition and reaction environment. Mainly for wet chemical methods like sol-gel, co-precipitation, and colloid emulsion technique, pH controller plays an important role. In this present work, eco-friendly NaOH is used to maintain pH. Particle size, crystalline nature, morphology, purity, and chemical composition were greatly influenced due to the physio-chemical properties of nanoparticles. Using chemical methods has been confirmed to efficiently control the morphology and chemical composition of prepared nanopowder. Among wet chemical techniques, sol-gel, hydrothermal, and colloid emulsions are time-consuming and involve highly unstable alkoxides and are difficult to maintain reaction conditions. Co-precipitation is one of the more successful techniques for synthesizing ultrafine nanoparticles, having a narrow particle size distribution. These advantages on the co-precipitation method motivated authors to synthesize $V_xMn_{1-x}Fe_2O_4$ [x=0.0, 0.2, 0.4, 1.0] nanoferrites by co-precipitation method. One mole of manganese (II) chloride tetrahydrate and two moles of iron (III) chloride were used as the precursors. They are mixed in stoichiometric ratio and added one by one on the basis of their electronegativity value. The mixture of aqueous solution is stirred vigorously at 338 K for 30 minutes. Meanwhile, NaOH is added to the precursor solution with the help of burette drop by drop until the pH of the solution reaches 10. The required composition of nanoferrites were formed from the conversion of metal salt into hydroxide and then transformed into ferrites. The precipitates obtained were thoroughly washed more than three times with the help of double distilled water and acetone. Then the colloidal sample was dried in sunlight for eight hours and which was then kept in a muffle furnace for 1,123 K at a ramping rate of 10°C. Samples were cooled and grinded and kept for further characterization techniques.

20.2.3 Characterization Techniques

The prepared samples were characterized for structural parameters by X-ray diffraction (XRD) using PANalytical X'Pert PRO powder X-ray diffractometer. 2θ values were taken from 10° to 80° using Cu K α[Å] radiation where l =1.54060 Å. Surface morphology and elemental composition are determined with the help of instrument named scanning electron microscope with energy dispersive X-ray analysis (SEM with EDAX) made by TESCAN and OXFORD instruments along with the software named VEGA3 and INCA. FTIR recorded the details about the positions settled by the ions and the spinel formation by Thermo Nicolet 380 FTIR Spectrophotometer. The electron-binding energy for the elements is measured by X-ray photoelectron spectroscopy (XPS) on a K-alpha surface analysis. The magnetic properties were measured at room temperature by CRYOGENIC (UK) vibrating sample magnetometer.

20.3 RESULTS AND DISCUSSION

20.3.1 X-Ray Diffraction Analysis

The XRD pattern of the prepared samples of vanadium-doped manganese ferrite for the concentration, X = 0.0, 0.2, 0.4, 1.0, is shown in Figure 20.1.

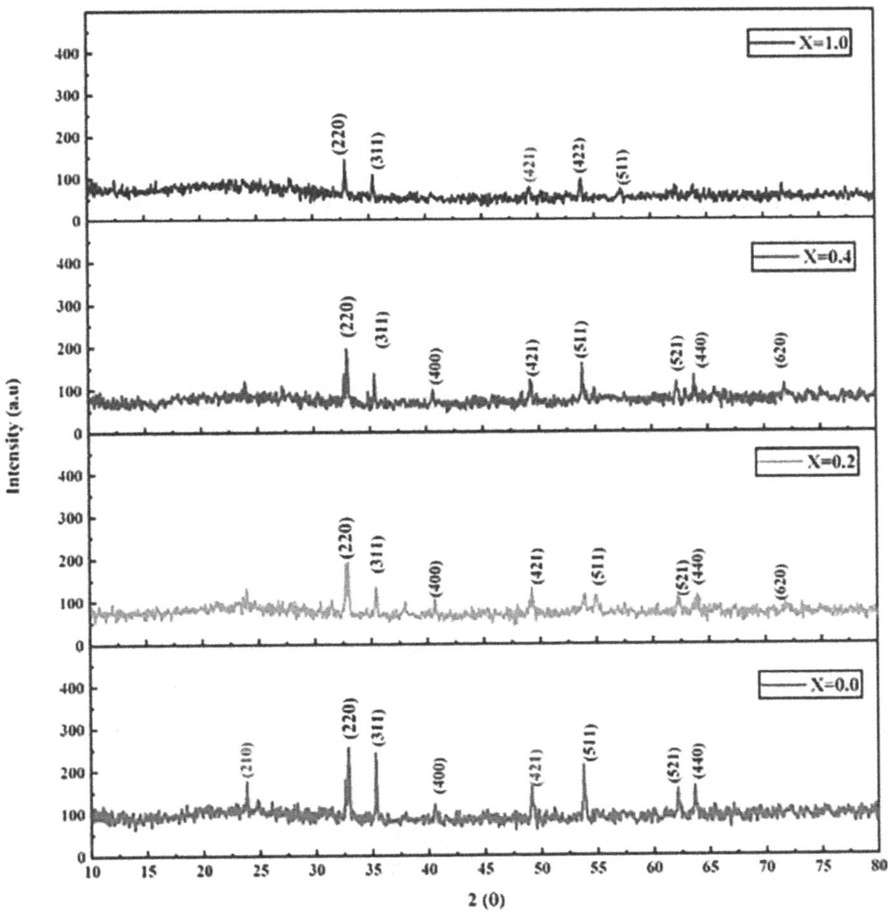

FIGURE 20.1 X-Ray diffraction pattern of $V_xMn_{1-x}Fe_2O_4$ (X = 0.0, 0.2, 0.4, 1.0) sintered at 1,123 K.

TABLE 20.1
Comparison of X-Ray Intensity

Vanadium Content "x"	Composition	I_{220}	I_{440}
0.0	$MnFe_2O_4$	100.00	44.42
0.2	$V_{0.2}Mn_{0.8}Fe_2O_4$	100.00	13.82
0.4	$V_{0.4}Mn_{0.6}Fe_2O_4$	100.00	49.56
1.0	VFe_2O_4	100.00	Nil

The peaks obtained are (220), (310), (311), (400), (422), and (440) confirmed the XRD pattern is in accordance with the inverse cubic spinel structure with the space group of Fd-3m, Fe^{3+} in tetrahedral (8c) and octahedral (12d). These peaks match with the JCPDS standard cards Ref. No. 10-0319 [7], Ref. No. 74-2403 [8], and Ref. No. 75-0035 [2] ($MnFe_2O_4$). In all the prepared samples, there exists extra peak (521); (421) denotes Fe_2O_3 are formed due to the presence of vanadium. Intensity of XRD pattern related to particle size and crystallinity of the samples. The intensities of (220) and (440) planes are more sensitive to cations in tetrahedral and octahedral sites shown in Table 20.1.

In all the prepared sample, minor reflection peak (422) is present. Heiba et al. [9] reported that vanadium can be substituted either as V^{5+} or V^{3+} ions. In the present study, VCl_3 was used as the source of vanadium +3 cations. Although the occupation of octahedral sites (B sites) by V^{3+} ions at low concentration has been reported. Maisnam et al. [10] reported that the size of Fe^{3+} and Co^{2+} ions is higher than V^{5+} so that V^{5+} ions tend to occupy the tetrahedral sites (A sites). However, it seems that the occupation of appropriate amount of Co^{2+} by V^{3+} cannot occur and V^{3+} ions occupy the Fe^{3+}, which results in the formation of secondary phase Fe_2O_3, which is confirmed by XRD patterns. The average crystallite size and lattice constant can be calculated from X-ray reflections indexed (220), (310), (311), (400), (422), (511), (440). The average particle sizes were calculated from the obtained peaks with the help of the Debye-Scherer formula:

$$D = \frac{n\lambda}{\beta \cos\theta} * 10^{-9} \text{ m},$$

where D is the crystallite size, β is the full-width half-maxima, λ is the incident Cu Kα [Å] of wavelength 1.54060 Å, and θ is the Bragg's angle of obtained peaks. Lattice constant has been calculated from equation $a = d\,(h_2 + k_2 + l_2)^{1/2}$, where a is the lattice constant, d' is the interplanar distance, and hkl is the miller indices. Lattice strain is calculated by Williamson-Hall formula $\varepsilon = \beta/4 \tan\theta$, where ε is the lattice strain of the structure. X-ray density can be calculated by $\rho_x = ZM/Na^3$, where Z is the number of molecules per unit cell, where, Z=8. M is the molecular weight, N is the Avogadro's number ($6.02214076*10^{23}$), and a is the lattice constant. Dislocation density is found by the formula $\delta = 15\varepsilon/aD$, where δ is the dislocation density. All these structural parameters are calculated and tabulated in Table 20.2.

TABLE 20.2
Structural Parameters of Vanadium-Doped Manganese Ferrite for Various Concentrations Sintered at 1,123 K

Vanadium Content "x"	Composition	Crystallite Size D (nm)	Lattice Constant a (Å)	Molecular Weight g/mole	X-ray Density g/cm3	Lattice Strain 10^{-3}	Dislocation Density 10^{15}
0.0	$MnFe_2O_4$	58.71	8.542	230.63	5.14	1.79	0.68
0.2	$V_{0.2}Mn_{0.8}Fe_2O_4$	39.47	8.469	229.83	5.22	2.64	1.92
0.4	$V_{0.4}Mn_{0.6}Fe_2O_4$	46.03	8.070	229.02	5.83	2.20	1.18
1.0	VFe_2O_4	37.75	8.256	226.62	5.39	2.66	1.61

The average crystallite size D calculated for $V_xMn_{1-x}Fe_2O_4$ nanoferrites for different X lies in between 37.75 nm to 58.71 nm. The prepared samples having lattice constant between 8.070 Å and 8.542 Å. The average crystallite size decreases and increasing on increasing vanadium doping. When the particle size decreases, the lattice strain, dislocation density, and X-ray density increases, as a result of doping of small amount of vanadium.

20.3.2 Scanning Electron Microscope (SEM) and Energy Dispersive Spectroscopic (EDS) Analysis

The morphological studies of the obtained $V_xMn_{1-x}Fe_2O_4$ nanoferrites sintered at 1,123 K have been investigated with the help of TESCAN along with the software named VEGA3 for all concentration X = 0.0, 0.2, 0.4, 1.0. Figure 20.2 shows morphology for $V_xMn_{1-x}Fe_2O_4$. As seen, the undoped $MnFe_2O_4$ agglomerated. The agglomeration decreasing by increasing vanadium concentration for X = 0.2 and X = 0.4 shows fine particle nature. Agglomeration is due to the magnetic interaction between particles and reduces the interface between nanoparticles [11]. The nanoferrite sample surface consists as many voids or pores that are attributed to the maximum amount of oxygen and chlorine that gets liberated during sintering process. Even though there is addition of higher ionic radius used as a dopant to the sample, the contraction of lattice occurs due to the presence of vacancies [12]. EDS spectrum for $V_xMn_{1-x}Fe_2O_4$ (X=0.0, 0.2, 0.4, 1.0) nanoferrite was recorded with OXFORD instruments along with the software named INCA, illustrated in Figure 20.3.

The results show that each peak corresponds to the element added in the prepared nanoferrite samples, confirming the presence of elements in respective concentration. Manganese, iron, and oxygen are the major constituents in the composition. Vanadium is the next major constituent of the composition. It is interesting to note that the preparation condition completely favours the formation of nano spinel ferrite and allows us to study the effect of increasing V content on the properties of $V_xMn_{1-x}Fe_2O_4$. The peak value variation is due to its stoichiometry, for all concentrations. The values of manganese vary with increasing vanadium concentration.

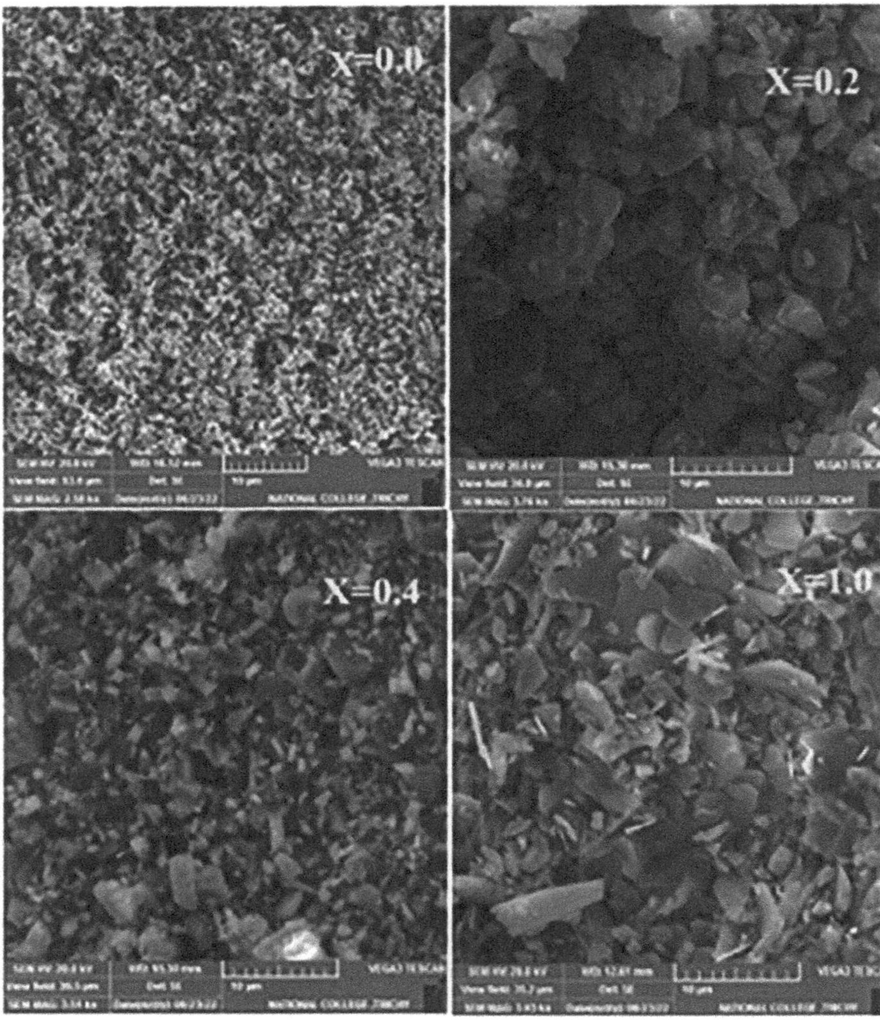

FIGURE 20.2 SEM micrograph of $V_xMn_{1-x}Fe_2O_4$ (X = 0.0, 0.2, 0.4, 1.0) sintered at 1,123 K.

20.3.3 Fourier Transform Infrared Spectroscopy (FTIR) Analysis

The FTIR spectrum of vanadium-doped manganese ferrite is shown in Figure 20.4.

The spectra are built with two bases and strong fundamental stretching vibration ranges from 400 cm^{-1} and 600 cm^{-1}, which is the common for all spinel structure [13]. The spinel structure assigned to the stretching vibrations of the unit cell occurs in the tetrahedral site (A) whereas the metal-oxygen vibration occurs in the octahedral site (B). These absorption bands are responsible for the changes in interactions between oxygen and cations, as well as to size of the obtained nanoparticles [14]. The statistical distribution of cations over A and B sites responsible for the broadening of

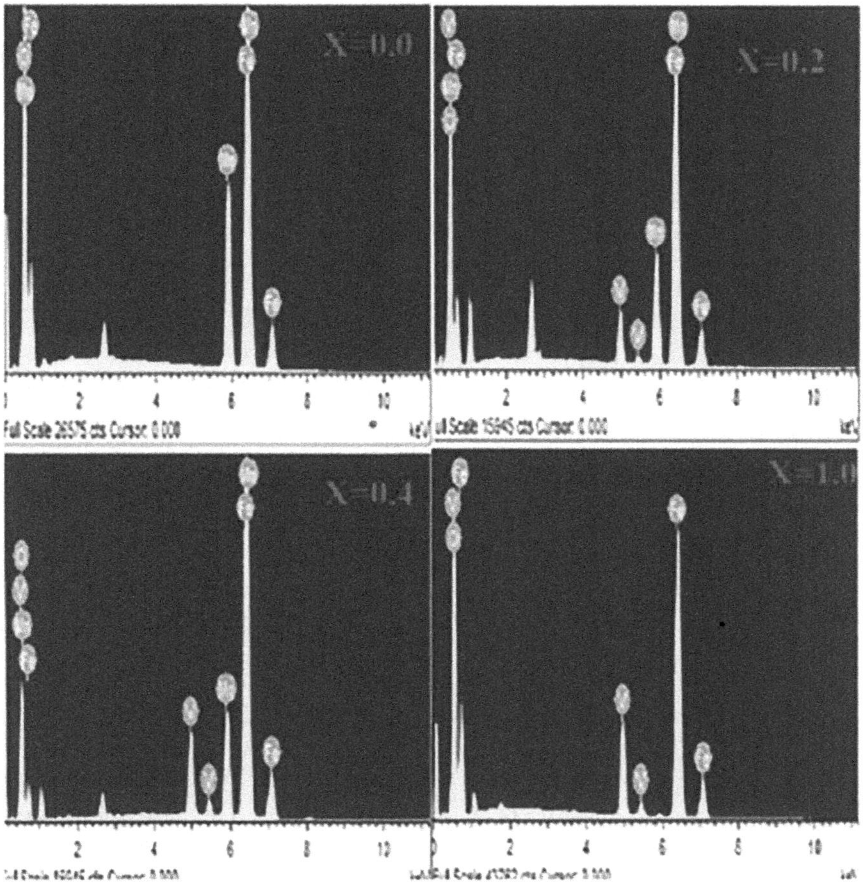

FIGURE 20.3 EDS of $V_xMn_{1-x}Fe_2O_4$ (X = 0.0, 0.2, 0.4, 1.0) sintered at 1,123 K.

the spectral bands whereas cation mass, cation-oxygen distance, and bending force responsible for vibration frequency [15]. From Table 20.3, intrinsic stretching vibration frequency of metal-oxygen at tetrahedral site from 559.28 cm^{-1} to 542.87 cm^{-1} decreases for concentration X = 0.0, and X = 0.2 represents the increase of vanadium concentration in the sample, which results as slight increase in metal oxygen bond length and decrease and increase in the wave number of octahedral and tetrahedral sites by increasing the substitution content [16].

The intensity of the peaks corresponds to tetrahedral and octahedral sites for the concentration X = 0.4 increases as compared to intensity at X = 0.2. This is because the intensity is a function of the change in dipole moment which further depends on the inter-nuclear distance. For X=1.0, the octahedral and tetrahedral intensity increases compared from X = 0.4 to X = 1.0. There is also an additional subsidiary peak observed at 475.47 cm^{-1} for X = 1; this value represents the contribution of ionic bond Fe-O in the lattice. So the observed increase and decrease in absorption band

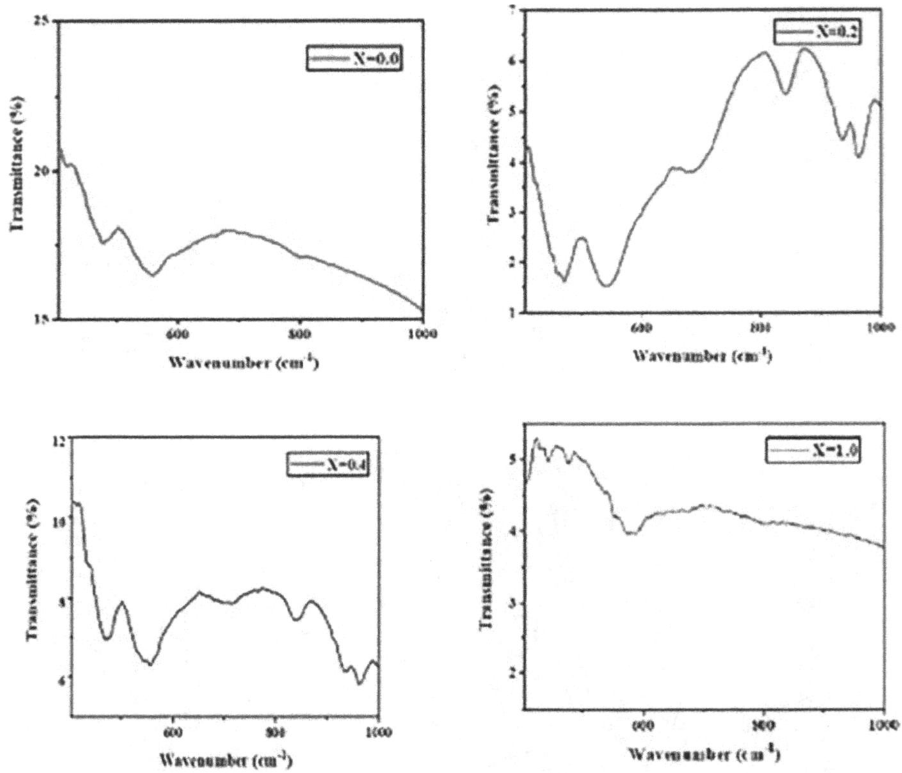

FIGURE 20.4 FTIR spectra of $V_xMn_{1-x}Fe_2O_4$ (X = 0.0, 0.2, 0.4, 1.0) sintered at 1,123 K.

TABLE 20.3
Vibrational Frequency of Tetrahedral and Octahedral Sites

"x"	Composition	$v_{1\ tetra}$ Cm^{-1}	$v_{2\ tetra}$ Cm^{-1}	$v_{2\ octa}$ Cm^{-1}
0.0	$MnFe_2O_4$	559.28	472.64	470.28
0.2	$V_{0.2}Mn_{0.8}Fe_2O_4$	542.87	468.22	470.23
0.4	$V_{0.4}Mn_{0.6}Fe_2O_4$	555.74	477.87	468.41
1.0	VFe_2O_4	587.06	475.47	442.09

intensity with the increase in vanadium content is due to perturbation occurring in Fe-O bond.

20.3.4 Vibrating Sample Magnetometer (VSM) Analysis

Vanadium-doped Manganese ferrite $V_xMn_{1-x}Fe_2O_4$ hysteresis loop recorded with CRYOGENIC vibrational sample magnetometer at 300 K are shown in Figure 20.5. By using the graph, magnetic saturation (M_s), remanence magnetization (M_r),

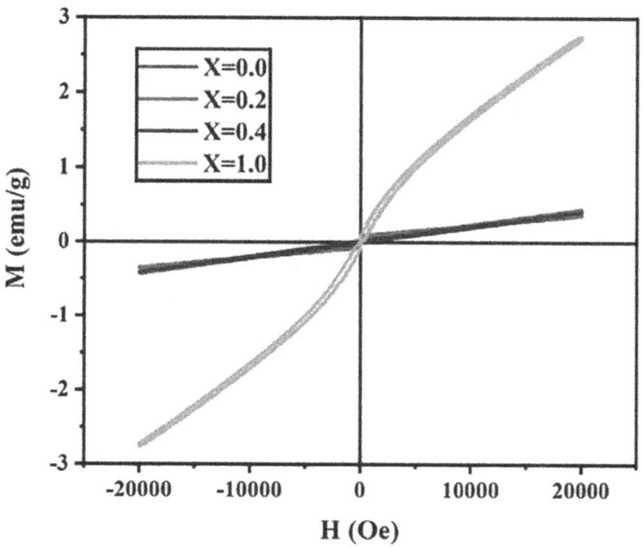

FIGURE 20.5 Magnetic hysteresis curves of $V_xMn_{1-x}Fe_2O_4$ (X = 0.0, 0.2, 0.4, 1.0) sintered at 1,123 K.

TABLE 20.4
Magnetic Parameters of Vanadium-Doped Manganese Ferrite for Various Concentrations Sintered at 1,123 K

Vanadium Content "x"	Composition	Ms (emu/g)	Mr (emu/g)	Hc (Oe)	Mr/Ms	K (erg/cm³)	nB
0.0	$MnFe_2O_4$	0.36	0.065	319	0.1805	114.84	0.014
0.2	$V_{0.2}Mn_{0.8}Fe_2O_4$	0.43	0.025	797	0.058	342.71	0.017
0.4	$V_{0.4}Mn_{0.6}Fe_2O_4$	0.414	0.015	957	0.036	396.198	0.01
1.0	VFe_2O_4	2.74	0	0	0	0	0.111

$\mathbf{M_s}$—saturation magnetization; $\mathbf{M_r}$—remanent magnetization; $\mathbf{H_c}$—coercivity; **K**—magneto-crystalline anisotropy; $\mathbf{n_B}$—magneton number

coercivity (H_c) were obtained. The value of magneto-crystalline anisotropy (K) was calculated from Stoner-Wohlfarth relation as follows: H_c = K/Ms. The magnetic moment in Bohr magneton is calculated using the relation n_B= (molecular weight*M_s)/5585 [17]. The magnetic parameters, such as saturation magnetization (M_s), magnetic remanence (M_r), coercivity (H_c), squareness ratio, and magnetic anisotropy constant (K), and magneton number are tabulated in Table 20.4.

Usually, the magnetic properties arise from coupling between spin and orbital angular momentum (L-S coupling) and electron spin (S-S coupling). In the case of Spinel structure nano magnetic ferrite materials, the parameters depend on cation distribution, collinearity, and non-collinearity (canting) of spins on their surface, crystallite size, and dopant.

The hysteresis loop shows that all the samples show paramagnetic nature having coercivity up to X = 0.4. At X=1.0, it shows diamagnetic. Generally, the magnetic saturation increases, and the coercive field decreases, but in our case, it is vice versa. Magnetic saturation decreases and coercive field increases [18]. In the work, vanadium-doped nickel-zinc-copper ferrites found that doping of a low amount of vanadium in the concentration results in a decrease in the magnetic saturation because the vanadium occupies the B-sites as V^{3+}. Increasing vanadium amount in the concentration, the magnetic saturation increases due to the migration of vanadium to A-sites [19]. As in the present case, the magnetic saturation increases on doping a low amount of Vanadium.

Increasing vanadium doping, the coercivity increases attributed to the change in anisotropy field which in turn the domain wall energy and also change in the crystallite size [20]. The coercivity (H_c) is the magnetic field that is responsible for overcoming the magneto-crystalline anisotropy to flip the magnetic moments. The K value increases on doping of low amount of vanadium and then decreasing. The change in the K value is due to the presence of Co^{2+} ions occupying the tetrahedral sites of the spinel structure [21].

20.4 CONCLUSION

Vanadium-doped manganese ferrite for different concentration $V_xMn_{1-x}Fe_2O_4$ (X = 0.0, 0.2, 0.4, 1.0) successfully prepared by co-precipitation method. The average crystallite size obtained in between 37.75 nm to 58.71 nm. All physical properties are studied for synthesized samples sintered at 1,123 K. The addition of V^{3+} ions influenced the structural properties, such as average crystallite size (D), lattice constant (a), X-ray density (ρ_x), lattice strain (ε), and dislocation density (δ) of the synthesized samples. From XRD, it is confirmed that the obtained sample is inverse cubic spinel structure. SEM shows morphology manipulated by V^{3+} ion in the sample and FTIR observation have fine match with results of XRD. From VSM, at x = 0.4, it has maximum coercivity and maximum magneto-crystalline anisotropy value.

REFERENCES

[1] J. Azadmanjiri, H.K. Salehani, "A study on the formation of MnFe2O4 nano-powder by coprecipitation method", *Phys Status Solidi (c)*. 2007;4(2):253– 255.

[2] M.J. Akhtar, M. Younas, "Structural and transport properties of nanocrystalline MnFe2O4 synthesized by co-precipitation method", *Solid State Sci.* 2012;14:1536–1542.

[3] Neda Akhlaghi, Ghasem Najafpour-Darzi, "Manganese ferrite (MnFe2O4) Nanoparticles: From synthesis to application: A review", *J Ind Eng Chem.* 2021;103:292–304.

[4] N.M. Deraz, A. Alarif, "Controlled synthesis, physicochemical and magnetic properties of nano-crystalline Mn ferrite system", *Int J Electrochem Sci.* 2012;7:5534–5543.

[5] E.V. Gopalan, K.A. Malini, S. Saravanan, D.S. Kumar, Y. Yoshida, M.R. Anantharaman, "Evidence for polaron conduction in nanostructured manganese ferrite", *J. Phys. D: Appl. Phys.* 2008;41:185005.

[6] S. Sukandhiya, B. Uthayakumar, S. Periandy, "Influence of Co^{2+} ions on structural and magnetic properties of Co-precipitated Ni-Cr nanoferrite for cancer cell diagnosis and treatment", *J Appl Phys (IOSR-JAP)*. 2017;9(4):4–12.

[7] Deivatamil D, John Abel M, Nancy Dyana P, Thiruneelakandan R, Joseph Prince J, "A comparative study on pure and cobalt doped manganese ferrite (Co:MnFe2O4) nanoparticles in their optical, structural, and gas sensing properties", *Solid State Commun*. 2021;339:114500.

[8] Chandunika R. Kalaiselvan, Nanasaheb D. Thorat, Niroj Kumar Sahu, "Carboxylated PEG-functionalized MnFe2O4 nanocubes synthesized in a mixed solvent: Morphology, magnetic properties, and biomedical applications", *ACS Omega*. 2021;6:5266–5275.

[9] Z.K. Heiba, M.B. Mohamed, S.I. Ahmed, "Cation distribution correlated with magnetic properties of cobalt ferrite nanoparticles defective by vanadium doping", *J Magn Magn Mater*. 2005;441:409–416.

[10] Mamata Maisnam, Sumitra Phanjoubam, H.N.K. Sarma, Chandra Prakash, L. Radhapiyari Devi, O.P. Thakur, "Magnetic properties of vanadium substituted lithium zinc titanium ferrite", *Mater Lett*. 2004;58:24122414.

[11] A. Hashhash, M. Kaiser, "Influence of Ce-substitution on structural, magnetic and electrical properties of cobalt ferrite nanoparticles", *J. Electron Mater*. 2015;45:462–472.

[12] S. Prabahar, M. Dhanam, "CdS thin films from two different chemical baths—Structural and optical analysis", *J Cryst Growth*. 2005;285(1–2):41–48.

[13] M. Hashim, Alimuddin, S.E. Shirsath, S. Kumar, R. Kumar, A.S. Roy, J. Shah, R.K. Kotnala, "Preparation and characterization chemistry of nano-crystalline Ni-Cu-Zn ferrite", *J Alloys Compd*. 2012;549:348–357.

[14] M. Farid, I. Ahmad, S. Aman, M. Kanwal, G. Murtaza, I. Alia, M. Ishfaq, "Structural, electrical and dielectric behavior of $Ni_xco_{1-x}NdyFe_2$-yO_4 nano-ferrites synthesized by sol-gel method", *Dig J Nanomater Biostructures*. 2015;10(1):265–275.

[15] A.M. Wahba, M.B. Mohamed, "Structural, magnetic, and dielectric properties of nanocrystalline Cr-substituted Co0. 8Ni0. 2Fe2O4 ferrite", *Ceram Int*. 2014;40(4):6127–6135.

[16] A.G. Bhosale, B.K. Chougule, "X-ray, infrared and magnetic studies of Al-substituted Ni ferrites", *Mater Chem Physics*. 2006;97(2–3):273–276.

[17] S. Singhal, K. Chandra, "Cation distribution and magnetic properties in chromium-substituted nickel ferrites prepared using aerosol route", *J Solid State Chem*. 2007;180(1):296–300.

[18] B. Abraime, K. El Maalam, L. Fkhar, A. Mahmoud, F. Boschini, M.A. Tamerd, A. Benyoussef, M. Hamedoun, E.K. Hlil, M.A. Ali, A. El Kenz, "Influence of synthesis methods with low annealing temperature on the structural and magnetic properties of CoFe2O4 nanopowders for permanent magnet application", *J Magn Magn Mater*. 2020;500:166416.

[19] M. Kaiser, "Magnetic and dielectric properties of low vanadium doped nickel–zinc–copper ferrites", *J Phys Chem Solids*. 2010;71(10):1451–1457.

[20] L. Kumar, P. Kumar, M. Kar, "Effect of non-magnetic substitution on the structural and magnetic properties of spinel cobalt ferrite (CoFe2– xAlxO4) ceramics", *J Mater Sci Mater Electron*. 2013;24(8):2706–2715.

[21] M.B. Mohamed, A.M. Wahba, M. Yehia, "Structural and magnetic properties of CoFe2– xMoxO4 nanocrystalline ferrites", *Mater Sci Eng B*. 2014;190:52–58.

21 Theoretical Analysis of Nuclear Properties of Pu-Isotopes to Synthesize Sustainable and Clean Fuels

Amandeep Kaur

21.1 INTRODUCTION

The Nuclear Energy Agency (NEA) conducted a sensitivity study to address the factors crucial for a cleaner and sustainable nuclear power generation and enlisted high priority for the most crucial nuclear isotopes and their pertinent quantities [1–2]. The understanding of these neutron-induced threshold reactions may be crucial for the management of nuclear waste. New reactor systems based on neutron-driven systems have been designed as a result of resurgence of interest in nuclear energy, utilizing the advancements in nuclear technology. These novel designs heavily rely on the transuranic nuclide ($Z > 92$) created in the nuclear fuel cycles by subsequent neutron capture. For the engineering and manufacturing of these reactor systems, reliable data compilations are crucial, notably for neutron-driven reaction cross-sections with these unstable transuranic isotopes occurring at energies lying in the range of MeV [3–4]. Four of the six designs being studied for the GEN-IV nuclear power reactors are based on a fast neutron energy spectrum and not the conventional thermal spread. To develop future nuclear energy with sustainable ideals, substantial research has been done on the advancement of nuclear fuel materials. Plutonium-238 is utilized as a thermal source to create thermoelectric power for heart implantable devices and telecommunication technologies in spacecraft. Nuclear weapons are indeed the primary application for plutonium-239. In addition to the radiation risk and problems with public acceptance, the amount of spent nuclear fuel and trash generated by nuclear power plants makes this sort of low carbon energy system unreliable for sustainability [5–6]. Additionally, the spent fuels' high heat load and long-lived radioactivity lead to the creation of a deep geological storage for nuclear waste, which is extremely dangerous for both the environment and people. The extraction and reprocessing of the valuable actinides from the spent fuels for peaceful purposes, rather than treating it as high level waste

 DOI: 10.1201/9781003367154-21

directly and disposing it off permanently, has been the subject of numerous studies to address this issue [7].

In reference to this, an effort has been made in the present paper to analyse the decay behaviour of 240Pu* formed in neutron-induced reaction, theoretically, at three different incident beam energies (Ebeam) 7.1, 7.7, and 9.0 MeV, within the collective clusterization approach of dynamical cluster-decay model (DCM) [8–17] in reference to an experimental data [18]. The calculations are carried out for hot and quadrupole (β2) deformed fragments, within non-sticking limits of moment of inertia (INS) in the centrifugal potential term to address the experimental data [18]. Note that the non-sticking limits of moment of inertia is favourable for light particle emission mode because it imparts lower angular momentum values to the compound system. The probability of fission decay at such lower values of angular momentum is very less. The experimentally available cross-sections for 2n-evaporation are nicely addressed within the framework of DCM by varying the neck-length parameter (ΔR) of the model, which in turn depends upon the limiting angular momentum value (lmax). We intend to analyse the role of nuclear properties such as deformations and angular momentum in the fragmentation of 240Pu* formed in n+239Pu reaction over the mentioned energy values. The decay channels 1n and 3n are also analysed in addition to 2n channel. The investigation of decay profile is done in terms of fragmentation and preformation profiles.

21.2 METHODOLOGY

The dynamical cluster-decay model (DCM) [8–17] approach is based on quantum mechanical fragmentation theory (QMFT) [19–21], which has been extensively applied to address various nuclear phenomenon such as cold fusion, cluster radioactivity, and spontaneous fission during the last five decades. This theory is worked out in terms of the collective coordinates of mass and charge asymmetries (ηA and ηZ) of the incoming/outgoing channel and relative separation *R*. For isotope production, the neutron evaporation leaves the system with high mass asymmetry values and the neutron evaporation is considered as the quantum tunnelling of the neutrons and the daughter isotope through the interaction barrier.

The input parameter to solve this wave equation is the temperature dependent total interaction potential existing between the interacting nuclei given as follows:

$$V_R(\eta,T) = \sum_{i=1}^{2} [V_{LDM}(A_i,Z_i,T)] + \sum_{i=1}^{2} [\delta U_i] exp\left(\frac{T}{T\frac{2}{0}} \right)$$
$$+V_c(R,Z_i,\beta_{\lambda i},\theta_i,T) + V_P(R,A_i,\beta_{\lambda i},\theta_i,T)$$

Here, VLDM(T) is the T-dependent liquid drop binding energies [22, 23], and δUi represents the shell-corrections of the nuclei [24], to which the temperature dependence is duly incorporated. The other terms represented in the equation are the

T-dependent Coulomb potential VC, nuclear proximity potential VP, and centrifugal potential Vl, the details of which can be taken from [25].

The angular momentum effects are defined by the centrifugal potential, which reads as follows:

$$\frac{\hbar \ell (\ell + 1)}{2}$$

$$V\ell(T) = 2I(T)$$

Here, I(T) is taken in non-sticking limits and defined as follows:

$$I(T)=INS=\mu R^2$$

In this case, the separation distance is assumed to be beyond the range of nuclear proximity forces, which is about 2 fm[26].

The preformation probability (P0) is calculated by solving the stationary Schrödinger wave equation in η-coordinates.

$$\left\{ - \rightarrow \frac{2\, \partial \rightarrow 1' \partial}{\hbar} + V_R (\eta, T) \right\} \psi^{\upsilon}(\eta) = E^{\upsilon} \psi^{\nu}(\eta)$$

$$2\sqrt{B_{\eta\eta}\, \partial_\eta \sqrt{B_{\eta\eta}\, \partial_\eta}}$$

Here, $\nu = 0, 1, 2, 3 \ldots$ refer to ground state ($\nu = 0$) and excited state solutions. The mass parameters $B\eta\eta(\eta)$, representing the kinetic energy part in the equation, are the smooth classical hydrodynamical masses since at high temperatures the shell effects are almost washed out [27]. The classical mass parameters of Kröger and Scheid [28] are used here, which are based on hydrodynamical flow. The preformation probability is given as follows:

$$P_0 = \sqrt{B_{\eta\eta}}\, |\psi[\eta(A_i)]|^2 \frac{2}{A}$$

The penetration probability is estimated by using the WKB approximation as shown here:

$$P = \exp\left[- \frac{2}{\hbar} \int_{Ra}^{Rb} \{2\mu [V(R) - Q_{eff}]\}^{1/2} dR \right]$$

where *Ra* and *Rb* represent the entry and exit points respectively and subsequently follow the condition

$$V(R_a) = V(R_t + \Delta R) = V(R_b) = Q_{eff} = TKE(T).$$

For the decay of a hot compound nucleus, we use the first turning point R = Ra, where P0 is calculated as follows:

$$R_a = R_1(a_1, T) + R_2(a_2, T) + \Delta R(\eta, T) = R_t(a, \eta, T) + \Delta R(\eta, T)$$

with radius vectors are taken from [29]

$$R_i(a_i, T) = R_{0i}(T)[1 + \sum_\lambda \beta_{\lambda i} Y_\lambda^{(0)}(a_i)]$$

The total cross-sections for isotope production and, alternatively, the neutron evaporation are given as follows:

$$\sigma = \frac{\pi}{k^2} \sum_{\ell}^{\ell_{max}} (2\ell + 1) P_0 P$$

21.3 RESULTS AND DISCUSSIONS

In the present work, the decay of the nuclear system $^{240}Pu^*$ populated via n-induced channel is investigated theoretically, for the synthesis of isotopes $^{237, 238, 239}Pu$ via 1n-, 2n-, 3n-evaporation and other actinides, such as ^{237}Np, ^{236}U via ^3H, and ^4He-emission in reference to the experimental data provided by V. Méot et al. [18], by employing the dynamical clusterization approach based upon the quantum mechanical fragmentation theory. The collective clusterization refers to the emergence of a particular exit channel from a hot-excited nuclear system, at the expense of all other possible decay modes. This allows us carry out a comparative analysis of the synthesis of various isotopes or other actinides formed via xn-evaporation or light particle emission (like ^3H or ^4He). All the nuclear decays are a result of the tug-of-war between the attractive nuclear forces and repulsive Coulomb forces and the total fragmentation potential for all the decay modes from $^{240}Pu^*$ is presented in Figure 21.1, for all the reported Ec.m. values at the steady state or zero angular momentum state. Note that zero angular momentum state indicates that the total fragmentation potential is the result of binding energies, shell-corrections, nuclear proximity potential and Coulomb potential and there is no contribution from centrifugal potential. The figure suggests that the synthesis of ^{238}Pu is highly favoured at this state as the fragmentation potential for ^{238}Pu+2n channel has lower magnitude.

The strong competition from the formation of ^{239}Pu isotope can be clearly seen because of the comparable fragmentation potential values of ^{239}Pu+1n channel at Ec.m. = 9.30 MeV. However, the formation of the lighter isotope ^{237}Pu and other competing actinide ^{236}U is further less as compared to heavier isotopes $^{238,239}Pu$ due to higher fragmentation potential values. In addition to these, certain heavier fragments tend to possess lower fragmentation potential values, depicted by minimas in the curve. These are identified as ^{48}Ca, $^{68,70}Ni$, ^{82}Ge, ^{84}Se, and $^{106,108}Ru$ as shown in the figure. The reason for the formation of dominant dips in the fragmentation curve

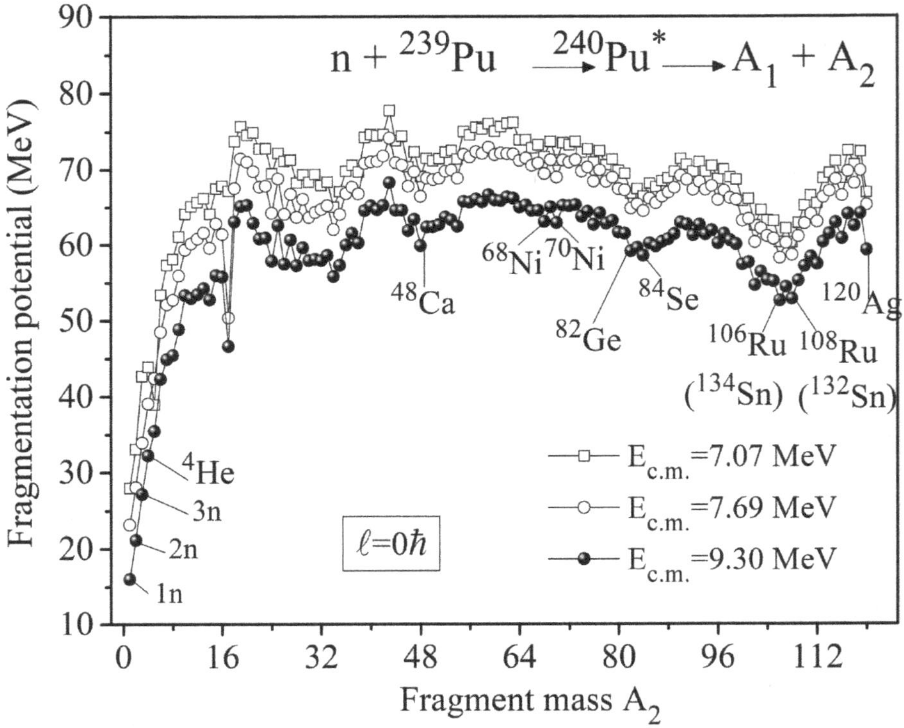

FIGURE 21.1 Fragmentation potential plotted in terms of light fragment mass A2 for the system $^{240}Pu^*$ populated via n-induced channel at all the three experimentally available energies (Ec.m.) for the s-wave ($l = 0\hbar$).

corresponding to these fragments is the closed magic shell structure. The mentioned fragments either correspond to magic Z = 28 ($^{68,70}Ni$), magic N = 50 (^{82}Ge, ^{84}Se) or are doubly magic Z = 20, N = 28 like ^{48}Ca. The emergence of $^{106,108}Ru$ in the decay channel of $^{240}Pu^*$ is attributed to the magic Z = 50 daughter nuclei $^{132,134}Sn$ formed in the complementary mass region. The probability of presence of these heavier fragments in the decay channel or equivalently, the synthesis of the daughter nuclei in the complementary mass region is relatively less as compared to the formation of Pu-isotopes or other actinides, attributed to their higher fragmentation potential values. The same behaviour is seen at all the three reported energy values, as seen from Figure 21.1, subjected to minor variations in the structure and notable change in magnitude.

Further the effect of energy on the decay profile is studied from the preformation probability (P0) profile, as shown in Figure 21.2. The figure is plotted for three different incident beam energy values (a) Ec.m. = 7.07 MeV, (b) Ec.m. = 7.69 MeV, and (c) Ec.m. = 9.30 MeV at best fit values of neck-length parameter, at the lowest and limiting value of angular momentum as mentioned in figure. This figure further clarifies that the probability of neutron emission or equivalently, the formation of isotopes $^{238,239}Pu$ is very high at lower values of angular momentum. Figure 21.2

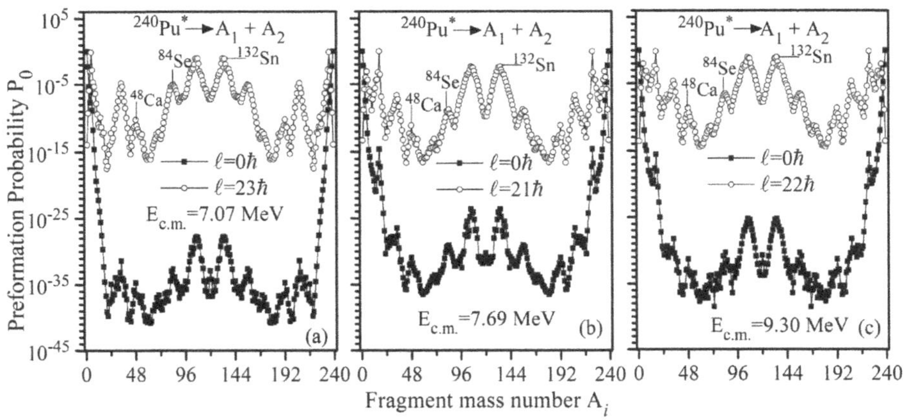

FIGURE 21.2 Probability of preformation of fragments in the decay of ^{204}Pu* populated via n-induced channel at all the three experimentally available energies: (a) Ec.m. = 7.07 MeV, (b) Ec.m. = 7.69 MeV, and (c) Ec.m. = 9.30 MeV for the lowest and limiting angular momentum.

confirms the results of Figure 21.1. At these lower angular momentum values, the decay channel purely consists of neutrons or other light-charged particles, synthesizing either the isotopes of Plutonium (^{237,238,239}Pu) or other neighbouring elements like ^{236}U. However, as higher angular momentum states are included, the structural changes come into picture, suggesting the enhancement in the probability of ^{48}Ca, ^{68,70}Ni, ^{82}Ge, ^{84}Se, and ^{106,108}Ru, as shown in the figure, attributed to the shell closure effects.

This also indicates the other possible reaction partners to synthesize the ^{240}Pu* compound nucleus. In simple words, the emergence of strong peaks at ^{48}Ca, ^{68,70}Ni, ^{82}Ge, ^{84}Se, and ^{106,108}Ru suggests that these nuclei or the corresponding nuclei lying in the complementary region may be used as the reaction partners (like ^{48}Ca + ^{192}W) to synthesize ^{240}Pu* nucleus in addition to the present neutron-induced channel. The comparative analysis of the three graphs depicts the variations in the structure and magnitude with energy, indicating that different fragments may be minimized in the fragmentation channel over different beam energies. However, dominant channels, such as neutron evaporation, light-charged particles evaporation or emergence of ^{48}Ca, ^{68,70}Ni, ^{82}Ge, ^{84}Se and ^{106,108}Ru, seems not to be affected to a greater extent. Our area of interest is neutron evaporation (synthesizing the plutonium isotopes) or light particle evaporation (synthesizing the neighbouring elements such as ^{236}U and is it quite interesting to explore the role of energy on these individual decay channels. For this purpose, the preformation probability (P0) of a particular channel is plotted in terms of the angular momentum imparted to the decay channel at all the three reported energies and presented in Figure 21.3. Note that Figure 21.3, parts (a), (b) and (c), represents the xn-evaporation channel where x = *1*, *2*, and *3*, respectively, or equivalently, it represents the synthesis of ^{239,238,237}Pu isotopes respectively, while part (d) represents the light-charged particle emission (^{4}He-emission) from ^{240}Pu*, thereby synthesizing ^{236}U. Figure 21.3 (a)

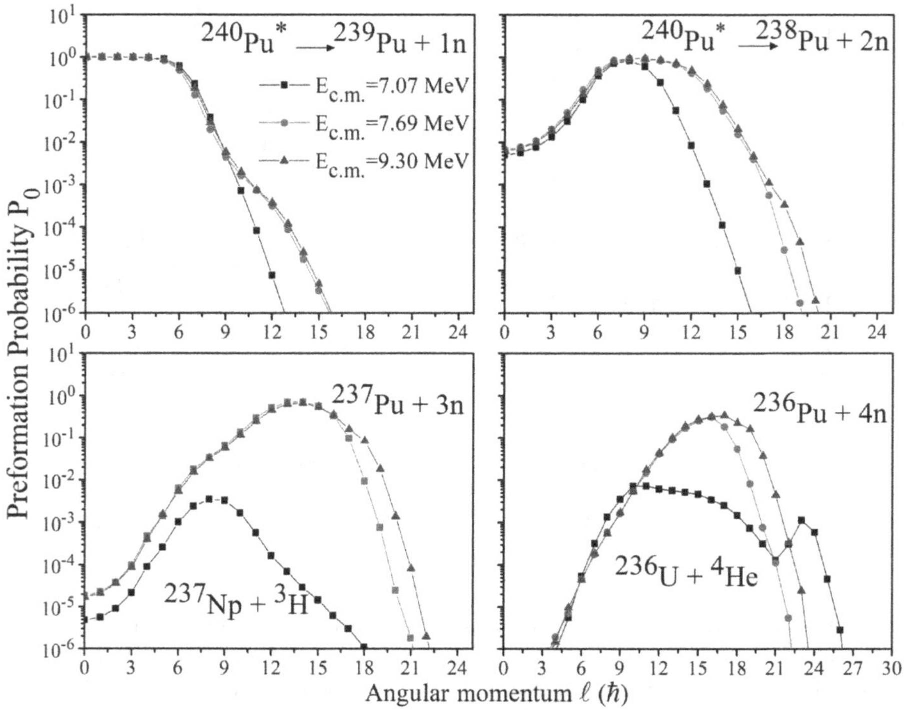

FIGURE 21.3 Probability of preformation of (a)-(b) isotopes $^{238,\,239}Pu$ via *1n*- and *2n*-evaporation, (c) isotope ^{237}Pu via *3n*-evaporation or actinide ^{237}Np via ^{3}H-emission, and (d) isotope ^{236}Pu via *4n*-evaporation or ^{236}U via ^{4}He-emission at all the three experimentally available energies (Ec.m.) up to the permissible angular momentum values.

and (b) clearly signifies that the role of energy of the incoming beam is evident for angular momentum values l > 9h because at angular momentum values less than 9h, the magnitude of *xn*-evaporation is almost equal at all the three energies and variation in magnitude is observed for l > 9h case. It is observed that as energy value increases, the limiting values of angular momentum (at which the P0→0 for *xn*-channel) also increases for 1n and 2n emission channels. In addition to this, the effect of energy seems to be more dominant for the synthesis of ^{237}Pu and ^{236}U, because a competition from other fragments corresponding to mass number A2 = 3 and 4 is also coming into picture.

The collective clusterization approach of the methodology used in the present case works in terms of mass minimization, which means that only those fragments are preferred in the decay channel, which possess minimum value of fragmentation potential corresponding to their respective isobars. It is evident from Figure 21.3(c) that for Ec.m. = 7.69 MeV and 9.30 MeV the preformation values are almost similar, while the values of P0 for the lowest energy Ec.m. = 7.07 MeV seem to be quite lower as compared to the other two energies. This is attributed to the fact that at this energy, the minimized fragment is ^{3}H leading to the formation of daughter nucleus ^{237}Np in place of ^{237}Pu that is formed at the other two energies via 3n emission.

It is quite interesting to explore this effect of energy because in nuclear reactors, ^{237}Np is always formed as a by-product of plutonium-based reactions, and it is a long living isotope of neptunium. This isotope holds significant importance in clean fuel technologies as ^{237}Np is transmuted into ^{238}Pu, which is used for the long-term operation of power sources that are used to propel satellites, spacecraft, and lighthouses. Hence, the incident beam energy may hold a significant position in deciding the daughter nucleus produced in a particular nuclear reaction. Similar analysis can be extended for the decay $^{240}Pu^* \rightarrow A1+A2$ with A2 = 4, the fourth fragment in the decay channel of $^{240}Pu^*$. In this case, it is observed that the fourth fragment is minimized as *4n* at the two higher energies Ec.m. = 7.69 MeV and 9.30 MeV, synthesizing ^{236}Pu as daughter nucleus, while, at the lowest energy Ec.m. = 7.07 MeV, the fragment is identified as ^4He synthesizing ^{236}U as daughter nucleus and it possesses much lower preformation values as compared to the other channels at two higher energies. A striking observation here is that at lowest energy, the fragment ^4He is identified at lower angular momentum values and towards the limiting angular momentum values, there is an emergence of shoulder in the P0 curve, which represents the minimization of *4n* instead of ^4He. Hence, for the A2 = 4 (the fourth fragment channel), the minimized fragment is identified as *4n* for lower angular momentum values, and as the limiting angular momentum values are approached, there is presence of ^4He in the decay channel. This is depicted by the emergence of a shoulder in preformation curve as plotted in Figure 21.3(d) for the lowest energy. This may be attributed to the role of quadrupole deformations coming into picture and is a subject of further analysis. An important point to be noted here is that within the framework of DCM, the total reaction cross-sections for a particular decay channel are a result of preformation probability and the barrier penetration probability. The high preformation probability values alone do not guarantee the presence of the fragment in the decay channel, but the role of penetration probability is equally important. The barrier penetration behaviour of the decaying fragments is also analysed and depicted in Figure 21.4 in terms of angular momentum states at all the three energies Ec.m. = 7.07, 7.69, and 9.30 MeV for *xn*-evaporation or light-charged particle emission. In part (a), the penetrability values of 1n-emission from $^{240}Pu^*$ synthesizing ^{239}Pu isotope at three energy values, evincing that penetration values saturate around l = 10h and saturates after that. The effect of energy on 1n-emission leading to the synthesis of ^{239}Pu isotope is also not very dominant; however, for other decay channels 2n, 3n, 4n, or other light-charged particle emission channels, the effect of energies is evident as change in magnitude is observed for different energy values (shown by black, blue, and red lines). In part (b), the penetrability values increase with angular momentum and becomes maximum for l > 15h. The effect of energy on penetrability P is similar to the effect of energy on preformation probability P0 as shown in Figure 21.3.

Moving further, Figure 21.5 presents the comparison of the theoretically calculated cross-sections for *2n*-emission from $^{240}Pu^*$ formed in *n*-induced channel, with the experimentally available data, at the three reported energies Ec.m. = 7.07 MeV, 7.69 MeV, and 9.30 MeV. Clearly, a nice agreement with the experimental results is observed, and the theoretical measurements are made by optimizing the neck-length parameter of the model, which is the only free parameter of the model. The values of neck-length parameter ΔR are also plotted in Figure 21.5, and the values of the

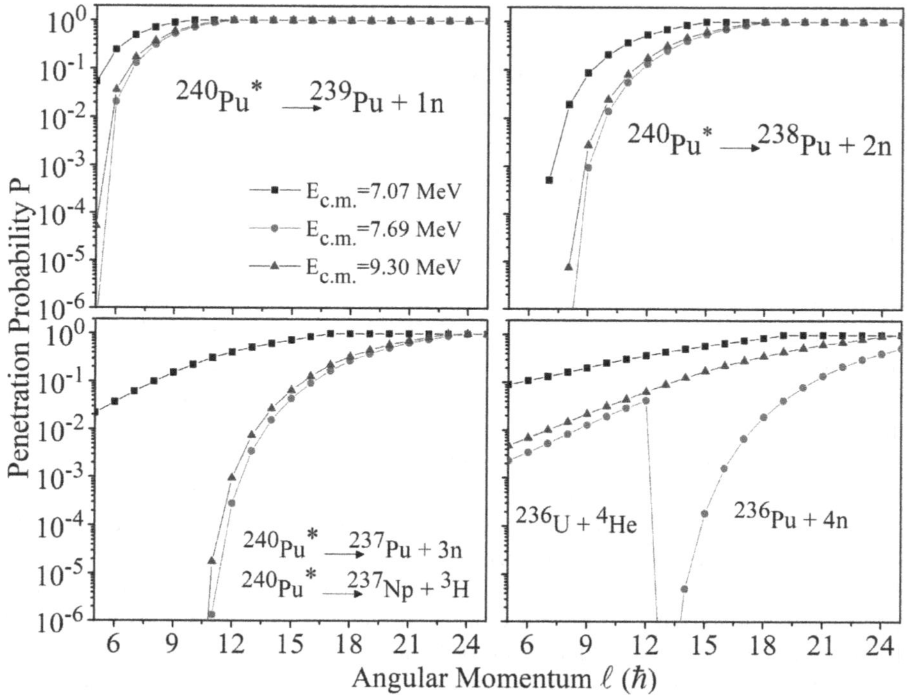

FIGURE 21.4 Probability of penetration of (a)-(b) isotopes $^{238,\,239}Pu$ via *1n*- and *2n*-evaporation, (c) isotope ^{237}Pu via *3n*-evaporation or actinide ^{237}Np via ^{3}H-emission, and (d) isotope ^{236}Pu via *4n*-evaporation or ^{236}U via *4He*-emission at all the three experimentally available energies (Ec.m.) up to the permissible angular momentum values.

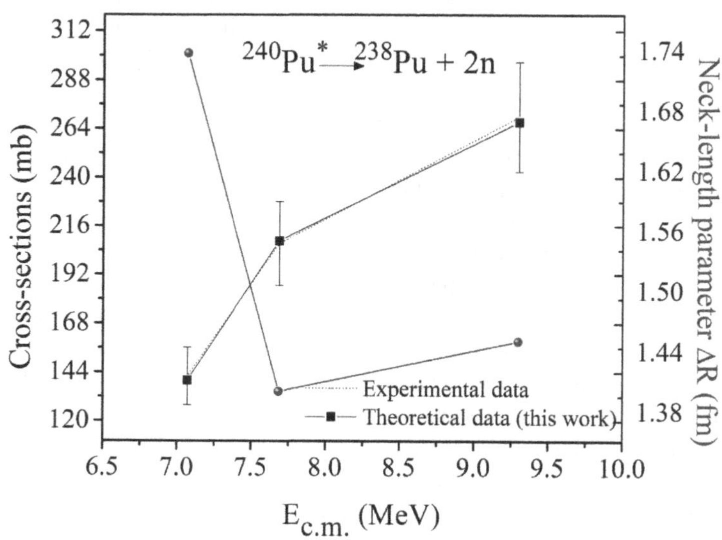

FIGURE 21.5 Theoretical cross-sections estimated in this work are plotted against the experimental data [] in terms of the available Ec.m. and the best-fit values of neck-length parameter; the free parameter of the dynamical clusterization approach is also shown in the figure.

TABLE 21.1
Theoretically Calculated Cross-Sections for the Synthesis of ^{238}Pu via *2n*-Evaporation from $^{240}Pu^*$ Formed in ^{n}-Induced Channel in Contrast to the Experimental Data along with Other Parameters of the Model

S. No.	Ec.m. (MeV)	T (MeV)	ΔR (fm)	lmax (h)	Theoretical Results (mb)	Experimental Data (mb)
1	7.07	0.816	1.737	16	140.78	142±22
2	7.69	0.834	1.4	19	206.89	207.3±32.4
3	9.30	0.878	1.45	20	270.56	269.9±40.3

neck-length parameter le between 1.4 fm to 1.75 fm. These results are also presented in Table 21.1.

21.4 CONCLUSIONS

In the present work, the decay of $^{240}Pu^*$ via xn-emission formed in n-induced reaction is analysed at three energies Ec.m. = 7.07, 7.69, and 9.30 MeV with an intent to comprehend the underlying dynamics governing the formation of the isotopes $^{240-xn}Pu$. In addition to this, the probable ^{3}H and ^{4}He channels are also investigated for the synthesis of other daughter nuclei, such as ^{239}Np or ^{238}U, so as to understand their nuclear properties justifying their candidature as nuclear fuels for the future sustainable reactors. The role of nuclear properties such as nuclear deformations, centre of mass energy and angular momentum of the decaying system is also analysed. It has been observed that the decaying system $^{240}Pu^*$ prefers to decay via neutron or light particle emission at lower angular momentum values. The DCM parameters, such as preformation probability and barrier penetration probability, seem to be dependent up on the centre of mass energy and, hence, show different variations at different energies. This results in competition between the emergence of neutrons or ^{3}H and ^{4}He channels and hence affecting the probability of formation of Pu-isotopes or ^{239}Np or ^{238}U.

REFERENCES

[1] M. Salvatores, R. Jacqmin, Uncertainty and target accuracy assessment for innovative system using recent covariance data evaluations. *Tech. Rep.*, NEA/WPEC-26, Argonne National Laboratory (2008).

[2] B. W. Brook, A. Alonso, D. A. Meneley, J. Misak, T. Blees, J.B. van Erp, Why nuclear energy is sustainable and has to be part of the energy mix. *Sustain. Mater. Technol.* 1–2, 8–16 (2014).

[3] I. Kodeli, Sensitivity analysis and uncertainty propagation from basic nuclear data to reactor physics and safety relevant parameters, in: *Workshop Proceedings of Evaluation of Uncertainties in Relation to Severe Accidents and Level-2 Probabilistic Safety Analysis*, November 7–9, 2005, Aix-En-Provence, Paris, France, pp. 180–233 (2007). https://www.oecd-nea.org/jcms/pl_18438

[4] Status and Advances in MOX Fuel Technology. *IAEA Technical Reports Series No. 415,* Vienna, Austria (2003).
[5] N. Colonna et al., *Energy Environ. Sci.* 3, 1910 (2010).
[6] N. Colonna et al., *Energy Environ. Sci.* 3, 1910 (2010).
[7] J. Lerendegui-Marc et al., *Phys. Rev. C* 97, 024605 (2018).
[8] M. K. Sharma et al., *J. Phys. G. Nucl. Part. Phys.* 38, 055104 (2011); R. K. Gupta et al., *J. Phys. G.* 31, 631 (2005).
[9] M. K. Sharma et al., *Phys. Rev. C* 85, 064602 (2012); G. Sawhney, M. K. Sharma, R. K. Gupta, *Phys. Rev. C* 83, 064610 (2011).
[10] R. K. Gupta, M. Balasubramaniam, R. Kumar, D Singh, C. Beck, W Greiner, *Phys. Rev. C.* 71, 014601 (2005); and earlier references therein.
[11] R. K. Gupta, M. Balasubramaniam, R. Kumar, D Singh, S. K. Arun, W. Greiner, *J. Phys. G: Nucl. Part. Phys.* 32, 345 (2006).
[12] B. B. Singh, M. K. Sharma, R. K. Gupta, W. Greiner, *Int. J. Mod. Phys. E* 15, 699 (2006).
[13] R. K. Gupta, S. K. Arun, R. Kumar, Niyti, *Int. Rev. Phys. (IREPHY)* 2, 369 (2008); R. Kumar, R. K. Gupta, *Phys. Rev. C* 79, 034602 (2009).
[14] S. Kanwar, M. K. Sharma, B. B. Singh, R. K. Gupta. W. Greiner, *Int. J. Mod. Phys. E* 18, 1453 (2009); R. K. Gupta, S. K. Arun, R. Kumar, M. Bansal, *Nucl. Phys. A* 834, 176c (2010).
[15] Amandeep Kaur, Manoj K. Sharma, *Mod. Phys. Lett. A* 33(1), 2050082 (2020).
[16] Amandeep Kaur, Manoj K. Sharma, *Nucl. Phy. A* 957, 274–288 (2017).
[17] Amandeep Kaur, Kirandeep Sandhu, Gudveen Sawhney, Manoj K. Sharma, *Eur. Phys. J. A* 58, 59 (2022).
[18] V. Méot et al., *Phys. Rev. C* 103, 054609 (2021).
[19] R. K Gupta, W. Scheid, W. Greiner, *Phys. Rev. Lett.* 35, 353 (1975).
[20] R. K. Gupta, W. Greiner, 'Two centre shell model in cold synthesis of superheavy elements' and 'Quantum Mechanical Fragmentation Theory for Cold Distribution of Masses and Charges in Fissioning Nuclei and Nuclei Formed in Heavy Ion Reactions.' *Heavy Elements and Related New Phenomena,* edited by W. Greiner and R. K Gupta, Vol. I, World Scientific, Singapore, p. 397; p. 536 (1999).
[21] S. S. Malik, R. K. Gupta, *Phys. Rev. C* 39, 1992 (1989).
[22] P. A. Seeger, *Nucl. Phys.* 25, 1 (1961).
[23] N. J. Davidson, S. S. Hsiao, J. Markram, H. G. Miller, Y. Tsang, *Nucl. Phys.* A570, 61C (1994).
[24] W. Myers, W. J. Swiatecki, *Nucl. Phys.* 81, 1 (1966).
[25] B. B. Singh, M. K. Sharma, R. K. Gupta, *Phys. Rev. C* 77, 054613 (2008).
[26] R. K Gupta, M. Balasubramaniam, C. Mazzocchi, M. La Commara, W. Scheid, *Phys. Rev. C* 65, 024601 (2002).
[27] R. A. Gherghescu, *Phys. Rev. C* 67, 014309 (2003).
[28] H. Kroeger, W. Scheid, *J. Phys. G: Nucl. Phys.* 6, L85 (1980).
[29] G. Royer, J. Mignen, *J. Phys. G: Nucl. Part. Phys.* 18, 1781 (1992).

22 Characterization of Plasma-Sprayed CNT-Reinforced Inconel 718 Coatings on Boiler Tube Steels

Rakesh Goyal, Punam, Hemender Yadav, Hitesh Singla

22.1 INTRODUCTION

Due to the ever-increasing complication of the requirements, it is becoming increasingly difficult, and in the current state of development, it may even be impossible, to incorporate all of the necessary properties into a single material. This is because the demands that are being placed on materials are continuing to rise. Therefore, a composite system consisting of a base material that offers the required mechanical strength and a protective surface layer that is distinct in structure and/or chemical composition may be the best option for combining material attributes (Sidhu et al., 2005; Singh et al., 2015). A composite system consists of a base material that offers the required mechanical strength and a protective surface layer that is distinct in structure and/or chemical composition.

Coatings that are applied using thermal spraying not only save money but also offer excellent protection against wear and corrosion. They are able to be produced through procedures that are not unduly complicated in any way. In addition to this, other advantageous features may be generated at the coating's or component's surface if either is given enough time. Because of this, these coatings are currently being used in a wide variety of different industrial applications (Wielage et al., 1998). This is a direct result of the previous point.

It is believed that plasma spraying is a versatile method that has been shown to be useful as a trustworthy and cost-effective solution for a number of problems that arise in the industrial sector (Deshpande et al., 2004). The plasma-spraying method is the sort of thermal spraying that offers the greatest degree of adaptability and variety in terms of the kinds of substances that can be sprayed. In plasma spray techniques, the high temperatures enable the deposition of coatings for applications in the fields of liquid and high temperature corrosion and wear protection, in addition to

DOI: 10.1201/9781003367154-22

specific applications for thermal, electrical, and biological purposes (Du et al., 2010). These coatings can also be used for liquid and high temperature wear protection.

Protective coatings like those made of Inconel 718 have only been around for a very short amount of time. In addition to being resistant to oxidation, it has been claimed that Inconel 718 is also capable of exhibiting high temperature mechanical strength (Malik et al., 1992). CNTs are long, cylinder-shaped allotropic forms of carbon that have high thermal conductivity, electrical conductivity, elasticity, and tensile strength (Iijima, 1991). (Note: CNTs are also referred to as carbon nanotubes.) In addition to that, the thermal expansion coefficient of these materials is quite low.

Plasma spray techniques are frequently used for the purpose of coating deposition because of the ease with which they can be applied and their capacity to deposit a wide variety of coating materials onto substrate alloys. In addition, plasma spray techniques are capable of depositing coatings onto a wide range of substrate alloys. It is one of the methods for changing the surface of a material that is used to improve the surface qualities of the material, such as resistance to corrosion and wear (Goyal et al., 2020; Goyal et al., 2018; Karthikeyan et al., 1988). It is one of the procedures that is used to increase the quality of the material's surface. The primary objective of the ongoing research is to develop a methodology for the characterization of coatings that are applied via plasma spraying on boiler steels. At this point in time, a study has been conducted into the microstructure, porosity, and microhardness of the coated samples. Analytical techniques like metallography and SEM/EDAX are typically employed for this purpose.

22.2 EXPERIMENTAL DETAILS

22.2.1 Development of Coatings

22.2.2 Substrate Materials

T22 and T91 are the two different types of substrate materials that were used. These materials are utilized in several of the power plants located in the northern region of India as boiler tube materials. American-made Thermo Jarrel Ash (TJA 181/81) Optical Emission Spectrometers were used to ascertain the actual chemical composition of these steels. The nominal and actual compositions of these steels are listed in Table 22.1. These steel samples were chopped to make specimens with dimensions of approximately 20 millimetres by 15 millimetres by 5 millimetres. Before the coating was applied, the specimens were grit blasted with Al_2O_3 (grit 60) and then polished.

22.2.3 Coating Materials

For plasma spraying, two different kinds of coating powders are utilized. The first powder mixture for coating is made up of Inconel 718 with a minimum assay of 99.5% and a size of 40 mesh. The second powder mixture for coating is made up of CNT-reinforced Inconel 718 with a minimum assay of 99.5%. Both of these powders

TABLE 22.1
Chemical Composition (wt%) for Various Steels

Type of Steel	ASTM Code	Composition	C	Mn	Si	S	P	Cr	Mo	Fe
T-22	SA213-T-22	Nominal	0.14	0.29–0.59	0.49	0.008	0.029	2.1–2.7	0.89–1.12	Bal.
		Actual	0.137	0.18	0.45	0.009	0.021	2.63	1.07	Bal.
T-91	SA213-T-91	Nominal	0.14	0.31–0.58	0.6–0.9	0.029	0.029	8.1–9.5	0.47–0.68	Bal.
		Actual	0.137	0.45	0.32	0.0024	0.014	8.664	0.928	Bal.

TABLE 22.2
Composition and Particle Size of Coating Powders

S. No.	Coating Powder	Composition (wt%)	Particle Size
1.	Inconel 718	Cr (21), Ni (55), Mo (3), Co (1), Cb (5), Ti (1)	–45μm + 10 μm
2.	CNT	C (99.99)	10 nm
3.	CNT-reinforced Inconel 718	CNT (7.5) Cr (21), Ni (55), Mo (3), Co (1), Cb (5), Ti (1)	–45μm + 10 μm

were mixed in a laboratory ball mill for eight hours to form a uniform mixture. Table 22.2 contains information regarding the chemical composition as well as the particle size of these powders.

22.2.4 Analysis of Coating Powders

SEM research on Inconel 718, carbon nanotubes (CNTs) and Inconel 718 that was reinforced with CNT has been carried out. Figures 22.1, 22.2, and 22.3 present the backscattered electron images (BSEIs) obtained from these samples. In the case of the Inconel 718 coating, the observation of spherical particles may be found in Figure 22.1. The CNT BSEIs display long tubes, as shown in Figure 22.2. These are long, cylinder-shaped allotropes of carbon that have high thermal conductivity, electrical conductivity, elasticity, tensile strength, and a low thermal expansion coefficient. Other desirable properties include high thermal conductivity and electrical conductivity.

The BSEIs of CNT-reinforced Inconel 718 are seen in Figure 22.3. The powder particles of Inconel 718 have been observed to have a dispersion of CNTs that is consistent throughout. This indicates that the CNTs will be evenly dispersed on the surface of the substrate during the coating process when they are sprayed onto the surface of the substrate. When reinforced into the coating powder, these nanotubes will boost the coating's resistance to deterioration, leading to improved performance.

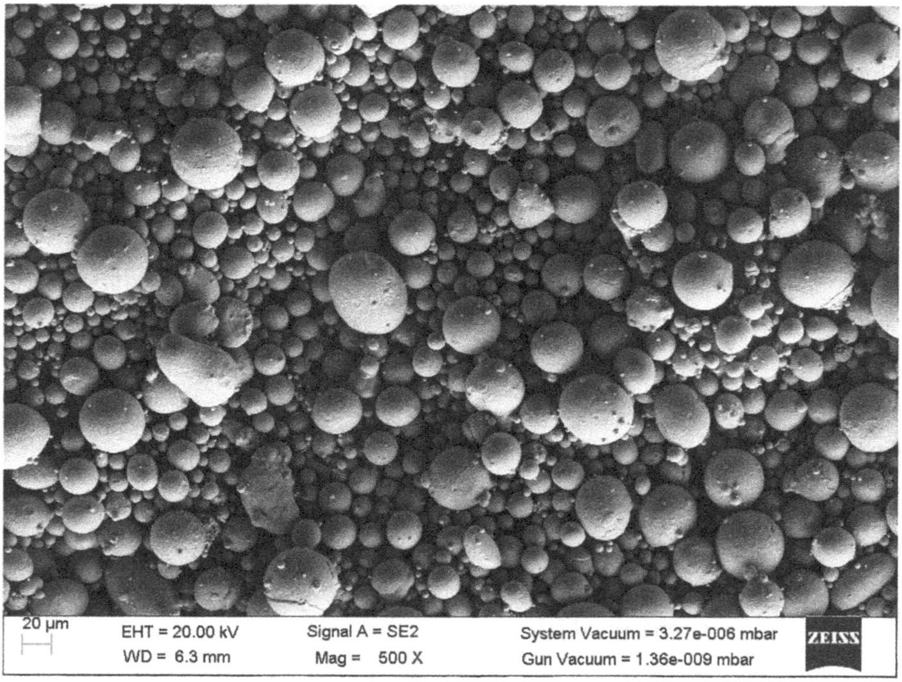

FIGURE 22.1 Morphology of Inconel 718 coating powder as seen by scanning electron microscopy, 500×.

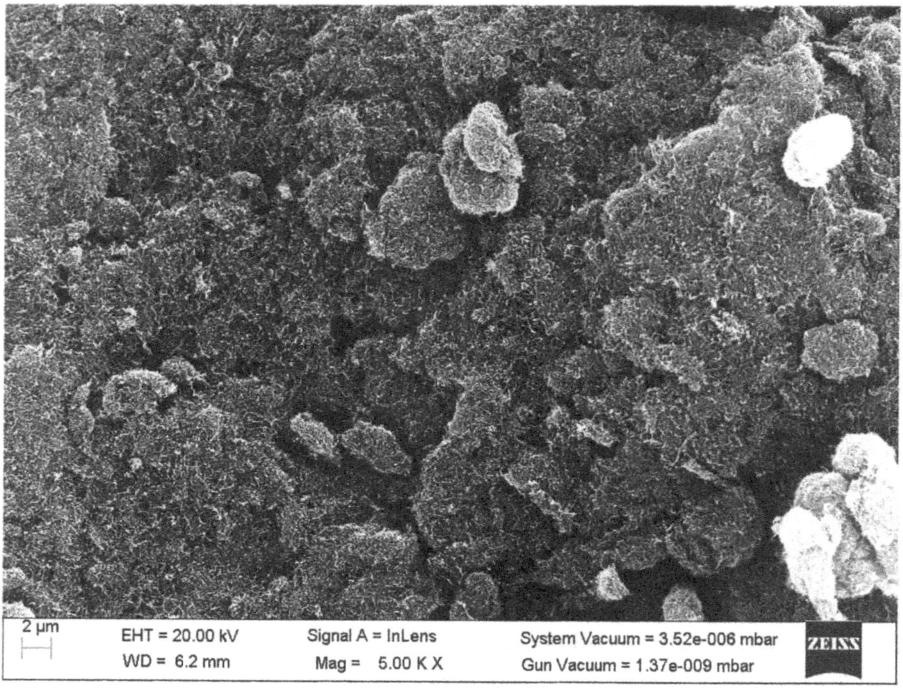

FIGURE 22.2 Morphology of CNT nanopowder as seen by scanning electron microscopy, 5K×.

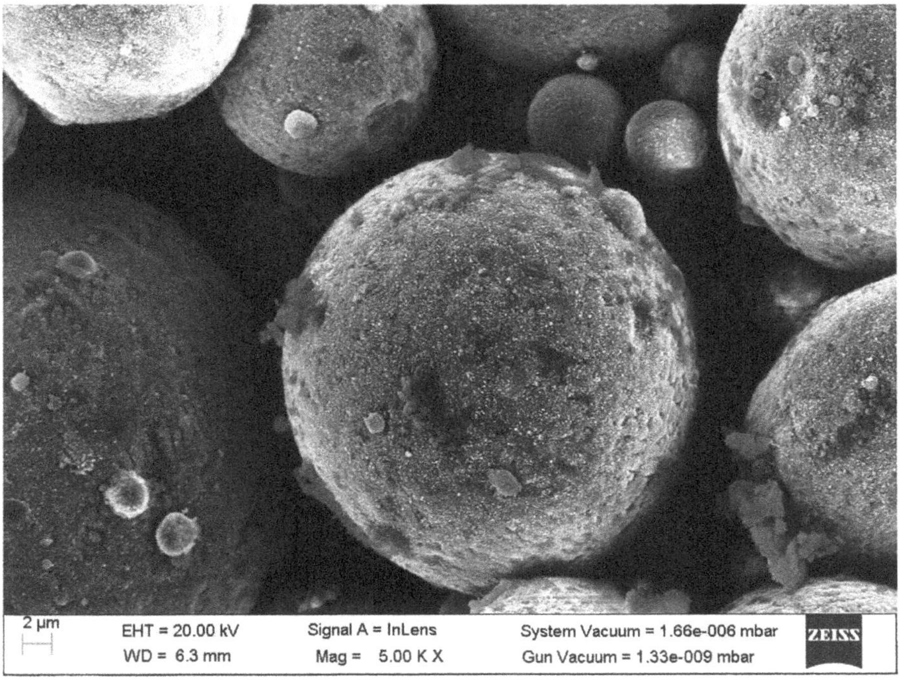

FIGURE 22.3 Morphology of CNT-reinforced Inconel 718 coating powder as seen by scanning electron microscopy, 5K×.

22.2.5 Formulation of Coating

Before plasma spraying the samples, grit blasting was performed. The coatings were applied using a 40 kW Miller Thermal Plasma Spray Apparatus. This apparatus was used to spray the coatings. The arc current never dropped below 700 A, and the arc voltage never rose beyond 35 V at any point during the experiment. Both a powder carrying gas (at a pressure of 59 psi) and a shielding gas (at a pressure of 40 psi) were accomplished with the help of argon. During the entirety of the coating process, all of the process parameters, such as the spray distance (90–110 mm), were maintained at a consistent level while the powder flow rate was preserved at 3.2 rev./min. In order to apply the final coats, Inconel 718 powder that had been enhanced with CNTs was utilized. The BSEIs of the CNT-reinforced Inconel 718–coated surface are displayed in Figure 22.4. The surface that has been coated is smooth. The particles of Inconel 718 have a consistent distribution of CNTs throughout the material. During the coating process, the CNTs are spread out in a uniform manner throughout the surface of the substrate. The coatings' hardness was evaluated using a microhardness tester, as well as by scanning electron microscopy (SEM) and electron backscatter X-ray diffraction (EDAX).

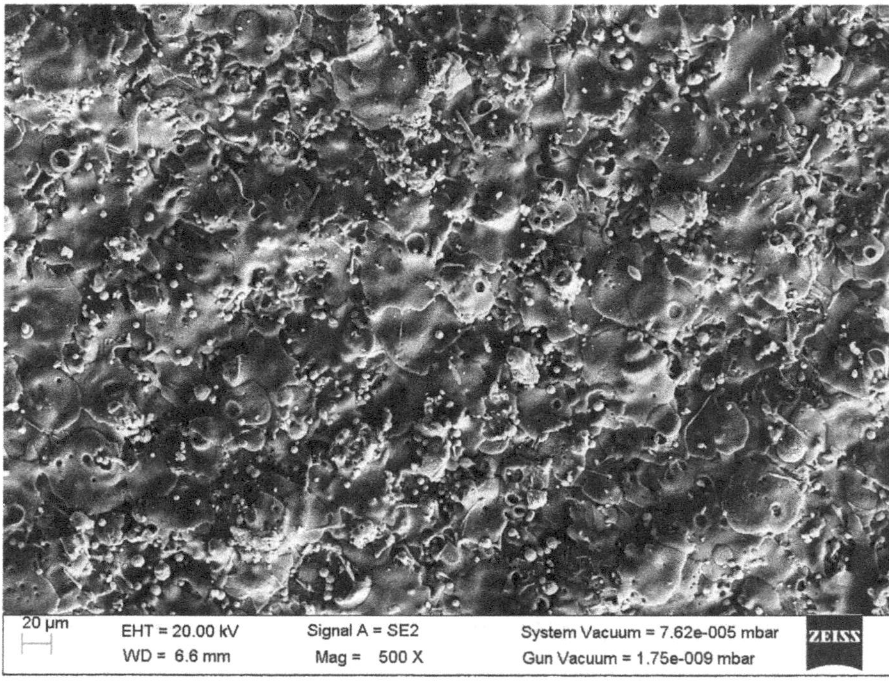

FIGURE 22.4 Morphology of CNT-reinforced Inconel 718–coated surface as seen by scanning electron microscopy, 500×.

22.3 RESULTS

22.3.1 Measurements of Coating Thicknesses

The plasma-sprayed Inconel 718 and the CNT-reinforced Inconel 718 coatings have both been subjected to examinations by the scanning electron microscope (SEM) along the cross-section. BSEIs of these samples may be seen in Figure 22.5 (a) and in Figure 22.5 (b), respectively. The outer layer in Figure 22.5 (a) illustrates the Inconel 718 coating that was applied by spraying it on. In the micrograph of CNT-reinforced Inconel 718 coatings shown in Figure 22.5(b), the outer coating is also composed of Inconel 718 with CNTs distributed uniformly throughout.

According to the micrographs, the coating layer is extremely smooth and dense. Table 22.3 contains a compilation of the measured thickness of the coatings obtained from the BSEI microscopy.

22.3.2 Porosity Analysis and Evaluation of Microhardness of Coatings

The image analyser with the programme Dewinter Material Plus 1.01, which was based on the ASTM B276 standard, was used to perform the porosity measurements for the as plasma-sprayed coatings. As-sprayed Inconel 718 coatings were found to

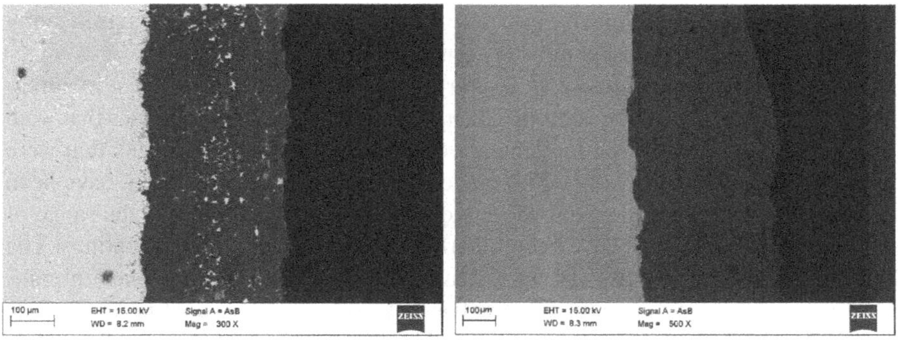

FIGURE 22.5 Morphology (cross-section) as seen by scanning electron microscopy (a) Inconel 718 coating, 300×, and (b) CNT-reinforced Inconel 718 coating on T91 steel, 500×.

TABLE 22.3
Average Coating Thickness

S. No.	Type of Coating	Coating Thickness (µm)
1	Inconel 718	250 ± 10 µm
2	CNT-reinforced Inconel 718	250 ± 10 µm

have a porosity in the range of 3.8%, according to the research. CNT reinforcement in the coating was observed to result in a sizable reduction in porosity, which was subsequently determined to be 3.2%.

Along the cross-section, the microhardness of coatings applied to a variety of substrate steels was analysed and measured. It was discovered that the critical hardness values of the substrate steels fell somewhere in the region of 200–220 Hv. The Inconel 718 coating has shown signs of reduced hardness, and its microhardness value was measured to be 204 Hv. This value is extremely close to the microhardness of steel substrates. Following the addition of CNT reinforcement, a rise in the microhardness value of the Inconel 718 coating was noticed. As a result of this, the microhardness of the coating reached 40% of that of the as-sprayed coating. It was determined that the microhardness value of the CNT-reinforced Inconel 718 coating was 285 Hv.

22.4 RESULTS AND DISCUSSION

The coating thickness can only go up to 250 micrometres because bigger coatings tend to self-degrade over time. During the process of applying Inconel 718 and CNTs reinforced Inconel 718 coating, the plasma torch became blocked numerous times due to the different melting points of nickel and aluminium as well as the nano size of CNTs; despite this, the necessary phases, which are Ni, Cr, Al, and C, were successfully coated. As a result of the incorporation of CNTs into the Inconel 718 coatings, gaps and pores in the structure of the coatings have been significantly reduced and, in some cases, completely removed.

While the measured porosity values for as-sprayed coatings are almost in agreement with the findings of the authors (Chen et al., 1993), the porosity values that were observed following CNT reinforcement are in agreement with the results that were reported by (Goyal et al., 2017). The range of microhardness values that have been recorded for plasma coatings (Goyal & Goyal, 2020) is comparable to the range of microhardness values that have been reported for sprayed Inconel 718 coatings. The CNT-reinforced coating has shown to have higher microhardness values in comparison to the Inconel coating, and the hardness values of this coating are in line with the hardness values that were published by the authors (Goyal et al., 2017). The structures that were obtained following CNT reinforcement are in agreement with the structures that were achieved by (Goyal et al., 2017) for CNT-reinforced–Al_2O_3 plasma-sprayed coatings. These structures are dense, homogeneous, and void-free.

22.5 CONCLUSIONS

Coatings made of CNTs and Inconel 718 were applied to T22 and T91 boiler tube steel with great success as part of this research project. The micrographs demonstrate that the surface of the coating after it has been sprayed is consistent, smooth, and free from fractures. Furthermore, there is a uniform dispersion of CNTs within the Inconel coatings. During the process of the plasma spray formation, the CNTs maintained their chemical stability. Even when subjected to temperatures that were very high during the processing, they did not react to generate oxides or carbides. The examination revealed that the elements nickel and chromium, in addition to carbon and oxygen, predominated in the composition of the coating, which was almost identical to the composition of the powder that was sprayed. The hardness of the coating was in the range of 204 Hv before the incorporation of CNTs into the Inconel 718 coating, and after the incorporation of CNTs, the hardness was raised by 40%. The coating thickness was measured to be in the region of 250 micrometres, and the porosity of CNTs–Inconel 718 coatings was found to have decreased due to the presence of CNTs.

REFERENCES

Chen, H. C., Liu, Z. Y., & Chuang, Y. C. (1993). Degradation of plasma-sprayed alumina and zirconia coatings on stainless steel during thermal cycling and hot corrosion. *Thin Solid Films*, *223*(1), 56–64. https://doi.org/10.1016/0040-6090(93)90727-7.

Deshpande, S., Kulkarni, A., Sampath, S., & Herman, H. (2004). Application of image analysis for characterization of porosity in thermal spray coatings and correlation with small angle neutron scattering. *Surface and Coatings Technology*, *187*(1), 6–16. https://doi.org/10.1016/j.surfcoat.2004.01.032

Du, L., Zhang, W., Liu, W., & Zhang, J. (2010). Preparation and characterization of plasma sprayed Ni3Al–hBN composite coating. *Surface and Coatings Technology*, *205*(7), 2419–2424. https://doi.org/10.1016/j.surfcoat.2010.09.036

Goyal, K., & Goyal, R. (2020). Improving hot corrosion resistance of Cr3C2–20NiCr coatings with CNT reinforcements. *Surface Engineering*, *36*(11), 1200–1209. https://doi.org/10.1080/02670844.2019.1662645

Goyal, R., Sidhu, B. S., & Chawla, V. (2017). Characterization of plasma-sprayed carbon nanotube (CNT)-reinforced alumina coatings on ASME-SA213-T11 boiler tube steel. *The International Journal of Advanced Manufacturing Technology*, *92*, 3225–3235. https://doi.org/10.1007/s00170-017-0405-z

Goyal, R., Sidhu, B. S., & Chawla, V. (2018). Improving the high-temperature oxidation resistance of ASME-SA213-T11 boiler tube steel by plasma spraying with CNT-reinforced alumina coatings. *Anti-Corrosion Methods and Materials*, *65*(2), 217–223. https://doi.org/10.1108/ACMM-03-2017-1778

Goyal, R., Sidhu, B. S., & Chawla, V. (2020). Hot corrosion performance of plasma-sprayed multiwalled carbon nanotube–al 2 o 3 composite coatings in a coal-fired boiler at 900° c. *Journal of Materials Engineering and Performance*, *29*, 5738–5749. https://doi.org/10.1007/s11665-020-05070-8

Iijima, S. (1991). Helical microtubules of graphitic carbon. *Nature*, *354*(6348), 56–58. https://doi.org/10.1038/354056a0

Karthikeyan, J., Sreekumar, K. P., Venkatramani, N., & Rohatgi, V. K. (1988). Preparation and characterization of plasma-sprayed thick ceramic coatings reinforced with metal pins. *High Temperatures. High Pressures (Print)*, *20*(6), 653–660.

Malik, A. U., Ahmad, R., Ahmad, S., & Ahmad, S. (1992). High temperature oxidation behaviour of Nickel aluminide coated mild steel/Hochtemperaturoxidationsverhalten von unlegiertem Stahl mit Nickelaluminid-Beschichtung. *Practical Metallography*, *29*(5), 255–267. https://doi.org/10.1515/pm-1992-290505

Sidhu, B. S., Puri, D., & Prakash, S. (2005). Mechanical and metallurgical properties of plasma sprayed and laser remelted Ni–20Cr and Stellite-6 coatings. *Journal of Materials Processing Technology*, *159*(3), 347–355. https://doi.org/10.1016/j.jmatprotec.2004.05.023

Singh, H., Sidhu, T. S., Karthikeyan, J., & Kalsi, S. B. S. (2015). Evaluation of characteristics and behavior of cold sprayed Ni–20Cr coating at elevated temperature in waste incinerator plant. *Surface and Coatings Technology*, *261*, 375–384. https://doi.org/10.1016/j.surfcoat.2014.10.060

Wielage, B., Hofmann, U., Steinhauser, S., & Zimmermann, G. (1998). Improving wear and corrosion resistance of thermal sprayed coatings. *Surface Engineering*, *14*(2), 136–138. https://doi.org/10.1179/sur.1998.14.2.136

23 Behavioural Study of Impact Energy

Nidhi Bansal Garg, Atul Garg, Mohit Bansal, Mohit Kakkar

23.1 INTRODUCTION

High-strength low-alloy (HSLA) steels are well-defined as a particular group of low-carbon steels with a specially designed chemical composition that imparts higher mechanical property values and some of these steels have a higher also significantly improved resistance to atmospheric corrosion. Carbon steel and HSLA steels are largely manufactured with importance on mechanical properties rather than chemical composition limitations. It is not considered an alloy steel, but with conventionally added alloy content it technically qualifies as an alloy steel [1][2]. HSLA steel can be gently processed into a fine-grained polygonal ferritic structure with an enhanced combination of properties [3] [4].

In general, for steels, the ductile-brittle transition temperature (DBTT) declines with declining grain size and increases with the existence of pearlite. The quenching and tempering processes further decrease the DBTT. The superiority of tempered and quenched steel is related with fine-grain extent and lower perlite content. Adequate alloy-carbide and carbonitride precipitates present in HSLA steels containing Nb and V act as precipitation strengthening agents, but DBTT increases by 0.3–0.5°C for every 1 MPa increase in yield strength[1][5]. Micro-alloy and thermo-mechanical processing techniques have been extensively developed over the last 50 years. These steels became more attractive with least processing and alloying cost, whereas its mechanical properties are more increased in comparison with the traditional carbon steel [6][7][8].

These so-called HSLA steels and thermos-mechanically processed (TMP) steels are an extension of C-Mn grades and fill the gap between 250 MPa yield strength of C-Mn steels and 700 MPa yield strength for tempered and quenched alloy steels. A combination of low C content, ferrite grain refinement, precipitation-hardening, and low alloy content are the results of these steels with the high mechanical properties of these steels.

Refining the ferrite grain size improves both the strength as well as toughness of HSLA steels. Optimal ferrite-grain-size reduction is generally achieved by tightly controlled rolling combined with accelerated cooling. For simple sheet-rolled steel compositions, this can reduce the ferrite grain size from 10μm for hot rolling and air cooling to 5μm for controlled rolling and water cooling, increasing the yield strength by about 80 MPa. At the same time, there was evidence that stress-induced

 DOI: 10.1201/9781003367154-23

transformation (that is, transformation during deformation rather than after deformation) can lead to significant refinement of ferrite. It has been found in literature that the strength and toughness of steel can be improved by applying transformation of structure like austenite to fine needle [9][10][11][12][13].

As a result, considerable effort has been expended to maximize the amount of acicular ferrite in the final microstructure. Some authors have shown that the presence of uniform layers of homomorphic ferrite along austenite grain boundaries induces the transformation of austenite to acicular ferrite rather than binate [3][5][10][11][14].

Authors in [15][16][17] contributed their knowledge and laid the foundation for continued growth in the field of metallurgy.

Rolled pearlitic steel typically contains manganese and carbon, although other elements may be added to increase strength. Duplex steels use micro-alloys to create a ferritic matrix with distributed martensite. The significance of these steels lies in their use in various commercial and industrial applications. As a result, they have served as key materials for achieving major advances in many areas, including rails, road cars, pipes, tanks for storage, and heavy-duty vehicles [6][18][14][19][20][21].

23.1.1 Material Selection and Metallographic Specification of HSLA X-65 Line Pipe Steel

300 mm Commercial X65 HSLA steel in sheet form of 300 mm × 24 mm was obtained. In this study, the line pipe steel HSLA X-65 was investigated. The current study is in widely used in the automotive and plumbing industries. For the purpose of metallographic examination, small pieces approximately 12 mm × 12 mm × 10 mm were cut from the as-supplied material using a hacksaw. Samples cut in this way are ground with grinding wheels, grinder belts, and so on.

Then, using freshly made 2% nital solution, the metallographic specimen was retrieved. To study the microstructure of the raw material, metallographic specimens (well-etched) were used. Then, using an optical microscope, they were studied from all three angles. Table 23.1 contains the chemical composition information.

TABLE 23.1
Chemical Composition (wt pct) of Steel

Elements	wt%	Elements	wt%	Elements	wt%
C	0.0616	Cr	0.109	Ti	0.0144
S	0.0046	Mo	0.258	V	0.0250
P	0.0087	Ni	0.256	N	0.0030
Mn	1.66	Al	0.0320	B	0.0016
Si	0.276	Co	0.0069		
Cu	0.0597	Nb	0.0594		

23.2 OPTICAL MICROSCOPY

For the purpose of analysing micro structural change, extensive optical microscopic exams were carried out. The specimens were manually polished using emery papers of 1/0, 2/0, 3/0, and 4/0 grit grades after undergoing various heat treatments. These polished specimens were afterwards polished with cloth using alumina paste to achieve a mirror surface.

Three per cent nital etchant was used to etch polished specimens in order to show the microstructure. The ingredients in the etchant are methanol (97%) and 3% nitric acid (HNO_3).

23.3 MEASURING HARDNESS

On a VM-50 Vickers Hardness Testing Machine, hardness measurements were made using a 30 Kgf force. Each specimen received at least five indentations, and the average was calculated.

23.4 CHARPY IMPACT TOUGHNESS TEST

To ascertain the impact energy of the X65 HSLA steel, Charpy-V test was performed. The energy engrossed up to breakage with impact loading was used to measure impact energy. The variance in impact toughness at different processing steps is depicted in Figure 23.1 (A–D). After treatment at 1,000°C, there has been a slight shift in impact energy noted. The impact toughness of the steel is significantly increased after 700°C of tempering and 400°C of ageing. Ageing after tempering, however, slightly reduces the impact energy values. The changes in impact energy levels following cold working are shown in Figure 23.1 (A–D). It can be seen that impact energy increases at first as degree of deformation improved.

Additionally, impact-energy degrades as the degree of deformation increases. On an Indian standard specimen with a 10 mm × 10 mm square cross-section and a 50 mm length, a 5 mm deep V-notch was provided for the Charpy impact toughness test.

The link between Charpy impact energy and impact toughness is as follows:

The specimen's toughness at ambient temperature can be ascertained using Charpy impact test. Calculations were made for the impact strength and toughness.

S. No.	Specimen Area (mm^2)	Impact Energy (joules)	Impact Strength (joule/mm^2)	Impact Toughness (Joule/mm^3)
1	8 × 8 = 64	270 − 75 = 195	195/64 = 3.05	195/5000 = 0.039

23.5 CALCULATIONS

Area of cross-section = 8 × 8 = 64 mm^2

Volume of specimen = 10 × 10 × 50 = 5,000 mm^3

$$\text{Impact strength} = \frac{Impact\ Energy}{Area\ of\ cross\ section} = \frac{195}{64} = 3.05\ \text{J/mm}^2$$

$$\text{Impact modulus} = \frac{\textit{Impact Strength}}{\textit{Volume of specimen}} == \frac{195}{5000} 0.039\ \text{J/mm}^3$$

23.6 RESULTS AND DISCUSSION

Various tests are used to determine the different mechanical parameters. This section tabulates and analyses numerous test findings from various tests that have been undertaken. This section describes the test findings for the Charpy impact test, hardness test, and tensile test.

Input Data (Specimen)		**Output Data**	
Shape	Line pipe	Load at yield	22 mm
Type	Sheet steel	At yield elongation	33.20 mm
Description	X65 HSLA line pipe	Yield stress	400 N/mm^2
Diameter	8 mm	Elongation at max.	40.00 mm
Gauge length (% elongation)	26.6 mm	Tensile strength	550.20 N/mm^2
Load value	1,000 KN	Load at break	12 KN
Cross-sectional area	310 mm^2	Elongation at break	60 mm
		(% elongation)	21.93%
		Area reduction (%)	21.57%

(A)

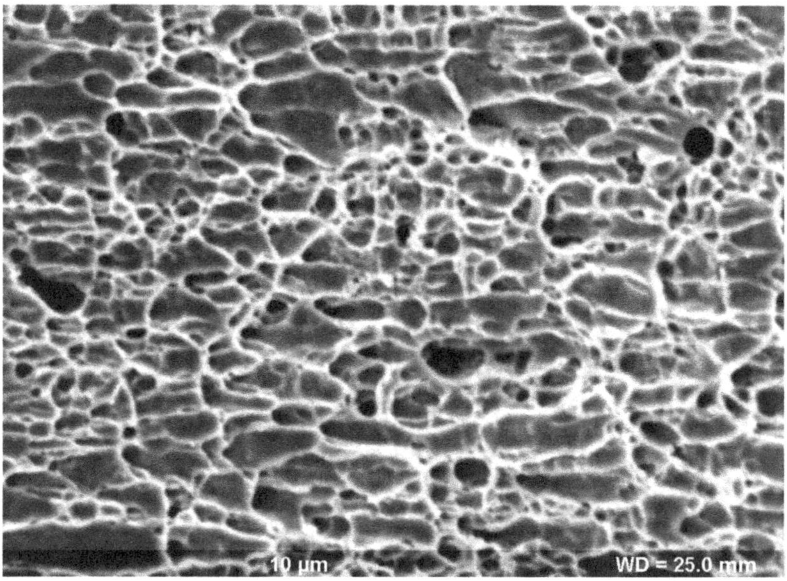

FIGURE 23.1 (A) SEM analysis in 3,000× magnification; (B) SEM analysis in 2,000× magnification; (C) SEM analysis in 1,000× magnification; (D) SEM analysis in 500× magnification.

(B)

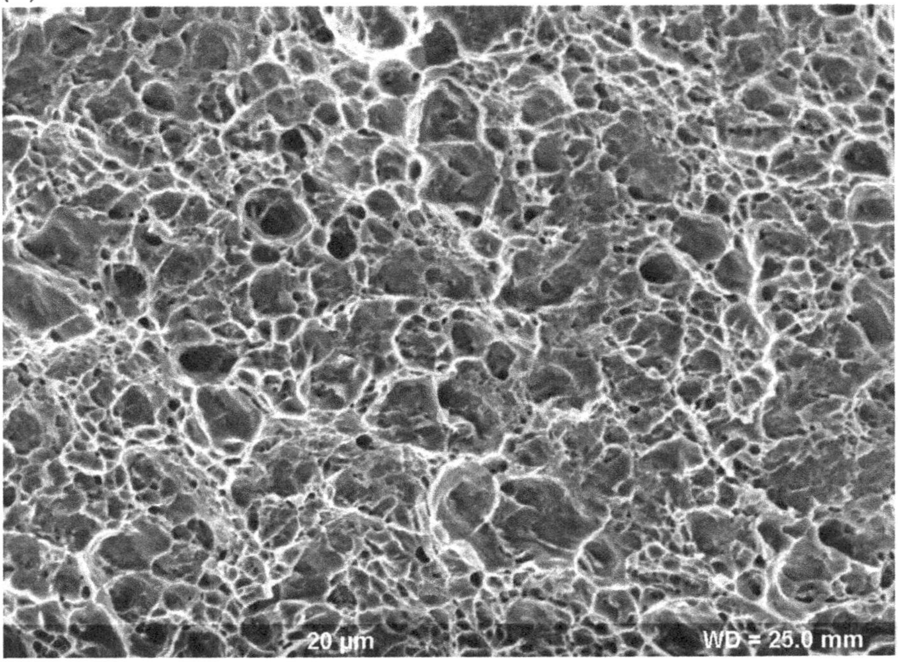

(C)

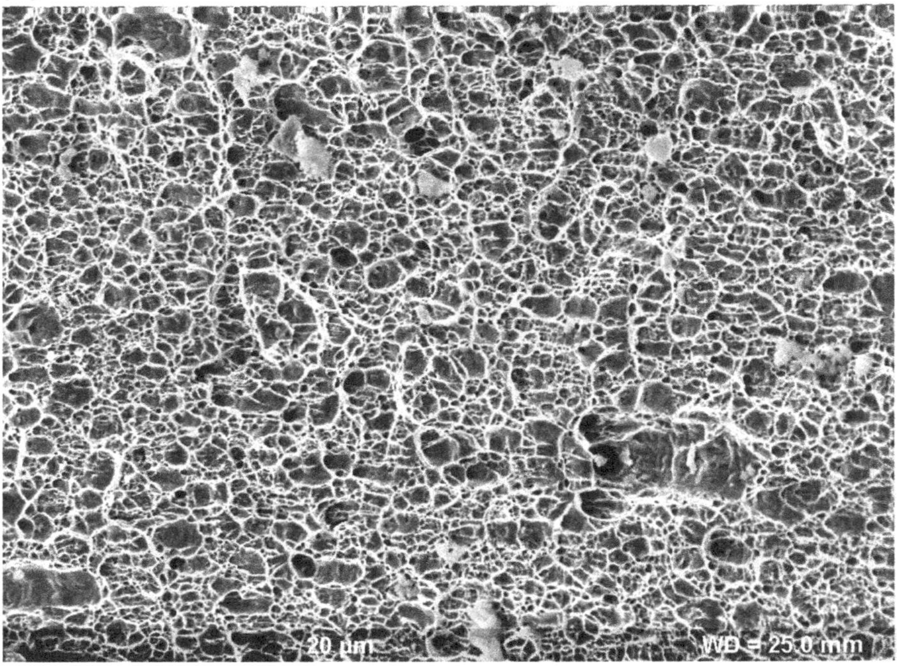

FIGURE 23.1 (Continued)

(D)

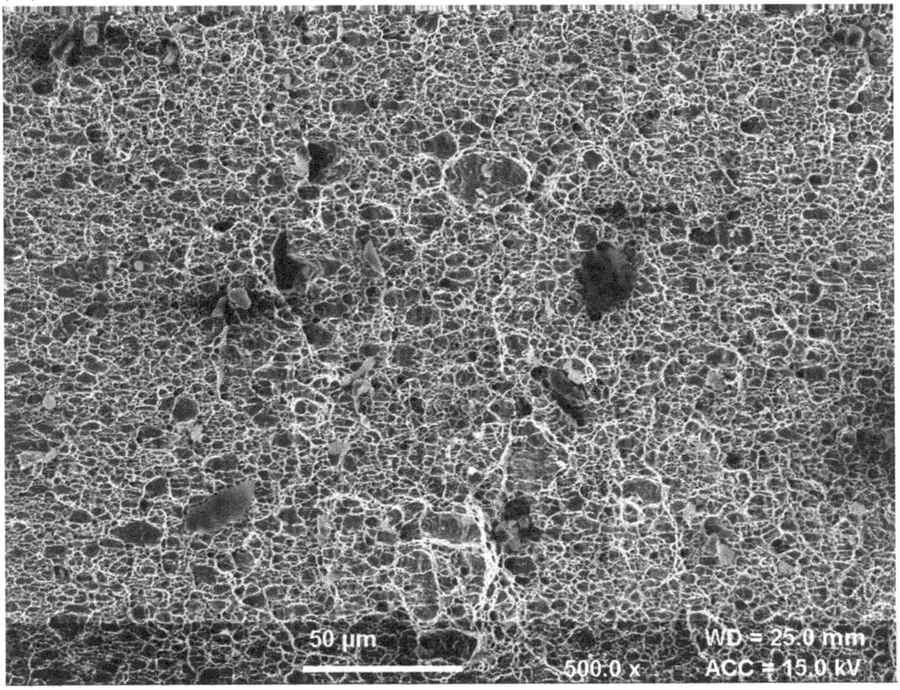

FIGURE 23.1 (Continued)

23.7 SEM ANALYSIS

Using an optical microscope, well-polished and etched metallographic specimens were examined. There were four magnifications used for the SEM investigation.

23.7 CONCLUSION

The study is conducted research on the behavioural study of impact energy of HSLA steel. This material can be used in its heat-treated state for a wide range of automotive parts and for general engineering applications involving small heat-treated parts and poor mechanical characteristics. Additionally, it can be used for components that naturally wear out. The mechanical properties of different steels are further impacted by the metallurgical transformation brought on by deformations, which is then followed by cooling or a final heat treatment. The number of thermal treatments used after hot/cold deformation can be minimized or even eliminated, which will lower the energy used in the production of steel. As a result, high strength steel production is enhanced. Additionally, steels can be toughened by quenching from a hot temperature. The main factors contributing to impact toughness are (1) increased dislocation density, (2) formation of thick precipitate-dislocation tangles, (3) an increase in the rate of carbonitride and/or carbide nucleation in micro-alloys, and (4) ongoing precipitate dissolution and new precipitate particle nucleation.

REFERENCES

[1] M. T. Miglin, J. P. Hirth, and A. R. Rosenfield, "Effects of microstructure on fracture toughness of a high-strength low-alloy steel." *Metall. Trans. A*, vol. 14A, pp. 2055–2061, October 1983.

[2] H. L. Dai, H. J. Jiang, T. Dai, W. L. Xu, and A. H. Luo, "Investigation on the influence of damage to springback of U-shape HSLA steel plates," *J. Alloys Compd.*, vol. 708, pp. 575–586, 2017, doi:10.1016/j.jallcom.2017.02.270.

[3] M. Talebi, M. Zeinoddini, M. Mo'tamedi, and A. P. Zandi, "Collapse of HSLA steel pipes under corrosion exposure and uniaxial inelastic cycling," *J. Constr. Steel Res.*, vol. 144, pp. 253–269, 2018, doi:10.1016/j.jcsr.2018.02.003.

[4] Y. Xia, X. Yan, T. Gernay, and H. Blum, "Experimental and numerical study of the behavior of HSLA and DP cold-formed high-strength steels at elevated temperature," *Ce/Papers*, vol. 4, no. 2–4, pp. 1264–1271, 2021, doi:10.1002/cepa.1420.

[5] O. Fatoba and R. Akid, "Uniaxial cyclic elasto-plastic deformation and fatigue failure of API-5L X65 steel under various loading conditions," *Theor. Appl. Fract. Mech.*, vol. 94, pp. 147–159, April 2018, doi:10.1016/j.tafmec.2018.01.015.

[6] V. Sharma, P. Kumar, Chinky, P. Malik, and K. K. Raina, "Preparation and electrooptic study of reverse mode polymer dispersed liquid crystal: Performance augmentation with the doping of nanoparticles and dichroic dye," *J. Appl. Polym. Sci.*, vol. 137, no. 22, 2020, doi:10.1002/app.48745.

[7] F. Huang, J. Chen, Z. Ge, J. Li, and Y. Wang, "Effect of heat treatment on microstructure and mechanical properties of new cold-rolled automotive steels," *Metals (Basel)*, vol. 10, no. 11, pp. 1–9, 2020, doi:10.3390/met10111414.

[8] E. Fan, Y. Li, Y. You, and X. Lü, "Effect of crystallographic orientation on crack growth behaviour of HSLA steel," *Int. J. Miner. Metall. Mater.*, vol. 29, no. 8, pp. 1532–1542, July 2022, doi:10.1007/S12613-022-2415-6.

[9] H. Zhao, J. Qi, G. Liu, R. Su, and Z. Sun, "A comparative study on hot deformation behaviours of low-carbon and medium-carbon vanadium microalloyed steels," *J. Mater. Res. Technol.*, vol. 9, no. 5, pp. 11319–11331, 2020, doi:10.1016/j.jmrt.2020.08.016.

[10] T. Xiao et al., "Effect of magnesium on microstructure refinements and properties enhancements in high-strength CuNiSi alloys," *Acta Metall. Sin. (English Lett.)*, vol. 33, no. 3, pp. 375–384, 2020, doi:10.1007/s40195-019-00953-9.

[11] D. Raabe et al., "Current challenges and opportunities in microstructure-related properties of advanced high-strength steels," *Metall. Mater. Trans. A Phys. Metall. Mater. Sci.*, vol. 51, no. 11, pp. 5517–5586, 2020, doi:10.1007/s11661-020-05947-2.

[12] B. Mintz and A. Qaban, "The influence of precipitation, high levels of Al, Si, P and a small B addition on the hot ductility of TWIP and TRIP assisted steels: A critical review," *Metals*, vol. 12, no. 3. 2022, doi:10.3390/met12030502.

[13] R. Bharadwaj, A. Sarkar, and B. Rakshe, "Effect of cooling rate on phase transformation kinetics and microstructure of Nb–Ti microalloyed low carbon HSLA steel," *Metallogr. Microstruct. Anal.*, vol. 11, no. 4, pp. 661–672, August 2022, doi:10.1007/S13632-022-00864-9.

[14] N. B. Garg and A. Garg, "Fractographic analysis of mechanical properties of microalloyed steel," *J. Phys. Conf. Ser.*, vol. 2070, no. 1, 2021, doi:10.1088/1742-6596/2070/1/012174.

[15] V. Aleksić, L. Milović, B. Aleksić, and A. M. Hemer, "Indicators of HSLA steel behavior under low cycle fatigue loading," *Procedia Struct. Integr.*, 2016, vol. 2, pp. 3313–3321, doi:10.1016/j.prostr.2016.06.413.

[16] J. Dhilipan, N. Vijayalakshmi, S. Suriya, and A. Christopher, "Prediction of students performance using machine learning," *IOP Conf. Ser. Mater. Sci. Eng.*, vol. 1055, no. 1, p. 012122, 2021, doi:10.1088/1757-899x/1055/1/012122.

[17] Y. Fukuda, M. Noda, T. Ito, K. Suzuki, N. Saito, and Y. Chino, "Effect of reduction in thickness and rolling conditions on mechanical properties and microstructure of rolled Mg-8Al-1Zn-1Ca alloy," *Adv. Mater. Sci. Eng.*, vol. 2017, 2017, doi:10.1155/2017/4065434.

[18] T. Coetsee and F. J. De Bruin, "Improved titanium transfer in submerged arc welding of carbon steel through aluminum addition," *Miner. Process. Extr. Metall. Rev.*, vol. 43, no. 6, pp. 771–774, 2022, doi:10.1080/08827508.2021.1945595.

[19] B. L. DeCost and E. A. Holm, "Characterizing powder materials using keypoint-based computer vision methods," *Comput. Mater. Sci.*, vol. 126, pp. 438–445, 2017, doi:10.1016/j.commatsci.2016.08.038.

[20] H. G. Hillenbrand, M. Gras, and C. Kalwa, "Development and production of high strength pipeline steels," *Niobium. Sci. Technol.*, pp. 543–569, 2001.

[21] V. Sharma, P. Kumar, and K. K. Raina, "Simultaneous effects of external stimuli on preparation and performance parameters of normally transparent reverse mode polymer-dispersed liquid crystals—a review," *J. Mater. Sci.*, vol. 56, 18795–18836, 2021.

24 Reliability of a Manufacturing Plant with Scheduled Maintenance, Inspection, and Varied Production

Reetu Malhotra, Harpreet Kaur

24.1 INTRODUCTION

Any industrial process must have well-equipped operating systems because if these systems are interrupted, the quality of the items produced will suffer and the infrastructure of the manufacturing plant will be affected. Thus, the system's dependability becomes significantly more crucial. Numerous authors [1–28] contributed to the reliability literature. Grewal et al. [1] created reliability models for non-identical units under various failure and repair assumptions. Further, Gupta et al. [2] discussed the dependability and availability of plastic-pipe manufacturing plant's serial operations to maintain constant demand. Afterward, Ajit et al. [3] addressed the reliability and importance of a single-unit boiler, which plays a vital role in garment industries. In their research, the authors explained the system analyser could not tolerate the failure of a boiler at any cost. They also discussed the primary reasons for the loss of the boiler. Further, Wang and Watada [4] developed redundancy allocation models for a parallel-series system by maximizing reliability and reducing costs. Moreover, Malik et al. [5, 8] established a reliability model with repair at various levels of damages subject to inspection and weather conditions for an operating system. Zhang et al. [7] discussed a diesel system's optimality and reliability modelling. Malik et al. [8] also considered priority and failure mechanisms in a non-identical system. Singh and Saini [14] studied the variables impacting the economy of steam by developing a model for a gas turbine system. In the model, the researchers addressed scheduled inspection. In these studies mentioned, the authors mainly considered demand fixed. Malhotra and Taneja [15, 19, 20] analysed that reliability and profit got affected according to the variation in demand/production. Suleiman et al. [16] dealt with the reliability of four dissimilar solar photovoltaic systems. Yusuf and Gatawa [21] developed a reliability model by taking three consecutive stages of deterioration. Malhotra and Taneja [17–20] compared two models to rectify the best model in a particular situation. Malhotra et al. [22, 29] discussed one working unit and one

 DOI: 10.1201/9781003367154-24

cold standby unit in which cold standby will operate if the demand for the product is high and the working unit cannot fulfil the need. But still, work has yet to be done on the autoclave, part of manufacturing ghee in industries. Taj et al. [23] discussed the six maintenance categories of a cable plant. Upasana et al. [27] analysed minor and significant failures as different types of losses sensitively affect production. Taneja et al. [28] worked on the redundancy failure possibly due to long-time non-use of standby unit non-operative mode. If it is found inoperable, then quickly go for repair by the repairman. Different researchers worked on reliability by considering various aspects. But only a few studies have been done using the concept of varied productions. Hence, the authors developed considering two different seasons winter (demand ≥ production) and summer (demand < production). The authors determine the reliability and availability of a single-unit autoclave system with varied production. Actual data is used from Markfed Vanaspati and Allied Industries, Punjab, India, to demonstrate the results.

The authors developed a reliability model for a single-unit autoclave that works with sufficient demand in the presented chapter. Failure in the autoclave will affect the production of the plant very severely. So the company user cannot tolerate any loss in the autoclave unit. The company hired one repairman to rectify such issues. Scheduled maintenance and inspection of the system was done to reduce any mishaps in the industry (like a blast). Scheduled maintenance is done after approximately five to six months while the repairman inspects regularly. The inspection technique determines the type of failure occurred if the autoclave stops working. After review, the repairman revealed three types of losses could occur: repair, replacement, or re-installation, according to whether it is a type I, II, or III failure. The authors analysed reliability and availability using the regeneration point technique and semi-Markov processes. Graphical study/method helps to draw various conclusions.

24.2 NOMENCLATURE

λ	Failure rate
λ_1/λ_2	Rate of decrease due to the summer season and the rate of increase in the winter season
λ_3/λ_4	Rate of change of states from upstate to downstate and vice versa
O_{SM}	Functioning unit when scheduled maintenance takes place
$Op_{d\geq p}/Op_{d<p}$	Functioning unit when demand ≥ production and when demand < production
D	Down unit
Pi, i = 1, 2, 3	Probability of repair/replacement/reconditioning or re-installation
Pi, i = 4, 5	Chances of repair when production ≤ demand and production > demand
Fi/Fr1/Fr2/Fr3	The failed unit under inspection/repair/replacement/reconditioning or re-installation
$h(t)/H(t)$, $g_i(t)/G_i(t)$, i = 1, 2, 3	p.d.f., c.d.f. of inspection/repair/replacement/reconditioning or re-installation of the unit

24.3 MODEL DESCRIPTION AND ASSUMPTIONS

Primarily, the authors consider the case of winter ("demand is greater than or equal to the production") to mean the autoclave is not working correctly or fails. Then inspection is done by a repairman responsible for detecting the type of failure. After the detection of failure, the system will be in a down state after the inspection state goes to the initial one (d ≥ p).

1. On failure of the unit, the repairman deals with inspection, repair, replacement, and re-installation.
2. Random variables are autonomous.
3. The breakdown time is considered to follow exponential, while the time distribution for repairs is taken arbitrarily.
4. The system works as new after every repair.

24.4 TRANSITION PROBABILITIES AND MEAN SOJOURN TIMES

The states S_0, S_1, and S_3 are regenerative states. States S_2, S_4, S_5, S_6, and S_7 are failed states. All system states at any time t are displayed in the state transition diagram shown in Figure 24.1. The transition probabilities are as follows:

$$q_{01}(t) = \lambda_1 e^{-(\lambda_1+\beta_1+\lambda)t}$$

$$q_{02}(t) = \lambda e^{-(\lambda_1+\beta_1+\lambda)t}$$

$$q_{03}(t) = \beta_1 e^{-(\lambda_1+\beta_1+\lambda)t}$$

$$q_{10}(t) = \lambda_2 e^{-(\lambda_2+\lambda_3+\lambda)t}$$

$$q_{14}(t) = \lambda e^{-(\lambda_2+\lambda_3+\lambda)t}$$

$$q_{15}(t) = \lambda_3 e^{-(\lambda_2+\lambda_3+\lambda)t}$$

$$q_{20}(t) = g(t)$$

$$q_{26}(t) = p_3 h(t)$$

$$q_{27}(t) = p_5 h(t)$$

$$q_{28}(t) = p_4 h(t)$$

$$q_{40}(t) = p_1 g_1(t)$$

$$q_{41}(t) = p_2 g_1(t)$$

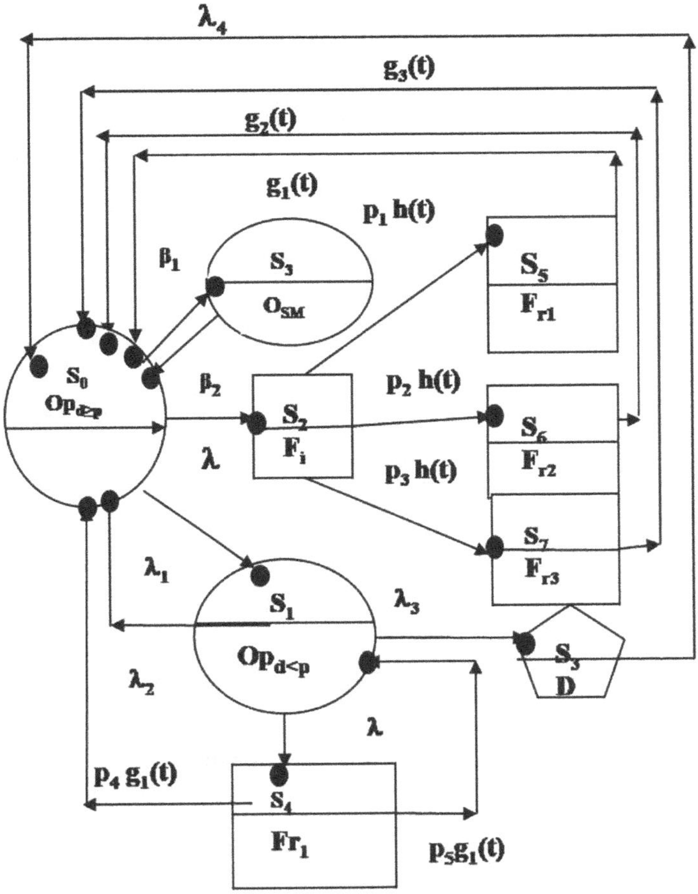

FIGURE 24.1 State transition diagram of the proposed model.

$$q_{50}(t) = \lambda_{34} e^{-(\lambda_4)t}$$

$$q_{60}(t) = g_1(t)$$

$$q_{70}(t) = g_2(t)$$

$$q_{80}(t) = g_3(t)$$

The non-zero p_{ij} are given by $p_{ij} = \lim_{s \to 0} q_{ij}^{*(s)}$.

$$p_{01} = \frac{\lambda_1}{\lambda + \lambda_1 + \beta_1}$$

$$p_{02} = \frac{\lambda}{\lambda + \lambda_1 + \beta_1}$$

$$p_{03} = \frac{\beta_1}{\lambda + \lambda_1 + \beta_1}$$

$$p_{10} = \frac{\lambda_2}{\lambda + \lambda_2 + \lambda_3}$$

$$p_{14} = \frac{\lambda}{\lambda + \lambda_2 + \lambda_3}$$

$$p_{15} = \frac{\lambda_3}{\lambda + \lambda_2 + \lambda_3}$$

$$p_{20} = p_{40} = p_1$$

$$p_{26} = p_3$$

$$p_{27} = p_4$$

$$p_{28} = p_5$$

$$p_{41} = p_2$$

$$p_{50} = p_{60} = p_{70} = 1$$

Since these are transition probabilities, it is validated that

$$\sum_{i=1}^{3} p_{0i} = 1 ; \quad \sum_{i=0,4,5} p_{1i} = 1 ; \quad \sum_{i=6}^{8} p_{2i} = 1 ; \quad \sum_{i=0}^{1} p_{4i} = 1$$

The mean sojourn time (μ_i) in state i is as follows:

$$\mu_0 = \frac{1}{\lambda + \lambda_1 + \beta_1}$$

$$\mu_1 = \frac{1}{\lambda + \lambda_2 + \lambda_3}$$

$$\mu_2 = \int_0^\infty \bar{H}(t)\,dt$$

$$\mu_3 = \frac{1}{\beta_2}$$

$$\mu_4 = \int_0^\infty \overline{G_1}(t)\,dt = \mu_6$$

$$\mu_5 = \frac{1}{\lambda}$$

$$\mu_{i,i=7,8} = \int_0^\infty \overline{G_j}(t)\,dt, \quad j = 2,3$$

The unconditional meantime is given by

$$m_{ij} = \int_0^\infty t q_{ij}(t)\,dt = q_{ij}^{**'}(0).$$

Therefore, $\sum_{i=1}^{3} m_{0i} = \mu_0;\quad \sum_{i=0,4,5} m_{1i} = \mu_1;\quad \sum_{i=6,7,8} m_{2i} = \mu_2$

$$m_{40} + m_{41} = \mu_4 \qquad m_{50} = \mu_5 \qquad m_{60} = \mu_6$$

$$m_{70} = \mu_7 \qquad m_{80} = \mu_8$$

24.5 MEASURES OF SYSTEM EFFECTIVENESS

24.5.1 MTSF (Mean Time to System Failure)

The failed states are considered as absorbing states while determining the MTSF of the system. The following recurring relations for $\phi_i(t)$ are acquired using probabilistic arguments:

$$\Phi_0(t) = Q_{01}(t)\ \S\Phi_1(t) + Q_{02}(t) + Q_{03}(t)\ \S\ \Phi_3(t)$$

$$\Phi_1(t) = Q_{10}(t)\ \S\Phi_0(t) + Q_{14}(t) + Q_{15}(t)\ \S\ \Phi_5(t)$$

$$\Phi_3(t) = Q_{30}(t)\ \S\Phi_0(t)$$

$$\Phi_5(t) = Q_{50}(t)\ \S\Phi_0(t)$$

Solving this equation for $\Phi_0^{**}(s)$ using Laplace-Steltjes transform (LST), we obtain

$$\Phi_0^{**}(s) = \frac{N(s)}{D(s)}$$

where $N(s) = q_{01}^{*}(s)q_{14}^{*}(s) + q_{02}^{*}(s)$

$$D(s) = 1 - q_{01}^{*}(s)q_{10}^{*}(s) - q_{03}^{*}(s)q_{30}^{*}(s) - q_{01}^{*}(s)q_{15}^{*}(s)q_{50}^{*}(s)$$

also,

$$MTSF = \lim_{s\to 0}\frac{\left(1-\Phi_0^{**}(s)\right)}{s}.$$

Applying the L'Hospital rule and substituting the value of ϕ_0**(s) from eq^n, we have

$$MTSF = \frac{N}{D}$$

$$N = \mu_0 + p_{01}\,\mu_1 + p_{03}\,\mu_{3+}\;\; p_{03}\,\mu_{3+} p_{01} p_{15}\,\mu_5$$

$$D = 1 - p_{03} - p_{01} p_{14}.$$

24.5.2 Availability of System in Winter (When Demand Is Greater Than Production)

By applying the probabilistic theory, the recurring relations obtained for availability $A_i^d(t)$ are as follows:

$$A_0^d = M_0(t) + q_{01}(t) \copyright A_1^d(t) + q_{02}(t) \copyright A_2^d(t) + q_{03}(t) \copyright A_3^d(t)$$

$$A_1^d = q_{10}(t) \copyright A_0^d(t) + q_{14}(t) \copyright A_4^d(t) + q_{15}(t) \copyright A_5^d(t)$$

$$A_2^d = q_{26}(t) \copyright A_{26}^d(t) + q_{27}(t) \copyright A_7^d(t) + q_{28}(t) \copyright A_8^d(t)$$

$$A_3^d = M_3(t) + q_{30}(t) \copyright A_0^d(t)$$

$$A_4^d = q_{40}(t) \copyright A_0^d(t) + q_{41}(t) \copyright A_1^d(t)$$

$$A_5^d = q_{50}(t) \copyright A_0^d(t)$$

$$A_6^d = q_{60}(t) \copyright A_0^d(t)$$

$$A_7^d = q_{70}(t) \copyright A_0^d(t)$$

$$A_8^d = q_{80}(t) \copyright A_0^d(t)$$

where $M_0(t) = e^{-(\lambda + \lambda_1 + \beta_1)t}, \; M_3(t) = e^{-(\beta_2)t}.$

Solve these equations for A_0^d *(s) by taking Laplace transformation. In the long run, availability is

$$A_0^d = \lim_{s \to 0}\left(sA_0^{d*}(s)\right) = \frac{N_1}{D_1}$$

where $N_1 = (1 - p_{41} p_{14})(\mu_0 + p_{03}\mu_3)$

$$D_1 = (1 - p_{41} p_{14})(\mu_0 + p_{02}\mu_2 + p_{02}\mu_2 + p_{03}\mu_3 + p_{02}(p_{26}\mu_6 + p_{27}\mu_7 + p_{28}\mu_8) + p_{01}\mu_1.$$

24.5.3 Availability of System in Summer (When Demand Is Less Than Production)

The recurring relations obtained for the system's availability $A_i^p(t)$ are as follows:

$$A_1^p(t) = q_{01}(t) \copyright A_1^p(t) + q_{02}(t) \copyright A_2^p(t) + q_{03}(t) \copyright A_3^p(t)$$

$$A_1^p = M_1(t) + q_{10}(t) \copyright A_0^p(t) + q_{14}(t) \copyright A_4^p(t) + q_{15}(t) \copyright A_5^p(t)$$

$$A_2^p = q_{26}(t)\copyright A_{26}^p(t) + q_{27}(t)\copyright A_7^p(t) + q_{28}(t)\copyright A_8^p(t)$$

$$A_3^p = q_{30}(t)\copyright A_0^p(t)$$

$$A_4^d = q_{40}(t)\copyright A_0^p(t) + q_{41}(t)\copyright A_1^p(t)$$

$$A_5^p = q_{50}(t)\copyright A_0^p(t)$$

$$A_6^p = q_{60}(t)\copyright A_0^p(t)$$

$$A_7^p = q_{70}(t)\copyright A_0^p(t)$$

$$A_8^p = q_{80}(t)\copyright A_0^p(t)$$

where $M_1(t) = e^{-(\lambda+\lambda_2+\lambda_3)t}$.

Solve these equations for A_0^P *(s) by taking Laplace transformation. In the long run, availability is

$$A_0^P = \lim_{s\to 0}\left(sA_0^{p*}(s)\right) = \frac{N_2}{D_1}$$

where $N_2 = p_{01}\mu_1$; D_1 is already defined.

24.6 GRAPHICAL REPRESENTATION

The repair rate (α_1), replacement rate (α_2), reconditioning/re-installation rate (α_3), and inspection rate (α) are distributed exponentially for the particular case. Therefore, we consider

$g_1(t) = \alpha_1 e^{-\alpha_1 t}$ $g_2(t) = \alpha_2 e^{-\alpha_2 t}$ $g_3(t) = \alpha_3 e^{-\alpha_3 t}$ $h(t) = \alpha e^{-\alpha t}$

$$\mu_0 = \frac{1}{\lambda+\lambda_1+\beta_1}, \quad \mu_1 = \frac{1}{\lambda+\lambda_2+\lambda_3}, \mu_2 = \int_0^\infty \bar{H}(t)dt \quad \mu_3 = \frac{1}{\beta_2}$$

$$\mu_4 = \int_0^\infty \overline{G_1}(t)dt = \mu_{6,}\mu_5 = \frac{1}{\lambda}$$

$$\mu_{i,i=7,8} = \int_0^\infty \overline{G_j}(t)dt, \ j = 2,3$$

$\alpha_1 = \alpha_2 = \alpha_3 = \alpha = 0.01$, $p_1 = 0.760$, $p_2 = 0.230$, $p_3 = 0.01$, $\lambda = 0.05$, $\lambda_1 = 0.01$, $\lambda_2 = 0.01$, $\lambda_3 = 0.001$, $\lambda_4 = 0.1$, $\beta_1 = 0.02$, $\beta_2 = 1$.

Using these particular values, MTSF and availability in two cases, winter and summer (when demand ≥ production and when demand < production), have been computed. The authors plot graphs to observe the behaviour concerning repair and failure rates as shown in Figure 24.2, Figure 24.3 and Figure 24.4.

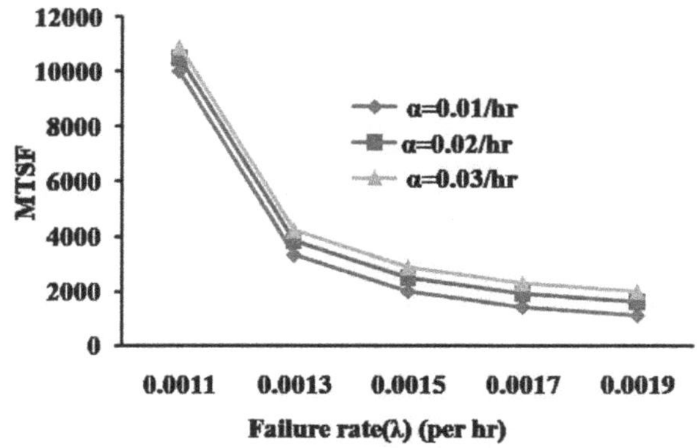

FIGURE 24.2 MTSF versus failure rate for different values of repair rate.

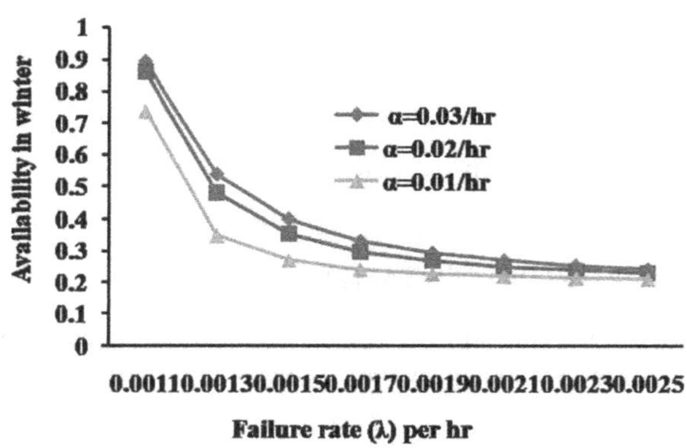

FIGURE 24.3 Availability in winter.

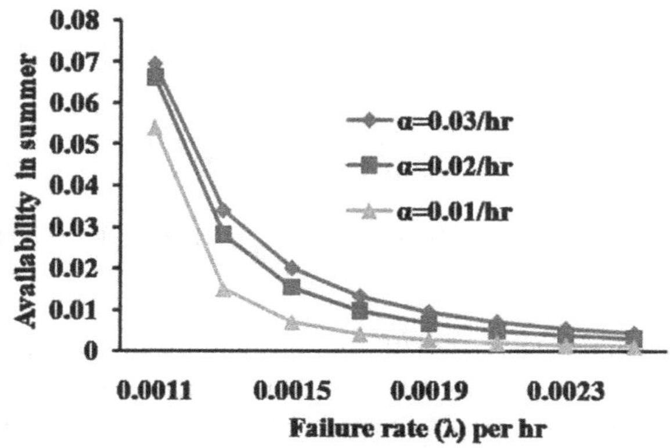

FIGURE 24.4 Availability in summer.

The authors draw similar other graphs for MTSF and availabilities by taking different parameters—inspection rate, repair rate, replacement rate, rate of re-installation, and schedule maintenance and productive results.

24.7 CONCLUSION

In the presented paper, the authors developed a reliability model that considered two seasons: winter and summer. Plotted graphs help to understand the variation in MTSF concerning the repair rate and failure rate of the autoclave unit. MTSF decreases, and for lower failure rate values, MTSF will be high. The graphs depict that the system's availability is more in winter than in summer for the exact discount of repair rate and failure rate, which is by the system analyst of the manufacturing plant. The system will be more available in winter for higher values of repair rate and lower values of failure rate. Availability in summer has a lower value for a high failure rate and a higher value for a more significant repair rate. Any company/plant can use the proposed general model.

REFERENCES

[1] M. S. Kadyan, A. Grewal, Stochastic analysis of non-identical units' reliability models with priority and different modes of failure, *Decision and Mathematical Sciences* (2004) 59–82.

[2] P. Gupta, A. Kumar, R. Kumar, J. Singh, Analysis of reliability and availability of serial processes of plastic-pipe manufacturing plant, *International Journal of Quality & Reliability Management* (2007) 404–419.

[3] A. Kumar, R. K. Agnihotri, Reliability analysis of a system of boiler used in readymade garment industry, *Journal of Reliability and Statistical Studies* (2008) 33–41.

[4] S. Wang, J. Watada, Modelling redundancy allocation for a fuzzy random parallel–series system, *Journal of Computational and Applied Mathematics* (2009) 539–557.

[5] S. C. Malik, S. Bahl, Steady state analysis of an operating system with repair at different levels of damages subject to inspection and weather conditions, *International Journal of Agricultural and Statistical Sciences* (2010) 225–234.

[6] W. Qingtai, W. Shaomin, Reliability analysis of two-unit cold standby repairable systems under Poisson shocks, *Applied Mathematics and Computation* (2011) 171–182.

[7] G. Wenke, Z. Yifan, Reliability modeling and maintenance optimization of the diesel system in locomotives, *Eksploatacja i Niezawodność* (2012) 302–311.

[8] S. C. Malik, S. Deswal, Reliability modeling and profit analysis of a repairable system of non-identical units with no operation and repair in abnormal weather, *International Journal of Computer Applications* (2012) 11.

[9] G. Taneja, R. Malhotra, Cost-benefit analysis of a single unit system with scheduled maintenance and variation in demand, *Journal of Mathematics and Statistics* (2013) 155–160.

[10] G. Taneja, R. Malhotra, Reliability and availability analysis of a single unit system with varying demand, *Mathematical Journal of Interdisciplinary Sciences* (2013) 77–88.

[11] G. Taneja, R. Malhotra, Cost-benefit analysis of a single unit system with scheduled maintenance and variation in demand, *Journal of Mathematics and Statistics* (2013) 155–160.

[12] G. Taneja, R. Malhotra, Reliability modelling of a cable manufacturing plant with variation in demand, *International Journal of Research in Mechanical Engineering and Technology* (2013) 162–165.
[13] G. Taneja, R. Malhotra, Reliability and availability analysis of a single unit system with varying demand, *Mathematical Journal of Interdisciplinary Sciences* (2013) 77–88.
[14] D. Singh, A. S. Saini, Computational analysis of parameters affecting economy of one gas and one steam turbine system with scheduled inspection, *International Journal of Statistics and Mathematica* (2014) 65–73.
[15] G. Taneja, R. Malhotra, Stochastic analysis of a two-unit cold standby system wherein both the units may become operative depending upon the demand, *Journal of Quality and Reliability Engineering* (2014) 13 pages.
[16] K. Suleiman, I. Yusuf, Ali, Analysis of reliability characteristics four dissimilar solar photovoltaic systems, *International Journal of Mathematics and Computation* (2015) 87–103.
[17] G. Taneja, R. Malhotra, Comparative study between a single unit system and a two-unit cold standby system with varying demand, *Springer Plus* (2015) 1–17.
[18] G. Taneja, R. Malhotra, Comparative analysis of two stochastic models subjected to inspection and scheduled maintenance, *International Journal of Software Engineering* (2015) 179–188.
[19] G. Taneja, R. Malhotra, Comparative analysis of two stochastic models with varying demand, *International Journal of Applied Engineering* (2015), 37453–37460.
[20] G. Taneja, R. Malhotra, Comparative analysis of two single unit systems with production depending on demand, *Industrial Engineering Letters* (2015) 43–48.
[21] I. Yusuf, R.I. Gatawa, Probabilistic models for reliability analysis of a system with three consecutive stages of deterioration, *Reliability theory and Application* (2016) 30–37.
[22] G. Taneja, R. Malhotra, A. Chitkara, Comparative profit analysis of two reliability models with varying demand, *Aryabhata Journal of Mathematics & Informatics* (2016) 305–314.
[23] S. Taj, S. Rizwan, B. Alkali, D. Harrison, G. Taneja, Reliability analysis of a single machine subsystem of a cable plant with six maintenance categories, *International Journal of Applied Engineering Research* (2017) 1752–1757.
[24] S. Gupta, R. Malhotra, A systematical study of reliability on some models, *International Journal of Advanced Search in Science and Engineering* (2017) 864–876.
[25] M. Kakkar, J. Bhatti, R. Malhotra, M. Kaur, D. Goyal, Availability analysis of an industrial system under the provision of replacement of a unit using genetic algorithm, *International Journal of Innovative Technology and Exploring Engineering* (2019) 1236–1241.
[26] R. Malhotra, T. Dureja, A. Goyal, Reliability analysis a two-unit cold redundant system working in a pharmaceutical agency with preventive maintenance, *Reliability and Risk Modeling of Engineering System* (2021) 1–10.
[27] U. Sharma, R. Kaur, Performance analysis of system where service type for boiler, *Depends Upon Major or Minor Failures. Reliability: Theory & Applications* (2022) 317–325.
[28] G. Taneja, A. K. Taneja, A. Manocha, Reliability and sensitivity analysis of two non-identical unit standby system subject to pre-operation random inspection of standby unit, *Reliability: Theory & Applications* (2022) 469–481.
[29] R. Malhotra, Reliability and availability analysis of a standby system with activation time and varied demand, *Engineering Reliability and Risk Assessment* (2022) 35–40.

25 Effect of A-Site Cation on the Specific Heat Capacity of $A_2Fe_2O_{6-\delta}$ (A = Ca or Sr)

Ram Krishna Hona, Tanner Vio, Mandy Guinn, Md. Sofiul Alom, Uttam S. Phuyal, Gurjot S. Dhaliwal

25.1 INTRODUCTION

Specific heat capacity is an intrinsic property of a material. It helps to know the thermal property, such as enthalpy and entropy. It can also help for solid state nuclear track of a material [1]. So it is important to study the specific heat capacity of energy-related materials. One of the energy-related materials is oxygen-deficient perovskite oxide. Because of the unique properties of oxygen-deficient perovskite oxides, they are applicable to the devices such as ceramic membranes for oxygen separation [2], gas diffusion membranes [3], solid oxide fuel cells (SOFCs) [4], electrodes [5], sensors [6], superconductors, and colossal magnetoresistance [7]. So attention has been paid to the specific heat capacity of these materials.

Oxygen-deficient perovskites are the oxides with the general formula ABO_{3-x} or $A_2B_2O_{6-\delta}$ where A and B are the alkaline earth metals and transition metals, respectively. X and δ represent a number of oxygen vacancies in the compound. Oxygen vacancies can arrange in an ordered or disordered fashion in the materials. The different arrangements and numbers of vacancies resulted in their structural flexibility, leading to great diversity and variation in the structure and properties. The vacancies can lead to different coordination geometries such as square pyramidal or tetrahedral. When the vacancies are ordered and tetrahedral geometries are formed, the tetrahedra share corners, forming chains and form alternate layers with octahedral layers. The tetrahedra are connected with the octahedra above and below through their apexes. This constitutes the brownmillerite-type structure. If square pyramids are formed, they form a unique crystallographic structure as will be discussed further in the structure of $Sr_2Fe_2O_{6-\delta}$ [8, 9].

The change in the A- or B-site ionic radius can lead to a variation of the structure and properties of materials. For example, the tetragonal crystal structure of $Sr_2Fe_2O_{6-\delta}$ transforms to the cubic crystal structure of $Sr_2FeMnO_{6-\delta}$ upon

DOI: 10.1201/9781003367154-25

the B-site cation replacement. Here, Fe in $Sr_2Fe_2O_{6-\delta}$ is substituted by Mn to get $Sr_2FeMnO_{6-\delta}$. This leads to the variation of their properties. The magnetic moments in the spin-density wave state of $Sr_2Fe_2O_{6-\delta}$ are modified to an inhomogeneous magnetic ground state in $Sr_2FeMnO_{6-\delta}$, where most of the sample at 4 K contains fluctuating spins with a magnetically ordered small fraction [10]. Similarly, the variation of a small amount of B-site between $Sr_2Fe_{1.9}Co_{0.1}O_{6-\delta}$ (orthorhombic) and $Sr_2Fe_{1.9}Cr_{0.1}O_{6-\delta}$ (cubic) also demonstrates the crystal structure change [11]. Different magnetic properties result in these two materials due to the B-site cation alteration [11].

A site cation substitution can also lead to structural transformation and property variation [12]. For example, $Ca_2FeGaO_{6-\delta}$ has *Pnma* space group and $CaSrFeGaO_{6-\delta}$ has *Ibm*2 space group [13]. The tetrahedra arrangements is different in these materials. All tetrahedra are oriented in the same direction through the crystal in $CaSrFeGaO_{6-\delta}$ (*Ibm*2), while the tetrahedral chains are oriented in opposite directions in alternating tetrahedral layers in the $Ca_2FeGaO_{6-\delta}$ (*Pnma*) crystal. They have different charge transport properties [13]. Similarly, The A-site cation variation $Ca_2FeCoO_{6-\delta}$ and $Sr_2FeCoO_{6-\delta}$ lead to structural difference and electrical property variation [9, 14]. $Ca_2FeCoO_{6-\delta}$ is oxygen-vacancy-ordered brownmillerite-type material with an orthorhombic crystal structure while $Sr_2FeCoO_{6-\delta}$ is vacancy-disordered material with a cubic crystal structure. The former demonstrates semiconductor-like behaviour, while the latter shows metallic-like behaviour.

Three well-known materials, $Ca_2Fe_2O_{6-\delta}$, $CaSrFe_2O_{6-\delta}$ and $Sr_2Fe_2O_{6-\delta}$, have been studied for their multiple properties and applications [15–17]. However, their specific heat capacities have not been reported yet. We report here their specific heat capacities and the effect of variation of the A-site average ionic radius on the specific heat capacity.

25.2 EXPERIMENTAL

The compounds $Ca_2Fe_2O_{6-\delta}$, $CaSrFe_2O_{6-\delta}$ and $Sr_2Fe_2O_{6-\delta}$ were prepared by solid state synthesis technique as usual using $CaCO_3$, $SrCO_3$, and Fe_2O_3 powders. Stoichiometric amounts of the powders were mixed in an agate mortar and a pestle. The uniformly mixed powders were used to make circular pellets of 3 mm diameter using a pellet press. The pellets were first calcined at 1,000°C for 24 hours and then sintered for 24 hours at 1,200°C. The calcination was followed by powdering and pelletizing before sintering. The ramp rate during calcination and sintering was 100°C/h. The phase purity was investigated for the materials using powder X-ray diffraction (PXRD) with Cu Kα1 radiation of wavelength λ = 1.54056 Å. Rietveld refinements with GSAS software[11] [18] and EXPEGUI interface[12] [19] were accomplished for structural analysis. A scanning electron microscopy (SEM) study was done for structural morphology. The specific heat capacities of these materials were investigated using a computer-controlled heat flow meter (HFM 446 Lambda from NETZSCH). Circular sintered pellets were used to study the specific heat capacity of these materials.

25.2.1 Specific Heat Capacity (Cp) Measurement Principle

The Cp represents the amount of energy required for a substance to alter the temperature of 1 kg of the substance by 1°C. It is, at constant pressure, represented by equation 25.1:

$$C_p =_{m^l} dQ_{dT} =_{mQ\,\Delta T} \tag{25.1}$$

where C_p, Q, m, and T represent specific heat capacity at a constant pressure (J/(kg·K)), heat energy (J), mass (kg), and temperature (K), respectively.

Though there is probability of thermal expansion, the expansion of a crystalline solid is negligibly small, and the expansion can be neglected [20–22].

The apparatus, heat flow meter, was used for Cp measurement. It follows the American standard ASTM C1784-13 (2013) [22] and is based on the principle that the quantity of heat absorbed by a sample from the heat flow meter is evaluated as in equation 25.2.

$$H = \sum_{i=1}^{n} \tau . [SUcal . (QUi - QUequil) + SLcal. (QLi - QLequil)] \tag{25.2}$$

where H = amount of heat energy per square meter (J/m2) τ = time interval (s)
S_{Ucal} = calibration factor of upper plate (W/(m^2·μV))
S_{Lcal} = calibration factor of lower plate (W/(m^2·μV))
Q_{Ui} = Heat Flow Meter signal value of upper plate (μV)
Q_{Li} = HFM signal value of lower plate (μV)
Q_{Uequil} = HFM signal at the final steady-state, upper plate (μV)
Q_{Lequil} = HFM signal at the final steady-state, lower plate (μV)

The calculated sum also involves the heat absorbed by the upper and lower two plates of the heat flow meter. Equation 25.3 shows the heat capacity of the heat flow meters. (These equations are explained in detail in references [22, 23]. It needs to be eliminated from the final results.)

$$C_p\,\rho = (_{\Delta T--}{}^{H} - C_p'\rho'2\delta x)._{x}{}^{1} \tag{25.3}$$

where $C_p\,\rho$ = specific heat of the specimen (J/(m^3 ·K)) (ρ is density (kg/m^3))
$C_p'\rho'$ = specific heat of the heat flow meters (J/(m^3·K))
x = thickness of the specimen (m)
2δx\` = thickness of the two heat flow meters (m)
ΔT = temperature change (K)

Finally, the specific heat capacity is calculated using the equation 25.4 [24].

$$Cp = (H——— -last-Hxh.fm\Delta T(T) . \Delta T.1\rho \tag{25.4}$$

where $H_{hfm}(T)$ = correction factor to remove the effect of the plates (J/(m^2·K)).

25.2.2 Crystal Structure

The crystal structures and phase purity of these materials were investigated with the help of their powder X-ray diffraction data. Rietveld refinement of the data demonstrated their crystal structures as reported before [15–17]. Figure 25.1 and Table 25.1 show the Rietveld refinement profile and the parameters for the refinement of $Ca_2Fe_2O_{6-\delta}$.

It has brownmillerite-type structure with orthorhombic unit cell and *Pnma* space group in which each tetrahedral layer is oriented in opposite direction of the other nearest tetrahedral layer (as seen in the inset of Figure 25.1). When one Ca is substituted in $Ca_2Fe_2O_{6-\delta}$ by Sr, the average ionic radius of A-site increases because of larger Sr. It changes the arrangements of oxygen vacancies resulting into different space groups. Figure 25.2 and Table 25.2 show the Rietveld refinement profile and the parameters for the refinement of $CaSrFe_2O_{6-\delta}$. It has also brownmillerite-type structure with orthorhombic unit cell but with *Ibm*2 space group in which all are oriented in the direction (as seen in the inset of Figure 25.2).

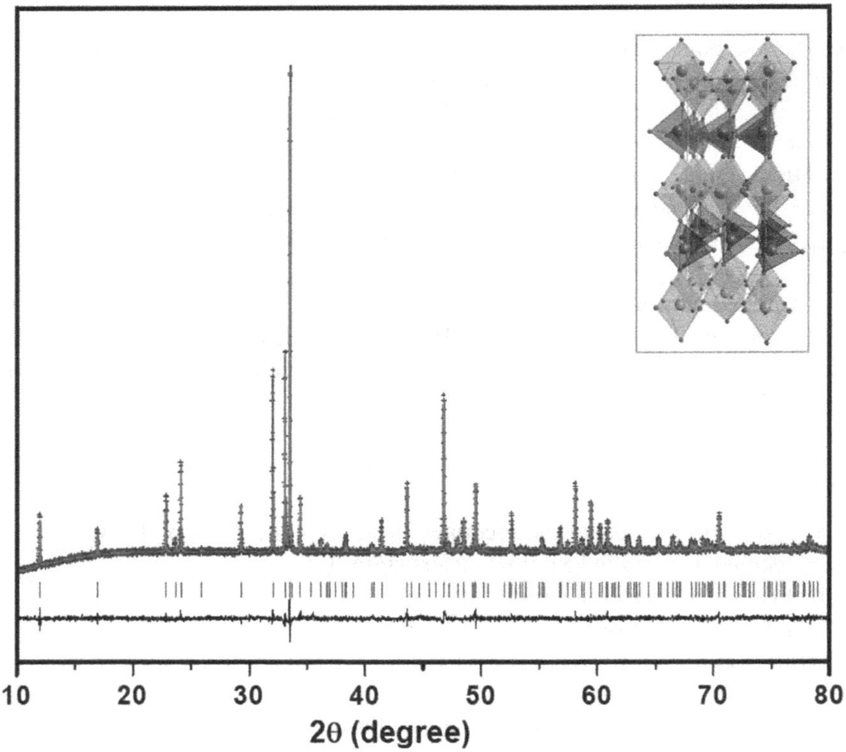

FIGURE 25.1 The Rietveld refinement profile of $Ca_2Fe_2O_{6-\delta.}$ The blue plus, red line, pink vertical lines, and black solid line represent the raw data, the model, Bragg peak positions, and difference plot, respectively. The inset shows the crystal structure with an orthorhombic unit cell and *Pnma* space group.

TABLE 25.1
Rietveld Refinement Parameters of $Ca_2Fe_2O_{6-\delta}$

Element	x	y	z	Uiso	Multiplicity	Occupancy
Ca	0.4809(1)	0.1079(4)	0.0247(7)	0.028(5)	8	1
Fe1	0.0000	0.0000	0.0000	0.028(2)	4	1
Fe2	−0.0550(7)	0.2500	−0.0671(1)	0.031(3)	4	1
O1	0.2565	−0.0153(6)	0.2338(6)	0.027(7)	8	1
O2	0.0302(9)	0.1420(1)	0.0774(5)	0.019(5)	8	1
O3	0.6119(1)	0.2500	−0.1250(3)	0.015(1)	4	1

Space group = Pnma; a = 5.4024(1) Å; b = 14.7017(3) Å; c = 5.5726(3) Å; wRp =0.0182; Rp = 0.0134

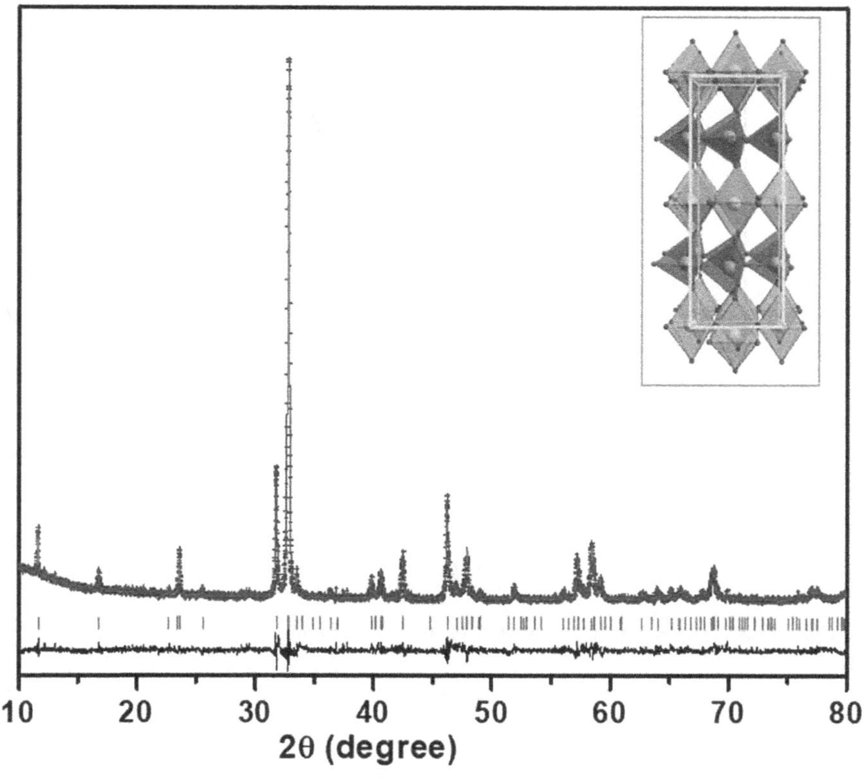

FIGURE 25.2 The Rietveld refinement profile of $CaSrFe_2O_{6-\delta}$. The blue plus, red line, pink vertical lines, and black solid line represent the raw data, the model, Bragg peak positions, and difference plot, respectively. The inset shows the crystal structure with an orthorhombic unit cell and *Ibm*2 space group.

TABLE 25.2
Rietveld Refinement Parameters of $CaSrFe_2O_{6-\delta}$

Element	x	y	z	Uiso	Multiplicity	Occupancy
Ca	0.5126(7)	0.1109(4)	0.0085(6)	0.029(8)	8	0.5
Sr	0.5126(7)	0.1109(4)	0.0085(6)	0.029(8)	8	0.5
Fe1	0.0760(2)	0.2500	−0.0039(4)	0.027(4)	4	1
Fe2	0.2500	0.0	0.0	0.038(6)	8	1
O1	0.2260(9)	0.0067(3)	0.2931(0)	0.028(7)	8	1
O2	−0.0815(6)	0.1490(7)	0.0016(2)	0.028(7)	8	1
O3	0.3834(3)	0.2500	0.8913(3)	0.028(7)	4	1

Space group = *Ibm*2; a = 5.6315(4) Å; b = 15.1809(9) Å; c = 5.4696(8) Å; wRp =0.0262; Rp = 0.0187

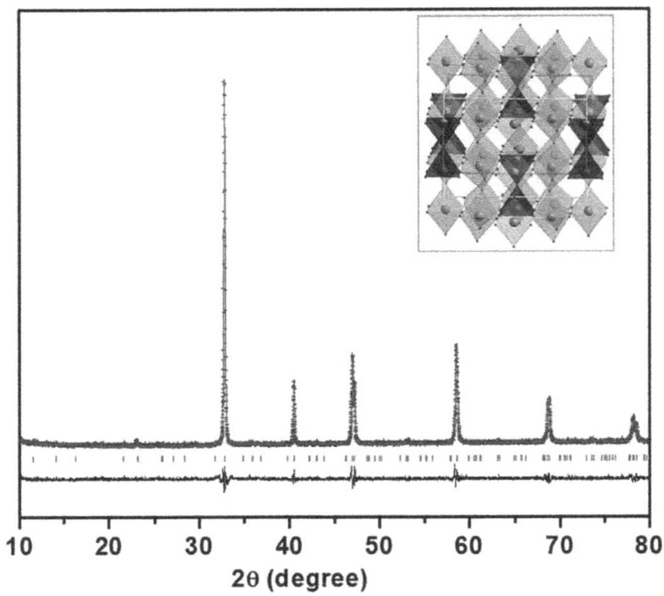

FIGURE 25.3 The Rietveld refinement profile of $Sr_2Fe_2O_{6-\delta}$. The blue plus, red line, pink vertical lines, and black solid line represent the raw data, the model, Bragg peak positions, and difference plot, respectively. The inset shows the crystal structure with a tetragonal unit cell and *I4/mmm* space group.

When all of the Ca is substituted—that is, one more Ca in $CaSrFe_2O_{6-\delta}$, by Sr—the resulting compound, $Sr_2Fe_2O_6\delta$, possesses square pyramidal coordination geometries around B cation instead of tetrahedra. Each square pyramid shares corners with octahedra on the sides and apex with other square pyramid former a dimer. The dimers are separated by octahedra. Figure 25.3 and Table 25.3 show the Rietveld refinement profile and the parameters for the refinement of $Sr_2Fe_2O_{6-\delta}$. It has tetrahedral unit cell with I4/mmm space group.

TABLE 25.3
Rietveld Refinement Parameters of $Sr_2Fe_2O_{6-\delta}$

Element	x	y	z	Uiso	Multiplicity	Occupancy
Sr1	0.2588(7)	0.0000	0.0000	0.015(4)	8	1
Sr2	0.2486(4)	0.0000	0.5000	0.018(2)	8	1
Fe1	0.0000	0.0000	0.2500	0.022(1)	4	1
Fe2	0.2500	0.2500	0.2500	0.010(0)	8	1
Fe3	0.5000	0.0000	0.2500	0.028(5)	4	1
O1	0.0000	0.0000	0.5000	0.041(0)	2	1
O2	0.1231(2)	0.1231(2)	0.2138(1)	0.010(0)	16	1
O3	0.2554(4)	0.2554(4)	0.5000	0.016(8)	8	1
O4	0.1251(7)	0.6251(7)	0.2500	0.002(8)	16	1
O5	0.5000	0.0000	0.0000	0.041(1)	4	1

Space group = *I4/mmm*; a = 10.9347(4) Å; b = 10.9347(4) Å; c = 7.6988(9) Å; wRp = 0.0304; Rp = 0.0212

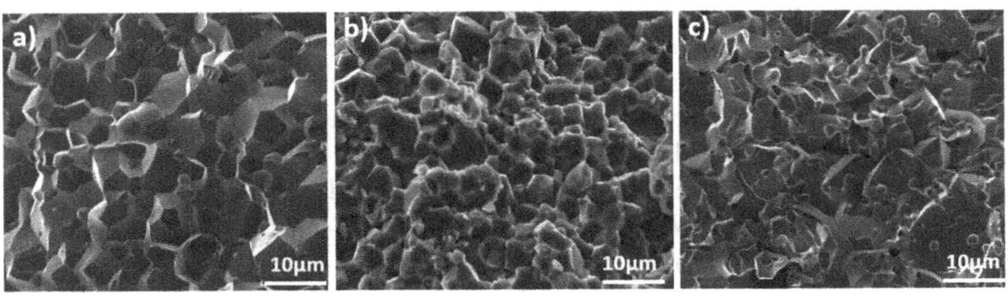

FIGURE 25.4 SEM images of (a) $Ca_2FeCoO_{6-\delta}$, (b) $CaSrFe_2O_{6-\delta}$, and (c) $Sr_2FeCoO_{6-\delta}$.

Figure 25.4 show the SEM images of the materials. The images reveal these materials' different microstructures, which are expected due to different crystal structures as evidenced by XRD analysis.

25.2.3 Specific Heat Capacity

The specific heat capacity (Cp) of materials is an important and intrinsic property. Cp is also used to calculate the entropy and enthalpy of a material [25]. Investigation of specific heat capacity for perovskite oxide is not uncommon [26]. Specific heat capacities of the vacancy ordered perovskite oxides, $Ca_2Fe_2O_{6-\delta}$, $CaSrFe_2O_{6-\delta}$, and $Sr_2Fe_2O_{6-\delta}$ were investigated on their circular pellets using computer-controlled heat flow meter. The Cp values for $Ca_2Fe_2O_{6-\delta}$, $CaSrFe_2O_{6-\delta}$ and $Sr_2Fe_2O_{6-\delta}$ at 50 °C are 0.288 $Jg^{-1}K^{-1}$, 0.191 $Jg^{-1}K^{-1}$ and 0.011 $Jg^{-1}K^{-1}$, respectively.

Specific heat capacity depends on different factors such as degree of freedom, physical state, molar mass, crystallinity, and temperature. Increasing the degree of freedom will lead to increase of specific heat capacity. Cp for amorphous phase is larger than that of crystalline phase [27]. Note that both of our materials are synthesized at the same temperature and environment with the same physical state. $Ca_2Fe_2O_{6-\delta}$, $CaSrFe_2O_{6-\delta}$, and $Sr_2Fe_2O_{6-\delta}$ all have crystalline structure. So the effect of degree of freedom of the atoms and physical states is ignored here. The specific heat capacity increases with increasing the molar mass [27, 28]. In contrast to the statement, $Sr_2Fe_2O_{6-\delta}$ with the highest molar mass has lower Cp value and $Ca_2Fe_2O_{6-\delta}$ with the lowest molar mass has the highest Cp value. Thus, the molar mass is also ignored for the analysis. The same material demonstrates the specific heat capacity variation with the phase or geometry change [29]. As discussed before, $Ca_2Fe_2O_{6-\delta}$, $CaSrFe_2O_{6-\delta}$, and $Sr_2Fe_2O_{6-\delta}$ have different crystal structures with different types of oxygen vacancy arrangements after A-site cation substitution. So the variation of specific heat capacities is expected. Note that Ca is smaller in size than Sr. When Ca in $Ca_2Fe_2O_{6-\delta}$ is substituted with Sr, the specific heat capacity is lowered. The low heat capacity value of $Sr_2Fe_2O_{6-\delta}$ is due to a bigger A-cation size, a different crystal structure, and vacancy arrangements. One report mentioned that the change of the specific heat capacity is related with nuclear track numbers after annealing some materials [1]. Another report discussed the effect of particle size on the specific heat capacity of carbon nanotubes [30]. Our SEM micrographs show the different grain structures and sizes for the materials. Thus, grain size may have contribution to the variation of the Cp values. An article discussed the effects of defects on the Cp variation [31]. Defects can be generated from impurities, broken bonds or vacancies resulting in the interruption in the regular or natural structure. The report mentions that the presence of the defects, characterized by the absence of broken C–C bonds generate the graphene blocks of various sizes [31]. which in turn affected in the Cp values. Grinding process of Ø18 nm MWCNTs increases the number of defects on the surface and the ends of the nanotubes and changes the parameters of defects. Thus, the observed significant increase in the specific heat of ground Ø18 nm MWCNTs are attributed to these changes [31]. $Ca_2Fe_2O_{6-\delta}$, $CaSrFe_2O_{6-\delta}$, and $Sr_2Fe_2O_{6-\delta}$ are compounds with different concentration of oxygen vacancies—that is, oxygen defects. We have already reported the oxygen vacancy concentration of these materials in the previous articles [16]. $Ca_2Fe_2O_{6-\delta}$ and $CaSrFe_2O_{6-\delta}$ have $\delta = 1$ and $Sr_2Fe_2O_{6-\delta}$ has $\delta = 0.75$ [16]. Not only the concentration of the oxygen defects but also the ways of their distribution are different. The oxygen vacancy forms tetrahedral coordination geometry around B cation in $Ca_2Fe_2O_{6-\delta}$ and $CaSrFe_2O_{6-\delta}$ but square pyramids in $Sr_2Fe_2O_{6-\delta}$.

25.3 CONCLUSION

Specific heat capacities of $Ca_2Fe_2O_{6-\delta}$, $CaSrFe_2O_{6-\delta}$, and $Sr_2Fe_2O_{6-\delta}$ are investigated. $Ca_2Fe_2O_{6-\delta}$ has the highest and $Sr_2Fe_2O_{6-\delta}$ has the lowest Cp values, respectively. The comparative study reveals that the A-site cation substitution in oxygen-deficient perovskite can lead to the variation of Cp values. The Cp value decreases with the increase of average A-site ionic radius demonstrating the inverse relation of the Cp with the average A-site ionic radius.

25.4 ACKNOWLEDGEMENTS

This work is supported in part by the National Science Foundation Tribal College and University Program Instructional Capacity Excellence in TCUP Institutions (ICE-TI) award number 1561004, and we express gratitude to the program managers and review panels for project support. A part of this work is also supported by NSF grant number HRD 1839895. Additional support for the work came from ND EPSCOR STEM grants for research. The authors also acknowledge the support of North Dakota EPSCoR for the purchase of thermal conductivity equipment and X-ray diffractometer. Permission was granted by United Tribes Technical Colleges (UTTC) Environmental Science Department to publish this information. The views expressed are those of the authors and do not necessarily represent those of United Tribes Technical College.

REFERENCES

1. Lu, J., et al., *A new method for determination solid state nuclear track by the change of the specific heat capacity*. Perspect Sci. 2019. **12**: p. 100403.
2. Kharton, V.V., et al., *Perovskite-type oxides for high-temperature oxygen separation membranes*. J. Membr. Sci., 1999. **163**(2): p. 307–317.
3. Leo, A., et al., *Oxygen permeation through perovskite membranes and the improvement of oxygen flux by surface modification*. Sci. Technol. Adv. Mater., 2006. **7**(8): p. 819–825.
4. Skinner, S.J., *Recent advances in Perovskite-type materials for solid oxide fuel cell cathodes*. Int. J. Inorg. Mater., 2001. **3**(2): p. 113–121.
5. Hona, R.K., A.K. Thapa, and F. Ramezanipour, *An anode material for lithium-ion batteries based on oxygen-deficient perovskite $Sr_2Fe_2O_{6-\delta}$*. ChemistrySelect, 2020. **5**(19): p. 5706–5711.
6. Gómez, L., et al., *Carbon dioxide gas sensing properties of ordered oxygen deficient perovskite $LnBaCo_2O_{5+\delta}$ (Ln=La, Eu)*. Sens. Actuators, B Chem., 2015. **221**: p. 1455–1460.
7. Maignan, A., et al., *A monoclinic manganite, $La_{0.9}MnO_{3-\delta}$, with colossal magnetoresistance properties near room temperature*. Solid State Commun., 1997. **101**(4): p. 277–281.
8. Hona, R.K., et al., *Transformation of structure, electrical conductivity, and magnetism in $AA'Fe_2O_{6-\delta}$, A = Sr, Ca and A′ = Sr*. Inorg. Chem., 2017. **56**(16): p. 9716–9724.
9. Hona, R.K., A. Huq, and F. Ramezanipour, *Unraveling the role of structural order in the transformation of electrical conductivity in Ca2FeCoO6−δ, CaSrFeCoO6−δ, and Sr2FeCoO6−δ*. Inorg. Chem., 2017. **56**(23): p. 14494–14505.
10. Ramezanipour, F., et al., *Local and average structures and magnetic properties of Sr_2FeMnO_{5+y}, y = 0.0, 0.5. comparisons with $Ca_2FeMnO5$ and the effect of the A-site cation*. Inorg. Chem., 2011. **50**(16): p. 7779–7791.
11. Ramezanipour, F., et al., *The effect of the B-site cation and oxygen stoichiometry on the local and average crystal and magnetic structures of $Sr_2Fe_{1.9}M_{0.1}O_{5+y}$ (M = Mn, Cr, Co; y = 0, 0.5)*. J. Mate. Chem., 2012. **22**(19): p. 9522–9538.

12. Alom, M.S., C.C.W. Kananke-Gamage, and F. Ramezanipour, *Perovskite oxides as electrocatalysts for hydrogen evolution reaction*. ACS Omega, 2022. **7**(9): p. 7444–7451.
13. Hona, R.K., A. Huq, and F. Ramezanipour, *Charge transport properties of $Ca_2FeGaO_{6-\delta}$ and $CaSrFeGaO_{6-\delta}$: The effect of defect-order*. Mater. Chem. Phys., 2019. **238**: p. 121924.
14. Ramezanipour, F., et al., *Intralayer cation ordering in a brownmillerite superstructure: Synthesis, crystal, and magnetic structures of Ca_2FeCoO_5*. Chem. Mater., 2010. **22**: p. 6008–6020.
15. Hona, R.K., et al., *Transformation of structure, electrical conductivity, and magnetism in $AA'Fe_2O_{6-\delta}$, A = Sr, Ca and A' = Sr*. Inorg. Chem., 2017. **56**(16): p. 9716–9724.
16. Hona, R.K. and F. Ramezanipour, *Remarkable oxygen-evolution activity of a perovskite oxide from the $Ca_{2-x}SrxFe_2O_{6-\delta}$ Series*. Angew. Chem. Int. Ed., 2019. **58**(7): p. 2060–2063.
17. Hona, R.K., G.S. Dhaliwal, and R. Thapa, *Investigation of grain, grain boundary, and interface contributions on the impedance of Ca_2FeO_5*. Appl. Sci., 2022. **12**(6): p. 2930.
18. Larson, A.C. and R.B. Von Dreele, *General structure analysis system (GSAS)*. Los Alamos National Laboratory Report LAUR, 1994: p. 86–748, University of California.
19. Toby, B.H., *EXPGUI, a graphical user interface for GSAS*. J. Appl. Crystallogr., 2001 April 1. **34**(2): p. 210–213.
20. Hagentoft, C.-E., *Introduction to building physics*. Studentlitteratur, 2003. **422**: s. ISBN 91-44-01896-7.
21. Young, H.D. and R.A. Freedman, *Sears and Zemansky's University Physics*, Tenth edition, 2000, Addison-Wesley Series in Physics, California, US: Pearson Publication, p. 460–586. ISBN 0-201-60322-5.
22. Ruuska, T., J. Vinha, and H. Kivioja, *Measuring thermal conductivity and specific heat capacity values of inhomogeneous materials with a heat flow meter apparatus*. J. Build. Engin., 2017. **9**: p. 135–141.
23. Tleoubaev, A., A. Brzezinski and L.C. Braga, *Accurate simultaneous measurements of thermal conductivity and specific heat of rubber, elastomers, and other materials*. Presented at the 12th Brazilian Rubber Technology Congress, April 22–24, 2008, Expocenter Norte, Sao Paulo, Brazil.
24. LaserComp, Inc. *Measurements of the volumetric specific heat Cpp and enthalpy of the PhaseChange materials (PCM) using the FOX heat flow meter instruments*. Application Note ANPCM, 2007–2013. 5 p.
25. Anderson, C.T., *The heat capacities at low temperatures of the oxides of strontium and barium*. J. Am. Chem. Soc., 1935. **57**(3): p. 429–431.
26. Chen, J.R., et al., *X-ray diffraction analysis and specific heat capacity of $(Bi_{1-x}La_x)FeO_3$ perovskites*. J. Alloys Compd., 2008. **459**(1): p. 66–70.
27. Borhani Zarandi, M., et al., *Effect of crystallinity and irradiation on thermal properties and specific heat capacity of LDPE & LDPE/EVA*. Appl. Radiat. Isot., 2012. **70**(1): p. 1–5.
28. Kokta, B.V., et al., *Effect of molecular weight of polystyrene on heat capacity and thermal transitions*. Thermochim. Acta., 1976. **14**(1): p. 71–86.
29. Kousksou, T., et al., *Effect of heating rate and sample geometry on the apparent specific heat capacity: DSC applications*. Thermochim. Acta, 2011. **519**(1): p. 59–64.

30. Hepplestone, S.P., et al., *Size and temperature dependence of the specific heat capacity of carbon nanotubes*. Surf. Sci., 2006. **600**(18): p. 3633–3636.
31. Bagatskii, M.I., et al., *Size effects in the heat capacity of modified MWCNTs*. Therm. Sci. Eng. Prog., 2021. **26**: p. 101097.

26 Recent Advancements in Iris Biometrics with Indexing
A Review

Preeti Gupta, Naveen Aggarwal, Renu Vig

26.1 INTRODUCTION

During the last decade, our society has become electronically joined to make one big worldwide community, but there is a need to bring out trustworthy personal recognition system remotely and through automatic ways. Biometrics is perceived as an extremely operative automatic system for individual identification **[1]**. It is like a pattern recognition which helps in finding the genuineness of particular physical or behavioural attributes owned by the individual person. Biometrics are unique, measurable, and enduring in every human and do not change over the course of one's whole life. They are intrinsically more consistent and proficient in distinguishing an authorized person from a fraudulent one than outdated approaches such as passwords and PINs **[2]**. Nowadays, precise and consistent security systems are a main area of researcher's interest.

Many of the physiological or behavioural attributes are present in biometric systems. Some of them are shown in Figure 26.1.

- Facial recognition systems include finding the distance between the eyes, measuring the length of nose, and calculating the angle of the nose, jaws, and so on by analysing the real-time or stored facial image. In general, the system, with the help of available information, creates an exclusive file normally called a template. The template is used for comparisons with stored image.
- A fingerprint includes the traits like orientation field, minutiae points, and pores. The pattern on fingertips is available in fingertips.
- Iris scan: Another physiological attribute is the iris. It is the area of the eye which surrounds the dark pupil of the eye. Generally, it is containing pigmented circle of colours like brown, green, or blue.
- Signature verification analyses various parameters of signature. It is used to identify the amount of pressure applied by pen while signing, shape of alphabets, and speed and timing info during the act of signing.

Every biometric attribute has its own pros and cons; depending upon the application and their performance will be considered in the research field. Researchers

DOI: 10.1201/9781003367154-26

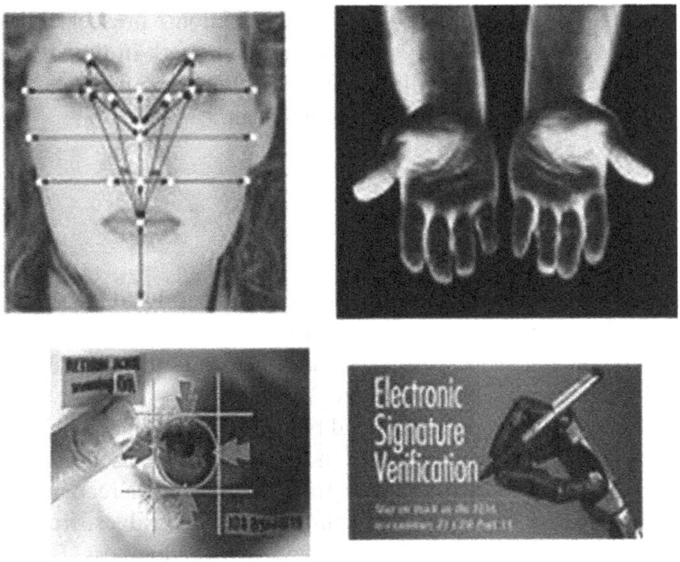

FIGURE 26.1 Different types of biometric traits [1].

Biometrics	Universality	Uniqueness	Permanence	Performance	Acceptability	Circumvention
Face	High	Low	Medium	Low	High	Low
Fingerprint	Medium	High	High	High	Medium	High
Hand geometry	Medium	Medium	Medium	Medium	Medium	Medium
Iris	High	High	High	High	Low	High
Retinal scan	High	High	Medium	High	Low	High
Signature	Low	Low	Low	Low	High	Low
Voice print	Medium	Low	Low	Low	High	Low

FIGURE 26.2 Comparison between various biometric traits [4].

in [3] proposed 7 criteria for the biometric application. They help in finding the appropriateness of a physical and behavioural attribute in the field of biometrics. A comparative analysis for the biometric traits, depending upon the seven factors is summarized in Figure 26.2.

The following sections compose this chapter: Section 2 presents a description of the iris biometric system, Section 3 reviews research work in the area of indexing, Section 4 discusses research gaps in this area, and Section 5 presents the interpretation and forthcoming option of indexing in iris scan.

26.2 IRIS BIOMETRIC SYSTEM

26.2.1 Iris Anatomy

In a human eye, the iris is a part that lies between the pupil and the sclera. It is a thin circular ring region. It presents a unique information, like rifts, rings, patterns, spots, coronal, colours, recess, stripes, furrows, minutiae, and other information of the of the infrared light, as represented in Figure 26.3.

Iris recognition is one of the automated biometric identifications. To identify and verify the people, the flowery patterns of the human iris are used. In different areas like access control and security at border, the demands of iris recognition are growing day by day because of its unique features like rings, ridges, freckles, furrows, and complex patterns. So it poses a great degree of randomness. Every individual has different iris patterns. Even both eyes of individual possess distinct and unique iris patterns. Few of the necessary features of the iris include stability over the whole life, uniqueness, being well protected and impossible to forge, all of which makes iris recognition an accurate and extremely reliable human identification system. One can misplace an identity card or forget a password, but one cannot lose one's physical characteristics **[5–6]**.

In 1987, Flom and Safir were the first to report the iris recognition as a reliable biometric **[7]**. Later, several researchers like K. W. Bowyer, Wildes, John Daugman, and A. K. Jain proposed multiple number of algorithms in this area. Iris is a visible

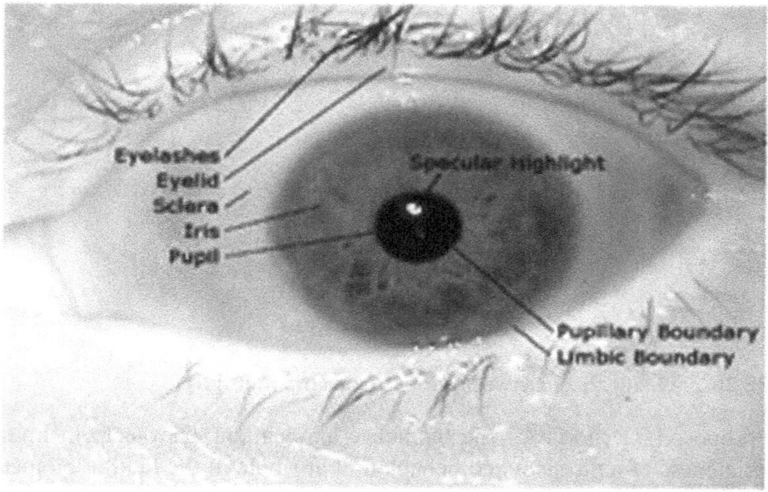

FIGURE 26.3 Structure of human eye **[1]**.

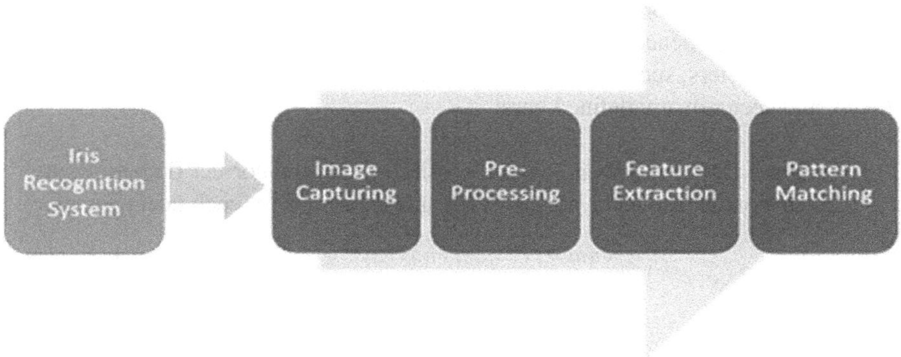

FIGURE 26.4 Modules in an iris recognition system.

part of human eye. The iris recognition system consists of four main sections. In Figure 26.4, the block diagram of iris recognition system is presented.

This deals with the picture capturing procedure of iris. This stride includes localization and isolation of iris and normalization and other pre-processing ventures to enhance the nature of picture. Feature extraction, an important process for analysis, is used to define the features set. A few transform methods can be utilized to concentrate local and global elements from the iris locale, such as SIFT, SURF, Gabor features, and so forth. After the iris code of an individual is created, this iris layout is contrasted and compared with stored template in database to check whether any matches happen. Hamming distance is by and large utilized for examination between bits.

26.2.2 Iris Indexing Techniques

A data structure which recognizes the locations where the index values occur is called an index. In a database, an index classifies which record contains which value. No natural order is defined by which the biometric data can be sorted. For a small-sized database, with respect to time and accuracy, the existing techniques work well. The size of the database is increasing enormously; for example, as the population of India increases, the demand of unique identification (UID) also increases.

With the increase in size on large databases, the computational time to perform iris identification is a thought-provoking task. To handle this issue, one of the useful ways is parallel computing **[8]**. Parallel processing is applicable on a large database, and moreover the computation time will be reduced. But again, there is a challenging task—the cost of operation. Further, to reduce the identification time and accurate finding, the demand of a general indexing technique is increased.

26.2.3 Performance Parameters

The effectiveness of any indexing technique for iris recognition systems depends upon its performance parameters, after extracting the features from each image.

Here, visual information from the images is explored and compiled in the form of a feature vector, and hence, a database is created. The analysis of indexing is done in terms of parameters, such as hit rate (HR), penetration rate (PR), false rejection rate (FRR), and accuracy are defined here:

1. **Hit rate (HR):** It determines the possibility that the correct ID belongs to at least one of the received subsets of images or IDs for each query image.

$$Hit\,Rate(HR) = \frac{Number\,of\,query\,images\,correctly\,identified\,(C)}{Number\,of\,query\,images\,in\,the\,probe\,set\,(P)}$$

2. **Penetration rate (PR):** It shows the fraction of the total number of database records retrieved by the system to identify a user or probe. It can also be defined as the amount of the total database scanned during each search (in %age). For a system to be more efficient, it should have a low value of the penetration rate.

$$Penetration\,Rate(PR) = \frac{1}{N}\sum_{i=1}^{N}\frac{L_i}{M}$$

where N : number of tests performed
L_i : number of images in the candidate set of ith input image
M : number of identities in the database

3. **False rejection rate (FRR):** It is defined as the number of times a genuine person is rejected because the criteria of reference threshold are not fulfilled. This number must be low as much as possible.

$$False\,Rejection\,Rate(FRR) = \frac{Rejected\,genuine\,attempts\,made\,for\,identity\,n}{All\,the\,genuine\,attempts\,made\,for\,identity\,n}$$

4. **Accuracy:** It is defined as the total number of correct predictions divided by the total number of predictions made for a dataset.

$$Accuracy = \frac{Total\,number\,of\,correct\,predictions\,for\,a\,dataset}{Total\,number\,of\,predictions\,made\,for\,a\,dataset}$$

26.3 LITERATURE SURVEY

There are several indexing systems that have been put out in the literature; however, each one has a specific application and a constrained scope. Iris indexing based on IrisCode was first introduced in **[9]** and is regarded as a groundbreaking achievement in this field. Three strategies were suggested in the methodology mentioned. The IrisCode and k-means algorithm are often used in the first technique to use principal component analysis (PCA) to optimize the dimension vector. The second method employs a technique known the local binary pattern (LBP), which separates each image into blocks before calculating a histogram for each block. By taking use of the close relationship between LBP histograms, the number of candidate hypotheses for matching is decreased. The third method examines the iris' inter-pixel relationships to identify pixel-level differences in block characteristics, and it calculates histogram entropy to assess similarity. It has been discovered that the indexing strategy using IrisCode's block-based features does not perform as well as LBP. Even for a small-scale database, the penetration rate is high and the hit rate is low. However, it has made room for more investigation into various indexing strategies.

Indexing noisy data has always been a difficult problem. One such method, **[10]**, eliminates eyelashes that are incorrectly categorized as texture by presuming that they grow in a single direction. In the subsequent stage, features are extracted using a 2D Gabor filtering method that combines data from many scales and directions. Lastly, ternary indexing coding is used to ensure both correctness and retrieval and is based on the Harris corner detection method, where undesirable corners are removed. The advantage of having a reduced penetration rate and miss rate as compared to the PCA method and SIFT key point approach is provided by the described method.

Getting the key value using a B-tree structure, pre-processing only with circular Hough transform, and transforming into rectangular pieces a feature vector using multi-resolution discrete cosine transformation are all part of an efficient indexing plan based on texture features using energy histogram that was introduced in **[11]**. The B-tree is implanted at the leaf level, but scalability is an issue because the bin and subband size is fixed, leading to misclassification when coarse texture is present.

Using a standardized feature vector projected to a lower vector space, an effective indexing method for a vast multi-modal biometric database was suggested in the literature **[12]**. To use a circular Hough transformation, the iris' inner border is discovered, and an intensity variation approach is used to retrieve the iris' outer part. The Haar wavelet transform and feature normalization were then employed to the feature vectors to build a kd-tree, and the Euclidean distance was calculated to identify the correct match. It provides excellent precision. However, because the structure must be explored for a minimal particular time frame on the subset, it is not appropriate for huge databases. It was described in **[13]** how to derive approximate inner and outer iris boundaries using an elliptical model and precise inner and outer iris borders to use a modified Mumford-Shah functional utilizing a feature-based iris-indexing approach for non-ideal insufficient iris images. For further classification or indexing, the topological match values and the textural match values are merged yielding an accuracy of 97.21% and a time of 2 seconds. The distribution of the iris patterns used a modified binary search algorithm, based on the spatial clustering property of the burrow-wheeler transform was proposed in **[14]** as a new method for indexing iris biometrics. With modest databases, the method works effectively.

A brand-new segmentation technique using the Hough circular transform was proposed **[15]** and is dependent on the iris' natural colour. Blue and red colour indices were used to distinguish the input image after it had been converted from RGB to YCbCr. The system works well for a small subset, but it requires a lot of inter-class variation, which affects search time and efficiency. A colour histogram indexing scheme was used to separate different bins of the histogram using the K-NN algorithm, and a kd-tree was created for identification and retrieval. The scheme is secure and generates a high hit rate, but the time required was extremely long. To locate ROI using the AdaBoost method on noisy datasets and colours, a unique iris indexing system was proposed **[16]**. To avoid the negative effects of luminance, the eyelids' and eyelashes' spectacular reflections were eliminated, and the average colour value was standardized. Analysis of hit and penetration rates was done both before and after colour correction, and it is clear that the HR is significantly higher after colour correction than without. After colour calibration, the system actually works well with low noisy datasets, but classification can get more difficult as the increases in size.

Using a sector-based approach and the circular Hough transformation, **[17]** is suggested as a method for indexing a large iris biometric database. The input iris image is first treated to obtain a noise-free image. The critical spots in the image that were resistant to scaling and rotation as well as to illumination, position, and occlusion were then found by extracting local features using SIFT. The precompiled image is utilized to extract key point descriptor using the SIFT. As a result, an improved technique for allocating hash table entries uniformly was described in **[18]**, using the rehashing function to generate regular occupancy. Each element in the hash table bin gets a vote, with the highest votes being deemed the greatest match and those above the threshold being evaluated prospective matches. Given that geometric hashing is naturally comparable and that the time required for hashing and rehashing is 29 seconds, a good accuracy of 98.5% has been achieved (approximate).

Various researchers are coming up with new techniques of iris indexing as the matter of concern is that computational efficacy is severely degraded with a surge in the database dimensions. An author in **[19]** did the investigation on four datasets and was able to show the effectiveness of the real-valued deep iris parameters binarized to IBCs (iris bar codes). Results concluded that M com loss can change the binary feature into an enhanced feature well-suited with the multi-index hashing system. This paper **[20]** discussed about the two approaches to reduce the problematic issue: the curse of dimensionality. It is to address the exact and approximation techniques and address the efficient algorithms for the nearest neighbour search. Authors clearly mentioned that universal solution is not possible for all biometric traits and also discussed that the penetration and hit rates are the best evaluation parameters to identify the precise accuracy.

The one of the goals in the field of iris identification is to decrease the time taken to identify a most appropriate match with the sample. In general, the database keeps on expanding, and it becomes difficult to identify a sample in less time. The author of this paper **[21]** discusses the problems and presents some of the indexing techniques for common biometric traits. It includes the four physiological traits, and one of them is the iris.

Researchers in this paper **[22]** discuss a new indexing technique. It efficiently lessens the exploration space for the database. Special CNN architecture was implemented for making index table using k-means and agglomerative clustering. Databases such as CASIA Interval and CASIA Lamp were mainly used. The projected methodology attained better results (approximately 99%). The authors in **[23]** present a retrieval system for a multi-biometric system by using a multidimensional spectral hashing scheme. The binary codes of face, iris, and palm prints were fused together, and promising results of penetration rate and hit rate were achieved as compared to the state of art techniques.

Another indexing space partitioning approach based on hashing was proposed by the authors in **[24]**, which gave desirable results of hit rate and penetration rate, when the applied on three different iris biometric datasets. Thus, it has been seen in the literature that many works have been proposed to reduce the search time for a high-dimensional biometric data. Drozdowski et al. **[25]** used a bloom-filter-based and binary search tree approach to the raw iris images for indexing of iris biometric systems. The author generated iris codes after segmentation and normalization of the images of the combined multi-iris dataset to propose an effective retrieval system.

26.4 COMPARATIVE ANALYSIS

This section gives an insight into the recent works proposed by various researchers for handling large biometric systems. The papers covered iris biometric systems and some works for multidimensional biometrics have also been included to give a broader view about the various challenges and research gaps associated with them. A comparison among the various indexing techniques is summarized in Table 26.1.

26.5 CONCLUSION

This chapter highlights the recent advancements and the different indexing techniques for iris recognition systems based on the feature vector, such as texture, energy histogram, binary codes, topology, colour, deep features, and so on. The performance of iris biometric systems can be evaluated by using hit rate, penetration rate, false rejection rate, and accuracy as the output parameters. From the exhaustive review, it has been found that the efficiency of these systems is greatly dependent on the number of classes, scale and rotation of images, noisy data, quality of images, and accurate segmentation. Various tree traversing techniques: B-tree and kd-tree as reported in the literature have a problem of dimensionality. Similarly, the problem of pure key generation is found in another technique of indexing—hashing. Further, in a real-life application, indexing with noisy iris data is rarely used in ongoing live projects. It has been concluded that researchers have to address these reported problems as it is the necessity of the current time by using a generic indexing technique to reduce the search time and improve the correct identification. The investigation found in this chapter is noteworthy and showed research potential in iris indexing for large noisy databases.

TABLE 26.1
Comparison of Various Indexing Techniques

Author Name	Name of Technique/ Algorithm Used	Features	Dataset Used	Performance Parameters	Remarks
Mukherjee et al. [9]	IrisCode (using Gabor wavelet with PCA) and texture (Signed Pixel Level Difference Histogram)	IrisCode and texture features	CASIA v3 iris dataset	PR = 8% (block avg.) PR = 17% (PCA) HR = 80% (row/col avg.) HR = 84% (SPLDH)	• Not evaluated for non-ideal poor-quality iris images
Si et al. [10]	Eyelash detection algorithm (using directional filters) and corner detection (2D Gabor filters)	Texture features	CASIA v1 and IITD iris datasets	PR <4%	• Performance measure is dependent on the block size and number of registered iris images.
Mehrotra et al. [11]	Energy histogram using DCT	Energy features	CASIA, BATH, and IITK Iris datasets	PR = 63% (CASIA) PR = 6% (BATH) PR = 20% (IITK)	• PR depends on the number of classes. • Variant to scale and rotation
Jayaraman et al. [12]	Haar wavelets and kd-tree with score-level fusion (with/ without PCA)	Texture features	IITK (iris, face, signature, ear)	FRR = 2.66 (no PCA) FRR = 0.66 (with PCA) (all databases)	• Feature-level fusion performs better than the score-level fusion.
Vatsa et al. [13]	Euler tech (topological features) and 1D log Gabor transform (textural features) with SVM fusion	Textural and topological features	ICE 2005, CASIA v3, and UBIRIS iris datasets	Accuracy = 92.39% (Euler) Accuracy = 96.57% (Euler + texture) Accuracy = 97.21% (Euler + texture + SVM fusion)	• Works only on greyscale iris • Sensitive to rotation • Requires accurate segmentation

Author Name	Name of Technique/ Algorithm Used	Features	Dataset Used	Performance Parameters	Remarks
Gadde et al. **[14]**	Burrow wheel transform indexing scheme, based on count of occurrence of 4- and 8-bit binary patterns	Texture and IrisCode features	CASIA v3 iris dataset	PR = 17.23% HR = 99.83%	• At least two normalized images required per eye for the implementation • Need to find an alternate searching mechanism which combines different index patterns (P1, P2) of the same length
Puhan et al. **[15]**	Semantic decision tree	Colour features	UBIRIS and UPOL iris datasets	PR < 25% HR > 98%	• Exclusive for colour data • Sensitive to parameter values (number indices)
Zhao et al. **[16]**	Colour compensation and quantization into three 1D feature spaces using bins	Colour features	UBIRIS v2 iris dataset	PR = 28.28% HR = 92.35% (with error correction) PR = 26.32% HR = 72.54% (without error correction)	• Exclusive for noisy colour data • Sensitive to parameters (interval length)
Mehrotra et al. **[17]**	SIFT key points to obtain indices of hash table	Texture features	CASIA, BATH, UBIRIS, and IITK iris datasets	PR = 24% Accuracy = 98.29% (BATH) Accuracy = 99.61% (IITK)	• Hr and Pr dependent on choice of the threshold • Speed and accuracy cannot be compromised.

TABLE 26.1 ***(Continued)***
Comparison of Various Indexing Techniques

Author Name	Name of Technique/ Algorithm Used	Features	Dataset Used	Performance Parameters	Remarks
Singh et al. [19]	Uses four strategies: ball tree, lamp interval, multi-index hashing (MIH), and optimized MIH methods with multi-index hashing	Real-valued deep features binarized to iris bar codes	CASIA Lamp, CASIA-v3 Interval, IITD-V1, and IITK iris datasets	HR > 98% PR range: (1–3)%	• Trade-off between accuracy and efficiency leads to performance deterioration. • The fuzziness of the iris dataset belonging to the same subject is high, making it challenging. • High computational costs • Needs large dataset
Arora et al. [22]	K-means and agglomerative clustering	Deep learning features	CASIA Interval and CASIA Lamp iris datasets	HR = 99% PR = 2.254% (CASIA Interval) PR = 0.008% (CASIA Lamp)	• Uses twin CNN (Siamese architecture) to construct compact feature vectors with low inter-class and high intra-class similarity in the latent representations • Shallow network • Employs larger-sized and more filters
Balasundaram and Sudha [23]	Multidimensional spectral hashing, cuckoo search algorithm, and Hamming affinity matrix (feature fusion)	Binary codes features	IITD (iris, face, and palm prints) datasets	HR = 90% PR = 7%	• Local features can be used to reduce the number of training samples and increase recognition accuracy.

Author Name	Name of Technique/ Algorithm Used	Features	Dataset Used	Performance Parameters	Remarks
Ahmed and Sarma [24]	Hash-based system to extract features using 1D Log-Gabor, DCT, and bins to store index keys	Texture features	IIT-D, CASIA-v3 Interval, and UBIRIS 2 datasets	PR range: 1–5% HR range: 95–99%	• Change in accuracy due to segmentation errors • EER between 1 and 5% depending upon the type of segmentation error
Drozdowski et al. [25]	1D Log-Gabor with Bloom filter and binary search trees	IrisCode features	CASIA v4 Interval, IITD v1, and ND iris template ageing datasets	HR = 98% PR = 10%	• Combined dataset was created. • Images having >70% usable iris segmented area were used. • High-quality images were considered and subjects wearing glasses were not covered completely.

REFERENCES

1. Bowyer KW, Hollingsworth K, Flynn PJ. Image understanding for iris biometrics: A survey. *Computer Visual Image Understanding*. 2008; 110 (2): 281–307. Available from: https://linkinghub.elsevier.com/retrieve/pii/S1077314207001373.
2. Jain A, Hong L, Pankanti S. Biometric identification. *Communication ACM*. 2000; 43 (2): 90–98. Available from: https://doi.org/10.1145/328236.328110.
3. Jain AK, Kumar A. Biometric recognition: An overview. In: Mordini E, Tzovaras D, editors. *Second Generation Biometrics: The Ethical, Legal and Social Context*. Dordrecht: Springer Netherlands; 2012, pp. 49–79. Available from: https://doi.org/10.1007/978-94-007-3892-8_3.
4. Jain A, Bolle R, Pankanti S. *Biometrics Personal Identification in Networked Society*. New York, Boston, Dordrecht, London, Moscow: Kluwer Academic Publishers; 1999, pp. 1–40.
5. Daugman JG. High confidence visual recognition of persons by a test of statistical independence. *IEEE Transactions on Pattern Analysis and Machine Intelligence*. 1993; 15 (11): 1148–1161. Available from: http://ieeexplore.ieee.org/document/244676.
6. Wildes RP. Iris recognition: An emerging biometric technology. *Proceedings IEEE*. 1997; 85 (9): 1348–1363. Available from: http://ieeexplore.ieee.org/document/628669.
7. Flom L, Safir A. *Iris Recognition System*. U.S. patent No.4641394 (1987). Washington, DC: U.S. Patent and Trademark Office.
8. Ma L, Tan T, Wang Y, Zhang D. Efficient iris recognition by characterizing key local variations. *IEEE Transactions on Image Processing*. 2004; 13 (6): 739–750.
9. Mukherjee R, Ross A. Indexing iris images. In: *2008 19th International Conference on Pattern Recognition* [Internet]. IEEE; 2008, pp. 1–4. Available from: http://ieeexplore.ieee.org/document/4761880/.
10. Si Y, Mei J, Gao H. Novel approaches to improve robustness, accuracy and rapidity of iris recognition systems. *IEEE Transactions on Industrial Informatics*. 2012; 8 (1): 110–117. Available from: http://ieeexplore.ieee.org/document/6011686/.
11. Mehrotra H, Srinivas BG, Majhi B, Gupta P. Indexing iris biometric database using energy histogram of DCT subbands. In: Ranka S, Aluru S, Buyya R, Chung Y-C, Dua S, Grama A, et al., editors. *Contemporary Computing*. Berlin, Heidelberg: Springer; 2009, pp. 194–204.
12. Jayaraman U, Prakash S, Gupta P. An efficient technique for indexing multimodal biometric databases. *International Journal of Biometrics*. 2009; 1 (4): 418–441. Available from: https://doi.org/10.1504/IJBM.2009.027304.
13. Vatsa M, Singh R, Noore A. Improving iris recognition performance using segmentation, quality enhancement, match score fusion, and indexing. *IEEE Transactions on Systems, Man, and Cybernetics, Part B*. 2008; 38 (4): 1021–1035.
14. Gadde RB, Adjeroh D, Ross A. Indexing iris images using the burrows-wheeler transform. In: *2010 IEEE International Workshop on Information Forensics and Security*. IEEE; 2010, pp. 1–6. Available from: http://ieeexplore.ieee.org/document/5711467/.
15. Puhan NB, Sudha N. A novel iris database indexing method using the iris colour. In: *2008 3rd IEEE Conference on Industrial Electronics and Applications*. IEEE; 2008, pp. 1886–1891. Available from: http://ieeexplore.ieee.org/document/4582847/.
16. Zhao Q. A new approach for noisy iris database indexing based on colour information. In: *2011 6th International Conference on Computer Science & Education (ICCSE)*. IEEE; 2011, pp. 28–31. Available from: http://ieeexplore.ieee.org/document/6028577/.
17. Mehrotra H, Majhi B, Gupta P. Robust iris indexing scheme using geometric hashing of SIFT keypoints. *Journal of Network and Computer Applications*. 2010; 33 (3): 300–313. Available from: http://doi.org/10.1016/j.jnca.2009.12.005.

18. Rigoutsos I, Hummel R. Implementation of geometric hashing on the connection machine. In: *Workshop on Directions in Automated CAD-Based Vision 1991 Jan 1.* IEEE Computer Society, pp. 76–77. http://doi.org/10.1109/CADVIS.1991.148760. Available from https://store.computer.org/csdl/proceedings-article/cadvis/1991/00148760
19. Singh A, Gaurav P, Vashist C, Nigam A, Yadav RP. IHashNet: Iris Hashing Network based on efficient multi-index hashing. In: *2020 IEEE International Joint Conference on Biometrics (IJCB).* IEEE Xplore; 2020, pp. 1–9.
20. Jaswal G, Kanhangad V, Ramachandra R, editors. *AI and Deep Learning in Biometric Security.* First edition. Boca Raton, FL: CRC Press, 2021. Available from: https://www.taylorfrancis.com/books/9781000291629.
21. Jayaraman U, Gupta P. Efficient similarity search on multidimensional space of biometric databases. *Neurocomputing.* 2021; 452: 623–652. Available from: https://www.sciencedirect.com/science/article/pii/S0925231220319159.
22. Arora G, Vichare S, Tiwari K. IrisIndexNet: Indexing on iris databases for faster identification. In: *5th Joint International Conference on Data Science & Management of Data (9th ACM IKDD CODS and 27th COMAD).* New York: ACM; 2022, pp. 10–18. Available from: https://dl.acm.org/doi/10.1145/3493700.3493715.
23. Balasundaram R, Sudha GF. Retrieval performance analysis of multi-biometric database using optimized multidimensional spectral hashing based indexing. *Journal of King Saud University-Computer and Information Sciences.* 2021; 33 (1): 110–117. Available from: https://doi.org/10.1016/j.jksuci.2018.02.003.
24. Ahmed T, Sarma M. Hash-based space partitioning approach to iris biometric data indexing. *Expert Systems with Applications.* 2019; 134: 1–13. Available from: https://doi.org/10.1016/j.eswa.2019.05.026.
25. Drozdowski P, Rathgeb C, Busch C. Bloom filter-based search structures for indexing and retrieving iris-codes. *IET Biometrics.* 2018; 7 (3): 260–268.

27 Modelling of the Wear Behaviour of AISI 4140 Alloy Steel Under the Nano Fly Ash Additive in SAE 10W-30 Engine Oil

Harvinder Singh, Yogesh Kumar Singla, Sahil Mehta, Abhineet Saini

27.1 INTRODUCTION

To increase the lubricated rubbing surfaces resistance to wear, researchers are searching for new lubricants and additives. New substances that could be added as lubricant additives have been discovered as a result of research. Numerous nanoparticles have been studied as lubricating oil additives [1–5]. These days, lubricating oil additives containing nanoparticles are widely recognized. When combined with the right surfactants, these nano-additives in colloidal form function as a lubricant's anti-wear, extreme pressure, and friction modifier [6, 7]. These nanoparticles can easily circulate in oil systems and access contact surfaces [8]. Some beneficial characteristics of the nano-additive include low reactivity with other nano-additives, ease in film production, a propensity to function even at high temperatures, and increased dependability [9–13]. Despite the fact that several nanoparticles have been investigated up to this point, picking one is challenging. The tribological behaviour of a material is influenced by the size, shape, compatibility, and concentration of nano-additives in lubricant [14]. Before determining the research gap and objectives in the area of nanoparticles as an additive in lubricants, the following should be reviewed: Another important element in reducing friction is the concentration of nanoparticles in lubricants [15–26]. The nanoparticles would always produce the best anti-wear results within a certain range [27]. According to a report, the nanoparticles' ideal concentrations vary depending on the medium. MoS_2 nanoparticles exhibit an optimal concentration of 0.58% by weight for mineral oil and 0.53% by weight for coconut oil [15]. In the case of nano CuO and nano ZnO in mineral and vegetable oils, 0.5% by weight was taken into consideration [28]. Depending on the characteristics of the employed lubricant and nano-additions, the ideal concentration would be determined. Nano fly ash was also used as additive in SAE 10W-30 engine oil for the tribological testing of AISI 4140 alloy steel which positive results [29–32].

 DOI: 10.1201/9781003367154-27

27.2 EMPIRICAL MODELLING OF COEFFICIENT OF FRICTION AND WEAR LOSS PERCENTAGE

Regression analysis was employed to develop a mathematical model to depict the relationship between input parameters and output factors—coefficient of friction and wear loss. The input factors were L (load), S_D (sliding distance), P_F, and S_V (sliding velocity), and COF (coefficient of friction) was one of the output parameters. The factor sliding velocity was not considered due to contribution in analysis of variance and SN ratio results. Multi-variable linear regression was employed to find the model of three factors: load, sliding velocity, and percentage of nano fly ash by weight.

A mathematical model having non-linear characteristics was build up to find the relation input parameters and coefficient of friction and wear loss.

$$C = Z(P_1)^{x1}(P_2)^{x2}\ldots\ldots\ldots(P_n)^{n} \tag{27.1}$$

Where C stands for coefficient of friction, P_1, P_2 . . . . P_n are the input parameters (L, S_D, and P_F) and the exponents were e_1, e_2 . . . . n

Equation becomes linear due the logarithmic conversion of equation 27.1 as shown in the following equation:

$$\log C = \log Z + e_1 * \log(P_1) + e_2 * \log(P_2) + \ldots\ldots\ldots + n * \log(P_n) \tag{27.2}$$

Further equation 27.2 can be rewritten in simple form as

$$K = \beta_0 + \beta_1 p_1 + \beta_2 p_2 + \ldots\ldots\ldots + \beta_n p_n \tag{27.3}$$

where the logarithmic value of coefficient of friction is denoted by K. $x_1, x_2, \ldots\ldots, x_n$ are the logarithmic values of input parameters and $\beta_0, \beta_1, \beta_3, \ldots\ldots\ldots, \beta_n$ denote the coefficients.

For simplicity, the equation is rewritten as

$$\hat{Y} = e_0 + e_1 * p_1 + e_2 * p_2 + \ldots\ldots\ldots\ldots + e_n * p_n + \varepsilon \tag{27.4}$$

where the predicted coefficient of friction is denoted by $\hat{Y}$ and ε is residual or error, e_0 is the intercept, and e_1, e_2. . . . . e_n are denoted as the estimates of P_1, P_2 . . . P_n parameters, respectively. Now, the data got from experimentation was transformed into logarithmic values per equation 27.4. The model estimation was performed using Minitab 16 software, as depicted in the Table 27.1.

Hence, the first-order empirical model for coefficient of friction was derived as equation 27.5:

$$\text{COF} = 0.0528 + 0.000694\ \text{P1} + 0.000012\ \text{P2} - 0.155\ \text{P3} \tag{27.5}$$

This could be also written as

$$\text{COF} = 0.0528 + 0.000694\ \text{L} + 0.000012\ \text{S}_\text{D} - 0.155\ \text{P}_\text{F} \tag{27.6}$$

TABLE 27.1
Coefficients of COF (First-Order Model)

Predictors	Coefficient
e_0	0.0528
P1	0.000694
P2	0.000012
P3	–0.155

TABLE 27.2
Regression Results in Minitab 16

Parameter (Predictors)	Coefficient	SE Coefficient	T	P
Constant	0.05276	0.01413	3.73	0.014
Load	0.0006943	0.0001010	6.87	0.001
Sliding distance	0.00001217	0.00000405	3.00	0.030
% Nano fly ash	−0.15481	0.01531	−10.11	0.000
S = 0.00992373		R Sq = 96.9%	R Sq (adjusted)= 95.1%	

27.3 RESULTS AND DISCUSSION

27.3.1 Regression Analysis

Table 27.2 gives the regression analysis results by Minitab software. In this regression model, the value of adjusted R^2 is .95, which means the model explains 95% of the total deviations.

Figure 27.1, Figure 27.2, Figure 27.3, and Figure 27.4 were the normal probability graph, versus fits, histogram, and versus order plots obtained from the regression analysis using Minitab Software.

27.3.2 Validation of Model

In a series of experiments for validating the model, equation 27.6 was used to perform the experiments using different arrangements of input factors, as depicted in Table 27.3.

As P_F was most significant parameter, three levels were selected. As L was a significant parameter, two levels were selected. In the case of the less significant parameter ($S_{D,}$), the optimum level was selected. So six experiments were conducted. The results in Table 27.4 depict the percentage error ranging from 8 to −9.2, and hence, the first-order empirical model was validated. Figure 27.5 shows the variation in the predicted and observed value for COF.

The same technique was employed for another output parameter—wear loss percentage by weight. Similarly, the first-order empirical model formed is as follows:

$$W_L = 9.03 + 0.107\,L + 0.00202\,S_D - 24.7\,P_F \tag{27.7}$$

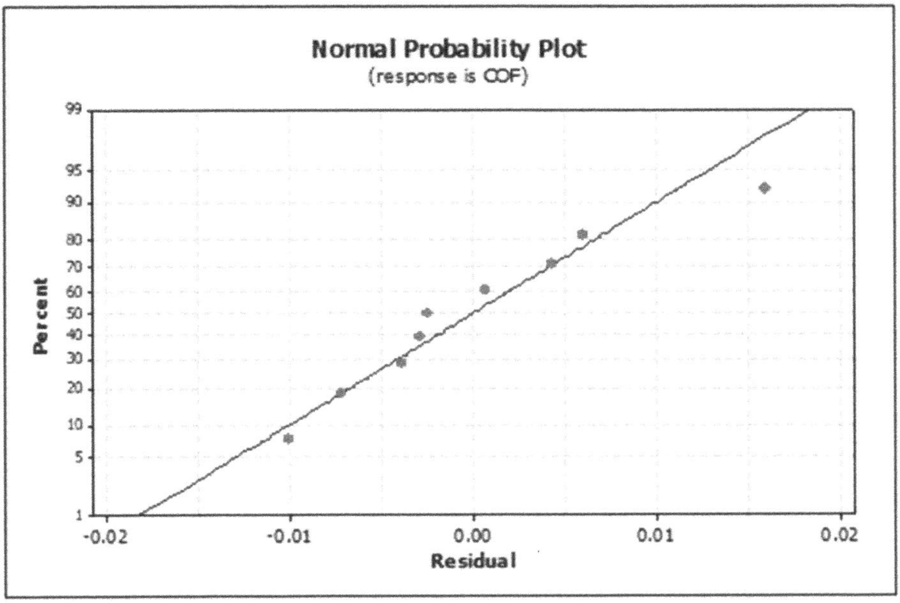

FIGURE 27.1 Normal probability graph for error of COF.

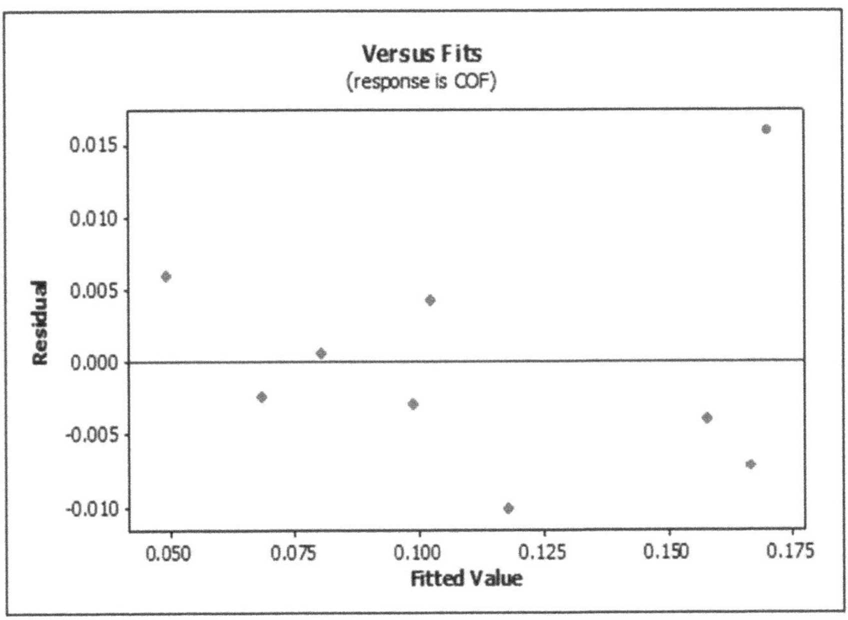

FIGURE 27.2 Graph of error/residual versus fits of COF.

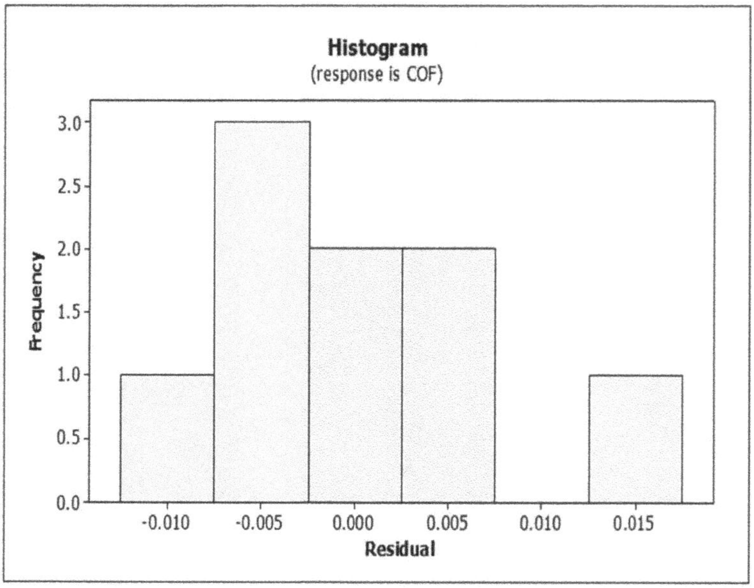

FIGURE 27.3 Histogram plot for residuals of COF.

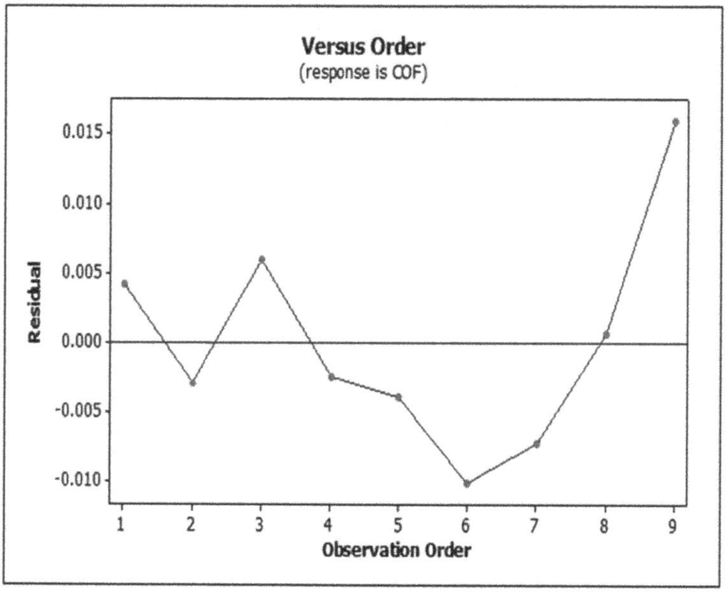

FIGURE 27.4 Versus order for residuals of COF.

TABLE 27.3
First-Order Empirical Model Validation of COF

S. No.	Input Factors L	S_D	P_F	Value Predicted	Average Observed Value	Percentage Error
1	90	1,500	0	0.1317	0.1211	8.0
2	125	1,500	0.1	0.1420	0.1316	7.3
3	90	1,500	0.5	0.0557	0.0519	6.7
4	125	1,500	0	0.1560	0.1650	–5.7
5	90	1,500	0.1	0.1177	0.1287	–9.2
6	125	1,500	0.5	0.0801	0.0750	6.3

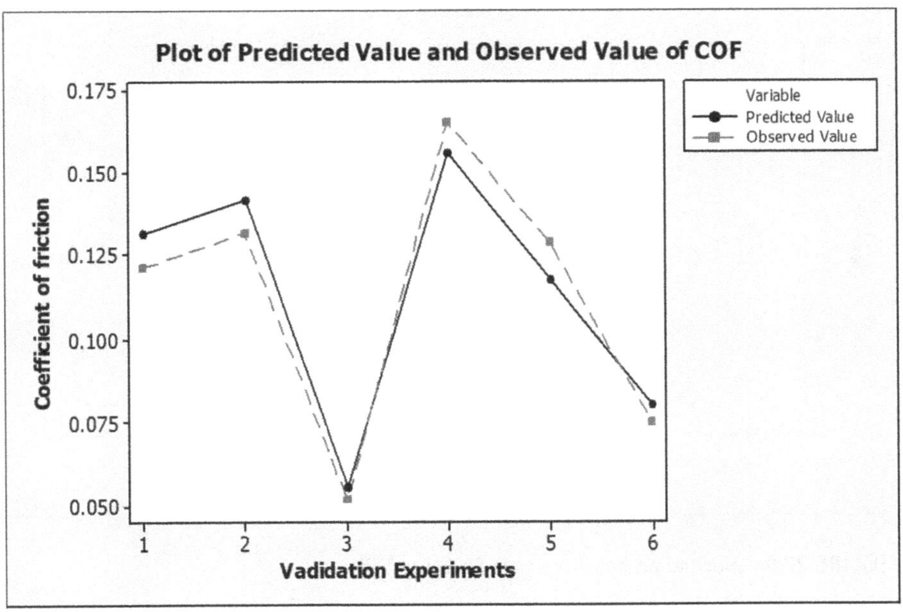

FIGURE 27.5 Variation in the predicted and observed value for COF.

Figure 27.6, Figure 27.7, Figure 27.8, and Figure 27.9 are the normal probability plot, versus fits, histogram, and versus order plots, respectively, obtained from the regression analysis using the Minitab software. Table 27.4 depicts the regression analysis results by Minitab software. In this regression model, the value of adjusted R^2 is 0.95, which means the model explains 95% of the total deviations or variations.

TABLE 27.4
Regression Results in Minitab 16

Predictor	Coefficient	SE Coefficient	T	P
Constant	9.034	1.500	6.02	0.002
Load	0.10731	0.01073	10.00	0.000
Sliding distance	0.0020197	0.0004303	4.69	0.005
% nano fly ash	−24.746	1.626	−15.22	0.000
S = 1.05396		R Sq = 98.6%	R Sq (adjusted) = 97.8%	

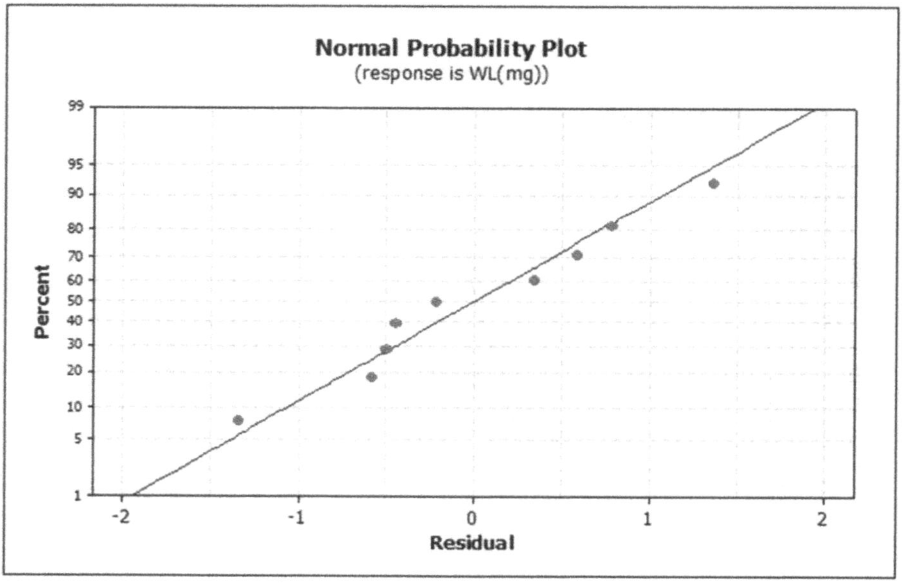

FIGURE 27.6 Normal probability graph for error of W_L.

The validation of the wear loss (W_L) model was also done experimentally for six combinations as done in the case of the coefficient of friction. Table 27.5 shows the combination and results, which show the percentage error ranging from 5.03 to −8.94, which validated the first-order empirical model. The variation in the predicted and observed value for wear loss in mg is depicted in Figure 27.10. This graph shows the goodness of the model developed for the wear loss response characteristic.

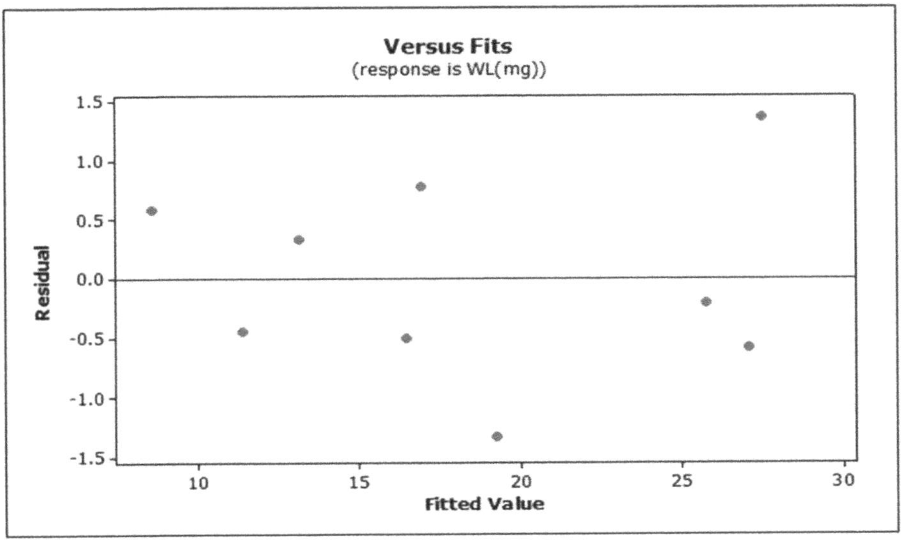

FIGURE 27.7 Graph of error/residual versus fits of W_L.

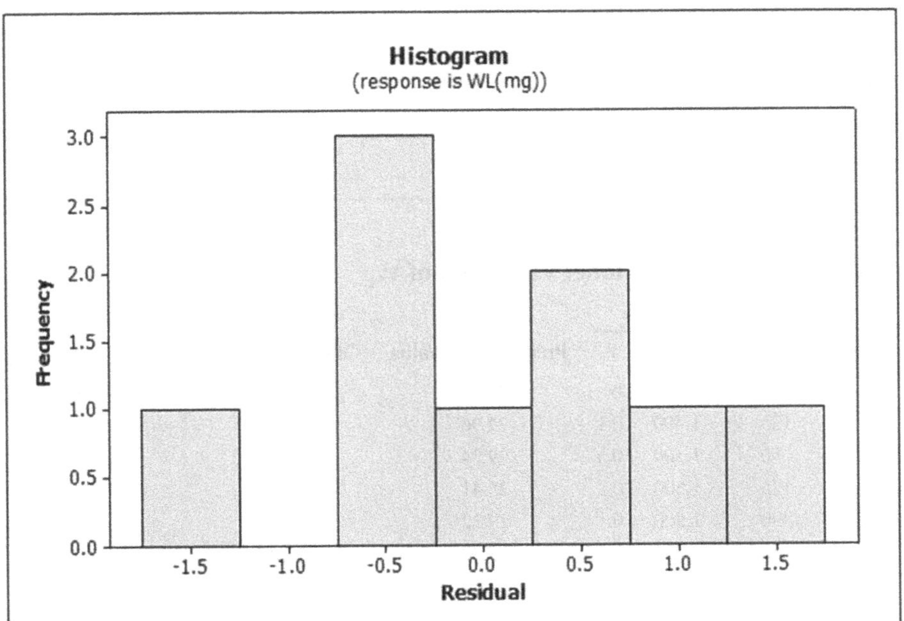

FIGURE 27.8 Histogram plot for residuals of W_L.

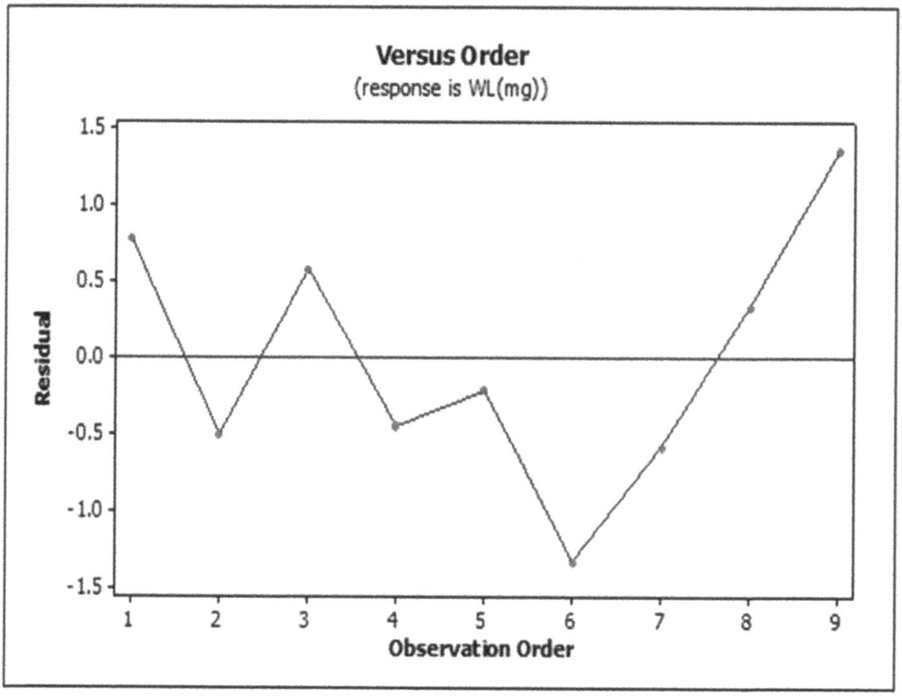

FIGURE 27.9 Versus order for residuals of W_L.

TABLE 27.5
First-Order Empirical Model Validation of W_L

S. No.	Input Parameters			Predicted W_L Value	Observed W_L Value	Error Percentage
	L	S_D	P_F			
1	90	1,500	0	21.66	20.15	6.97
2	125	1,500	0.1	22.96	24.64	–7.31
3	90	1,500	0.5	9.34	8.87	5.03
4	125	1,500	0	25.41	23.78	6.41
5	90	1,500	0.1	19.22	20.94	–8.94
6	125	1,500	0.5	13.085	12.14	7.222

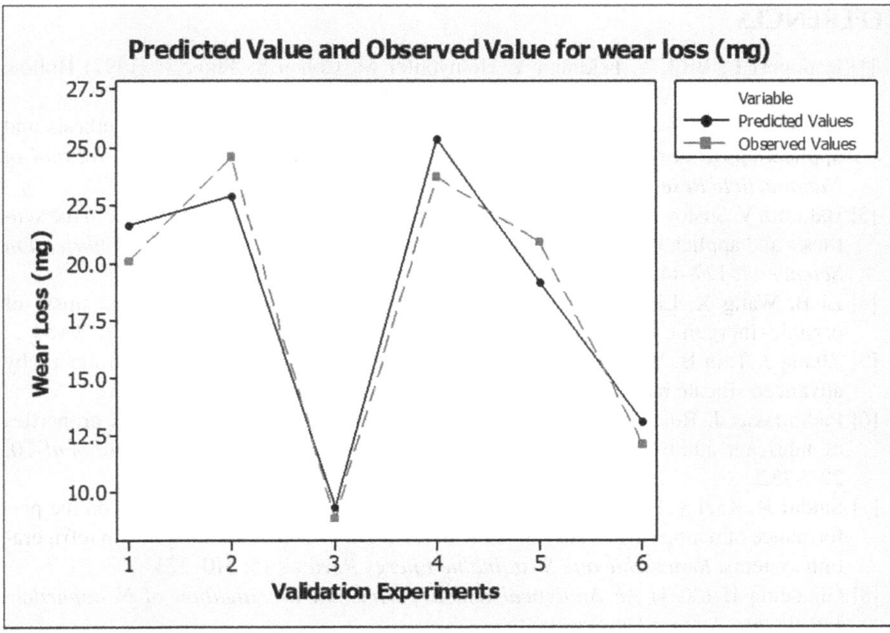

FIGURE 27.10 Variation in the predicted and observed value for wear loss (W_L).

27.4 CONCLUSION

The first-order empirical model for coefficient of friction and wear percentage loss was developed in this present study.

1. The input factors were L (load), S_D (sliding distance), P_F, and S_V (sliding velocity). COF (coefficient of friction) and wear percentage loss were the output parameters.
2. The mathematical model was validated by conducting validation experiments. The range of deviation in error percentage for the developed multivariate linear regression models was within the acceptable limit of ±10% for both coefficients of friction and wear loss by weight.
3. During validation of the six selective combinations, the percentage errors ranged from 8% to −9.2% in the case of the coefficient of friction, and in the case of percentage wear loss, the error ranged from 5.03% to −8.94%.

REFERENCES

[1] Rapoport L, Bilik Y, Feldman Y, Homyonfer M, Cohen S, Tenne R (1997) Hollow nanoparticles of WS_2 as potential solid-state lubricants, *Nature* 387: 791–793.

[2] Bakunin V, Suslov AY, Kuzmina G, Parenago O, Topchiev A (2004) Synthesis and application of inorganic nanoparticles as lubricant components–a review, *Journal of Nanoparticle Research* 6: 273–284.

[3] Bakunin V, Suslov AY, Kuzmina G, Parenago O (2005) Recent achievements in the synthesis and application of inorganic nanoparticles as lubricant components, *Lubrication Science* 17: 127–145.

[4] Li B, Wang X, Liu W, Xue Q (2006) Tribochemistry and antiwear mechanism of organic–inorganic nanoparticles as lubricant additives, *Tribology Letters* 22: 79–84.

[5] Zhang J, Tian B, Wang C (2013) Long-term surface restoration effect introduced by advanced silicate based lubricant additive, *Tribology International* 57: 31–37.

[6] Padgurskas J, Rukuiza R, Prosyčevas I, Kreivaitis R (2013) Tribological properties of lubricant additives of Fe, Cu and Co nanoparticles, *Tribology International* 60: 224–232.

[7] Saidur R, Kazi S, Hossain M, Rahman M, Mohammed H (2011) A review on the performance of nanoparticles suspended with refrigerants and lubricating oils in refrigeration systems, *Renewable and Sustainable Energy Reviews* 15: 310–323.

[8] Ghaednia H (2014) *An Analytical and Experimental Investigation of Nanoparticle Lubricants*, Auburn University.

[9] Spikes H (2015) Friction modifier additives, *Tribology Letters* 60: 1–26.

[10] Nallasamy P, Saravanakumar N, Nagendran S, Suriya E, Yashwant D (2014) Tribological investigations on MoS_2-based nanolubricant for machine tool slideways, *Proceedings of the Institution of Mechanical Engineers, Part J: Journal of Engineering Tribology* 229(5): 559–567.

[11] Verma A, Jiang W, Abu Safe H, Brown W, Malshe A (2008) Tribological behavior of deagglomerated active inorganic nanoparticles for advanced lubrication, *Tribology Transactions* 51: 673–678.

[12] Srinivas V, Thakur R, Jain A (2016) Antiwear, antifriction, and extreme pressure properties of motor bike engine oil dispersed with molybdenum disulfide nanoparticles, *Tribology Transactions*, 60(1): 12–19.

[13] Chou R, Battez A, Cabello J, Viesca J, Osorio A, Sagastume A (2010) Tribological behavior of polyalphaolefin with the addition of nickel nanoparticles, *Tribology International* 43: 2327–2332.

[14] Peña-Parás L, Taha-Tijerina J, Garza L, Maldonado-Cortés D, Michalczewski R, Lapray C (2015) Effect of CuO and Al_2O_3 nanoparticle additives on the tribological behavior of fully formulated oils, *Wear* 332–333: 1256–1261.

[15] Koshy C, Rajendrakumar P, Thottackkad M (2015) Evaluation of the tribological and thermophysical properties of coconut oil added with MoS_2 nanoparticles at elevated temperatures, *Wear* 330–331: 288–308.

[16] Kumar Dubey M, Bijwe J, Ramakumar S (2013) PTFE based nano-lubricants, *Wear* 306: 80–88.

[17] Yu W, Xie H (2012) A review on nanofluids: Preparation, stability mechanisms, and applications, *Journal of Nanomaterials*: 1–17.

[18] Song X, Zheng S, Zhang J, Li W, Chen Q, Cao B (2012) Synthesis of monodispersed $ZnAl_2O_4$ nanoparticles and their tribology properties as lubricant additives, *Materials Research Bulletin* 47: 4305–4310.

[19] Luo T, Wei X, Huang X, Huang L, Yang F (2014) Tribological properties of Al_2O_3 nanoparticles as lubricating oil additives, *Ceramics International* 40: 7143–7149.

[20] Chen S, Liu W (2006) Oleic acid capped PbS nanoparticles: Synthesis, characterization and tribological properties, *Materials Chemistry and Physics* 98: 183–189.
[21] Ye W, Cheng T, Ye Q, Guo X, Zhang Z, Dang H (2003) Preparation and tribological properties of tetrafluorobenzoic acid-modified TiO_2 nanoparticles as lubricant additives, *Materials Science and Engineering A* 359: 82–85.
[22] Jiao D, Zheng S, Wang Y, Guan R, Cao B (2011) The tribology properties of alumina/silica composite nanoparticles as lubricant additives, *Applied Surface Science* 257: 5720–5725.
[23] Ma S, Zheng S, Cao D, Guo H (2010) Anti-wear and friction performance of ZrO_2 nanoparticles as lubricant additive, *Particuology* 8: 468–472.
[24] Sui T, Song B, Zhang F, Yang Q (2015) Effect of particle size and ligand on the tribological properties of amino functionalized hairy silica nanoparticles as an additive to polyalphaolefin, *Journal of Nanomaterials*: 9.
[25] Demas NG, Timofeeva EV, Routbort JL, Fenske GR (2012) Tribological effects of BN and MoS_2 nanoparticles added to polyalphaolefin oil in piston skirt/cylinder liner tests, *Tribology Letters* 47: 91–102.
[26] Thottackkad MV, Perikinalil RK, Kumarapillai PN (2012) Experimental evaluation on the tribological properties of coconut oil by the addition of CuO nanoparticles, *International Journal of Precision Engineering and Manufacturing* 13: 111–116.
[27] Azman S, Zulkifli N, Masjuki H, Gulzar M, Zahid R (2016) Study of tribological properties of lubricating oil blend added with graphene nanoplatelets, *Journal of Materials Research* 31(13): 1932–1938.
[28] Alves SM, Barros BS, Trajano MF, Ribeiro KSB, Moura E (2013) Tribological behavior of vegetable oil based lubricants with nanoparticles of oxides in boundary lubrication conditions, *Tribology International* 65: 28–36.
[29] Singh H, Singh A, Singla Y, Chattopadhyay K (2020) Design & development of a low cost tribometer for nano particulate lubricants, *Materials Today: Proceedings* 28: 1487–1491.
[30] Singh H, Singh A, Singla Y, Chattopadhyay K (2020) Tribological study of nano fly ash as lubrication oil additive for AISI 4140 steel for automotive engine applications, *International Journal of Mechanical and Production Engineering Research and Development* 10(3): 5939–5946.
[31] Singh H, Singh AK, Singla YK, Chattopadhyay K, Saini A, Singh K (2022) Interpretation of the wear characteristics of AISI 4140 under nano-fly ash based engine lubricant, *Materials Today: Proceedings* 50: 1683–1689.
[32] Singh H, Singla Y, Singh A, Chattopadhayay K (2020) Effect of nanofly ash as lubricant additive on the tribological properties of SAE 10W-30 oil: A novel finding. *Transactions of the Indian Institute of Metals* 73(9): 2371–2375.

28 Machine Learning Forecasting Model for Recycled Aggregate Concrete's Strength Assessment

Amruthamol N. A., Kanish Kapoor

28.1 INTRODUCTION

There are several tests performed to assess concrete's effectiveness, and yet CS is often considered as the much crucial. Concrete's CS is linked to diverse mechanical and durability properties, in either a direct or indirect manner (Singh et al., 2019). Another vital consideration when constructing structural concrete is FS. The water-to-cement ratio (w/c), replacement ratio, moisture content, mechanical properties, and physical characteristics of the RA all have an impact on the strength of RAC. (Bui et al., 2017; Saha & Rajasekaran, 2016; Xiao, Li, & Poon, 2012). The strength of the RAC was significantly impacted by the scrap concrete's age used to create the RA. The parent concrete's strength, from which the RA are generated, also has an influence on the RAC's strength. As a matter of fact, a number of factors affect the strength of RAC, making it difficult to investigate their joint effect through field studies. Making use of computational tools allows researchers to investigate the combined impact of these variables on RAC strength (Awoyera, 2016; Cao et al., 2018). Machine learning is a great approach for predicting the properties of RAC since it is faster and cost efficient.

In this chapter, ANN, which is one of the best algorithms in machine learning is used to predict RAC flexural and compressive strength. Correlation coefficients (R) were utilized to assess the efficiency of the ANN technology. The ANN approach was chosen since the literature indicated that it outperformed alternative machine learning methods (A. Ahmad et al., 2021; Chou et al., 2014; Farooq et al., 2021). It is challenging to investigate the combined influence of many factors on RAC strength utilizing an experimental method. Machine learning algorithms can more easily assess the combined impact of many factors. The goal of this study is to find the efficiency of suggested machine learning approach for estimating RAC compressive and flexural strength depending on the outcomes forecast and the impacts of varying aspects on RAC strength.

 DOI: 10.1201/9781003367154-28

28.2 LITERATURE REVIEW

W. Ahmad et al. (2021) looked at how the RA's physio-mechanical characteristics used and the resultant matrix's microstructure affected the FS and CS of RAC. Experimental data by Kou et al. (2007) and Xiao, Li, Fan, et al. (2012) suggests that while w/c is maintained constant, the RA replacement rate significantly affects the strength of RAC. Butler et al. (2011) and Deshpande et al. (2011) have found out that the CS of RAC could be lowered by as much as 30% when natural aggregate is supplanted with RA. Etxeberria et al. (2007) and Rahal (2007) found that using 100% RA reduced CS by 12 to 25%. A. Ahmad et al. (2021), Chou et al. (2014), Farooq et al. (2021) found out that ANN outperformed alternative machine learning methods for finding out the FS and CS of RAC.

28.3 METHODOLOGY

28.3.1 Data Extraction

Dataset was made from different literatures (Dutta et al., 2018; Ilyas et al., 2021; Olalusi & Awoyera, 2021). In this research, machine learning approaches were utilized to estimate CS using 638 data samples and FS using 139 sets of data (Yuan et al., 2022). The range of input and output parameters for the prediction of FS and CS is represented in Table 28.1. As input parameters, the following 12 variables were considered, and the output variables were also chosen as compressive and flexural strength:

TABLE 28.1
Input and Output Parameters of RCA

Parameters	Min.	Max.
Effective water-to-cement ratio (w_{eff}/c)	0.19	0.87
Aggregate cement ratio (a/c)	1.5	6.5
Ratio of RCA replacement (RCA%)	0	100
Parent concrete strength (MPa)	0	100
Max. nominal aggregate size of RCA (mm)	8	32
Max. nominal aggregate size of NA (mm)	10	38
Bulk density of RCA (kg/m^3)	0	2,880
Bulk density of NA (kg/m^3)	0	2,970
Water absorption of RCA (%)	0	11.9
Water absorption of NA (%)	0	3
Los Angeles (LA) abrasion of RCA	0	42
LA abrasion of NA	0	32
Compressive strength of RCA (MPa)	13.4	108.5
Flexural strength of RCA (MPa)	1.9	10.2

With the use of MATLAB code (ML toolbox), the machine learning approach (ANN) was employed to achieve the goals of this study. Machine learning algorithms are commonly used to predict target output depending on the given variables. Temperature effects, material strength qualities, and material durability can all be predicted using this method. Data splitting for training, testing, and validating models (75% for training, 15% for testing, and 15% for validating) and picking the best model based on R-value are some of the ways for finding optimum parameters of the model. R-value is a measure of machine learning algorithms' performance and validity. The variance of a model's output variable is determined using the R-value. To put it another way, it quantifies how well the model fits the data.

28.3.2 ANN

NN are non-linear parallel processing computational models (Gupta, 2013). A programming tool similar to human behaviour is referred to as an ANN. The human nervous system is composed of a massive parallel network. The neurons are just active for a few seconds at a time. By using an appropriate teaching approach and data, ANN discovers correlations between input and output data. It is used in a variety of situations (Gupta, 2013; U. Khan et al., 2013).

There seem to be three different layers: input, hidden, and output in the preceding simple example of ANN. Neurons from the hidden units are mixed with input data. The output layer, on the other hand, contains the target data that the hidden layer needs (U. Khan et al., 2013). Overall learning operation is performed inside the hidden layers. Weights are used to link neurons that are usually chosen randomly at first (BKA et al., 2021). In a training task, particular epochs are used to raise or drop weights in order to produce a perfect NN which can predict accurately. A typical architecture of ANN is shown in Figure 28.1.

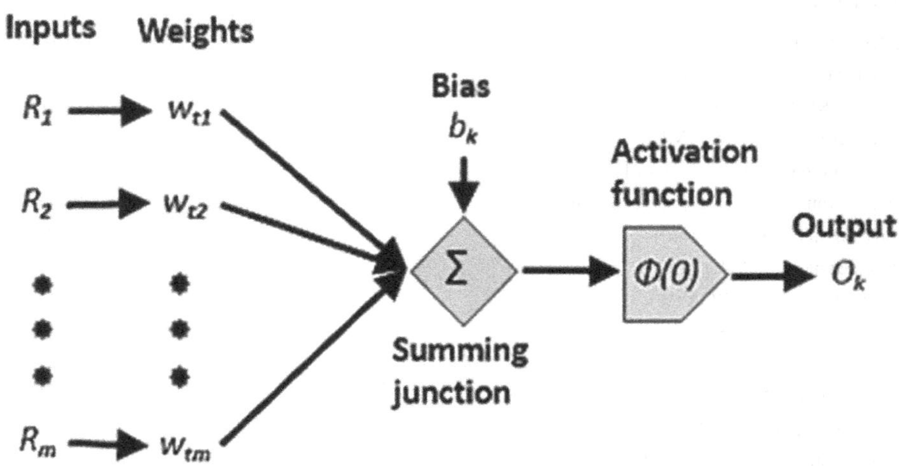

FIGURE 28.1 Depiction of an ANN structure.

28.4 RESULTS AND DISCUSSIONS

28.4.1 Compressive Strength

Figures 28.2, 28.3, 28.4, and 28.5 demonstrate the ANN model's outcomes for RAC compressive strength. The ANN method produced accurate results with a smaller difference contrasting experimental with forecasted values. The ANN model can reasonably estimate the RAC's compressive strength with 0.935 R-value. Figures 28.3 and 28.4 show the ANN validation performance and gradient, respectively. Figure 28.5 demonstrates the connection between the experimental results and the predicted outcomes. These differences show that the anticipated results of the ANN differed little from the experimental findings. As a result, the ANN predicts the compressive strength of RAC pretty accurately.

The ANN's gradient used in this research work is single-layer feedforward network with log-sig function in the hidden layer and the tan-sig function in the output layer. The network architecture is 12-60-1. The aspects for the ANN model are given in Table 28.2. The generality of the dataset will help us evaluate concrete's reliability. The test has terminated at 26th epoch on a slope 73.6435, and Figure 28.3 represents the ANN training state. The error is repeated starting at 27th epoch, as there are six iterations of the errors, suggesting the data being overfitted. Consequently, the weights of epoch 26 are utilized as both the beginning point and the final weights.

The validation performance with respect to the MSE of the network is depicted in Figure 28.3. The training, validation, and test were all depicted on the graph. The best validation performance after six error iterations is 81.9786 at epoch 26, and the program stops at the 26th epoch, as displayed on the x-axis of the plot.

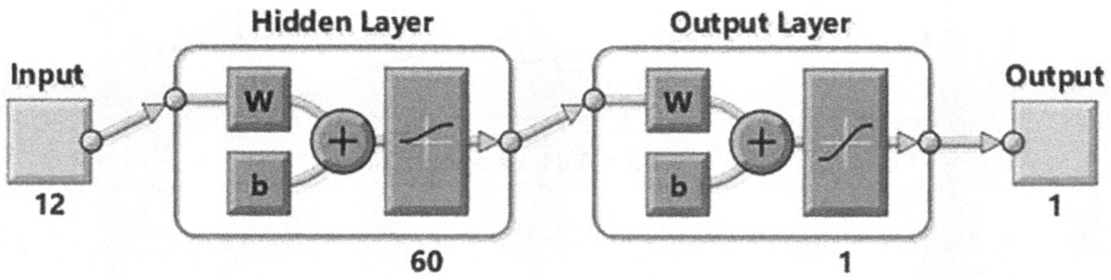

FIGURE 28.2 ANN architecture of compressive strength prediction model.

TABLE 28.2
Aspects of Neural Network

No. of inputs	12
Training algorithm	Levenberg-Marquadt
No. of hidden layers (HL)	1
No. of neurons in HL	60
Activation function of HL	Log-sig

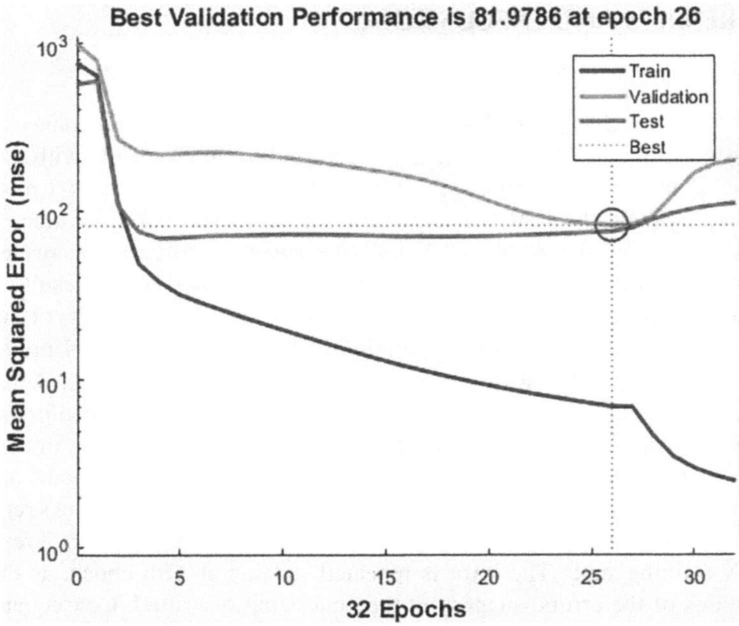

FIGURE 28.3 Validation performance.

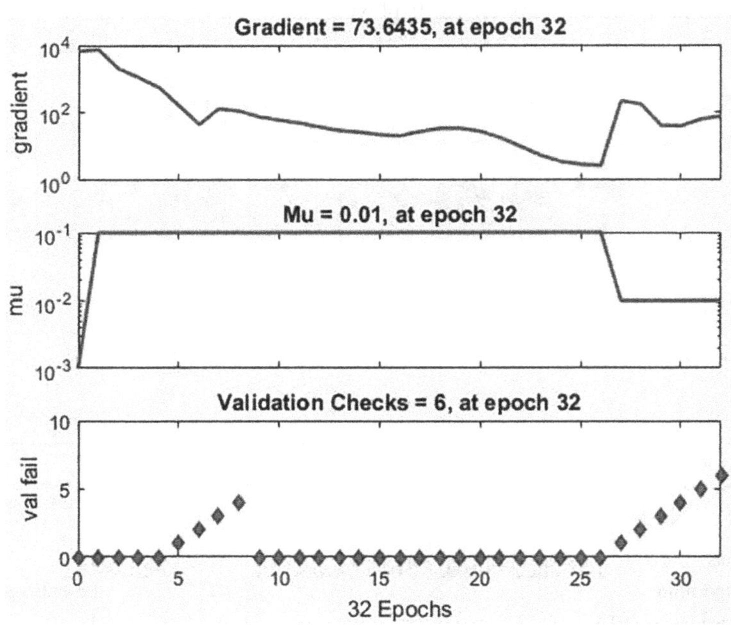

FIGURE 28.4 Training state.

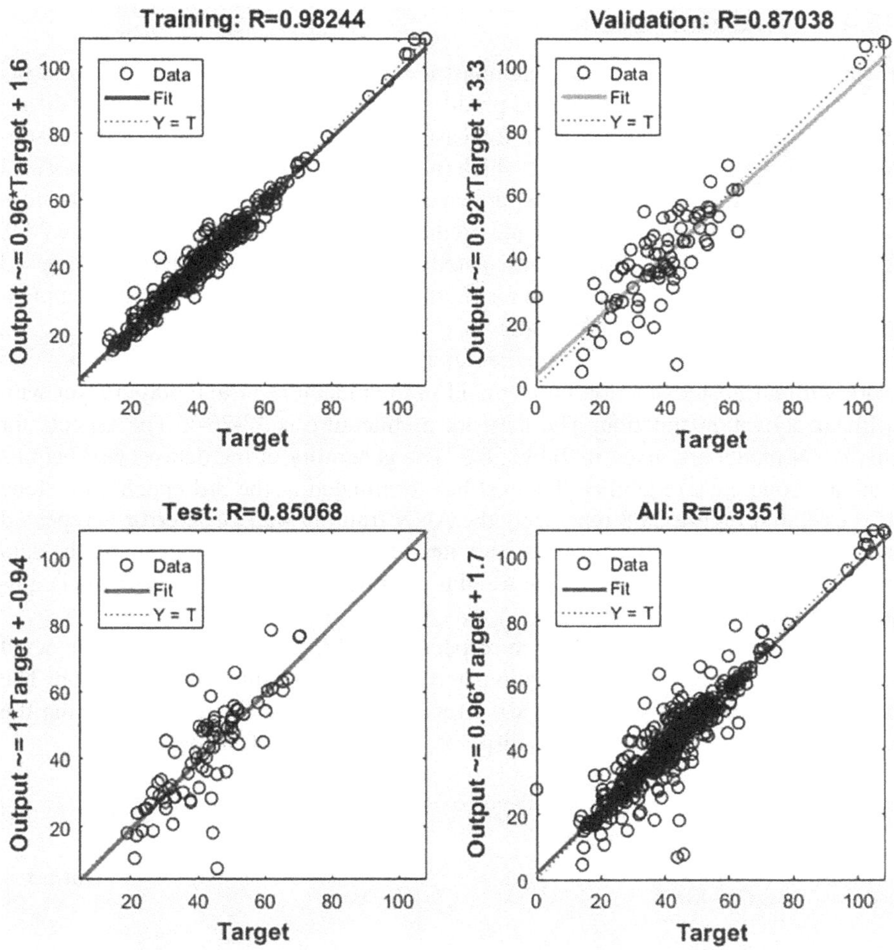

FIGURE 28.5 Plot of regression for training, validation, and test.

Figure 28.5 shows a neural network training regression that matches the test dataset on a slope of 0.935. It also demonstrates the correlation coefficients for training, validation, and test processes, which is the degree of similarity between the target and outcome. R-values serve as primary measures of how well a dataset resembles the best fit line. The "Measured" values are represented by the "Target" in the regression plots, whereas the "Predicted" values are represented by the "Output." Both during the phases of training and validation, the R-values demonstrated the satisfactory accuracy rate of model in the regression plot.

28.4.2 Flexural Strength

Figures 28.6, 28.7, 28.8, and 28.9 demonstrate the ANN model's outcomes for RAC flexural strength. The ANN method produced accurate results with a smaller difference contrasting experimental with forecasted values. The ANN model can reasonably estimate the R-value of 0.952, which indicates the RAC's strength. Figures 28.7 and 28.8 show the ANN validation performance and gradient, respectively. The connection between the experimental and predicted results are indicated in Figure 28.9. These differences show that the anticipated results of the ANN differed little based on the results of experiments. As a result, the ANN approach predicts the compressive strength of RAC pretty accurately.

The ANN's gradient used in this research work is a single-layer feedforward network with a transfer function tan-sigmoid in the hidden layer and output layer with a linear activation function. The network architecture is 12-26-1. The aspects for this ANN model are given in Table 28.3. The generality of the dataset will help us evaluate concrete's reliability. The test has terminated at the 3rd epoch on a slope 1.73 e^{-08}, and Figure 28.8 represents the ANN training state. The error is repeated starting at the 4th epoch, as there are six iterations of the errors, suggesting the data being overfitted. Consequently, the weights of epoch 3 is utilized as both the beginning point and the final weights.

The validation performance with respect to the MSE of the network is depicted in Figure 28.7. The training, validation, and test were all depicted on the graph. The best validation performance after six error iterations is 0.5733 at epoch 3, and the program stops at the 3rd epoch, as displayed on the x-axis of the plot.

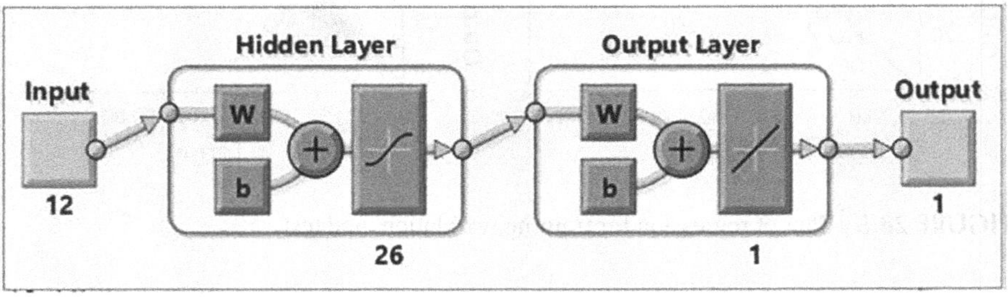

FIGURE 28.6 ANN architecture of flexural strength prediction model.

TABLE 28.3
Aspects of Neural Network

No. of inputs	12
Training algorithm	Levenberg-Marquadt
No. of hidden layers (HL)	1
No. of neurons in HL	26
Activation function of HL	Tan-sigmoid

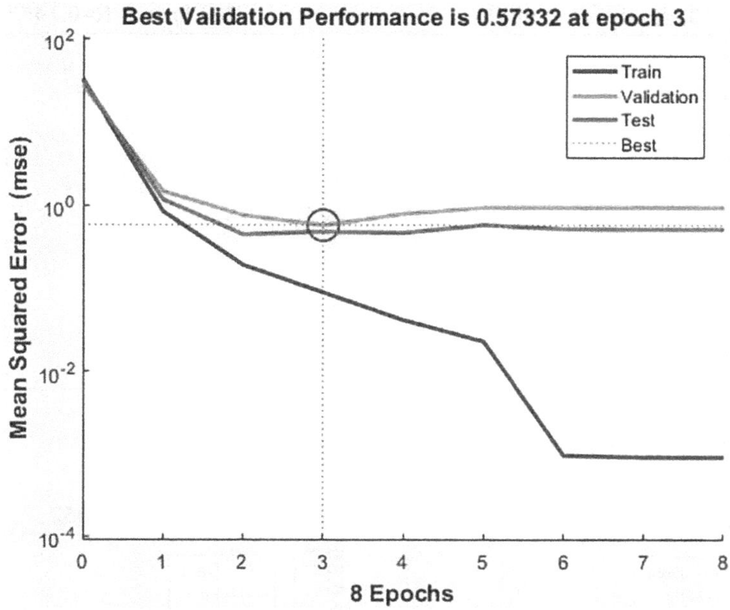

FIGURE 28.7 Validation performance.

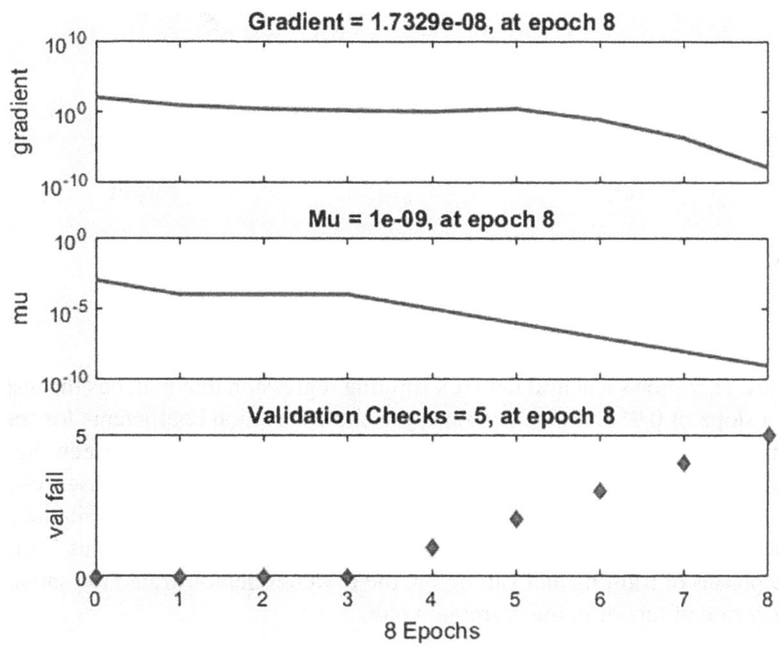

FIGURE 28.8 Training state.

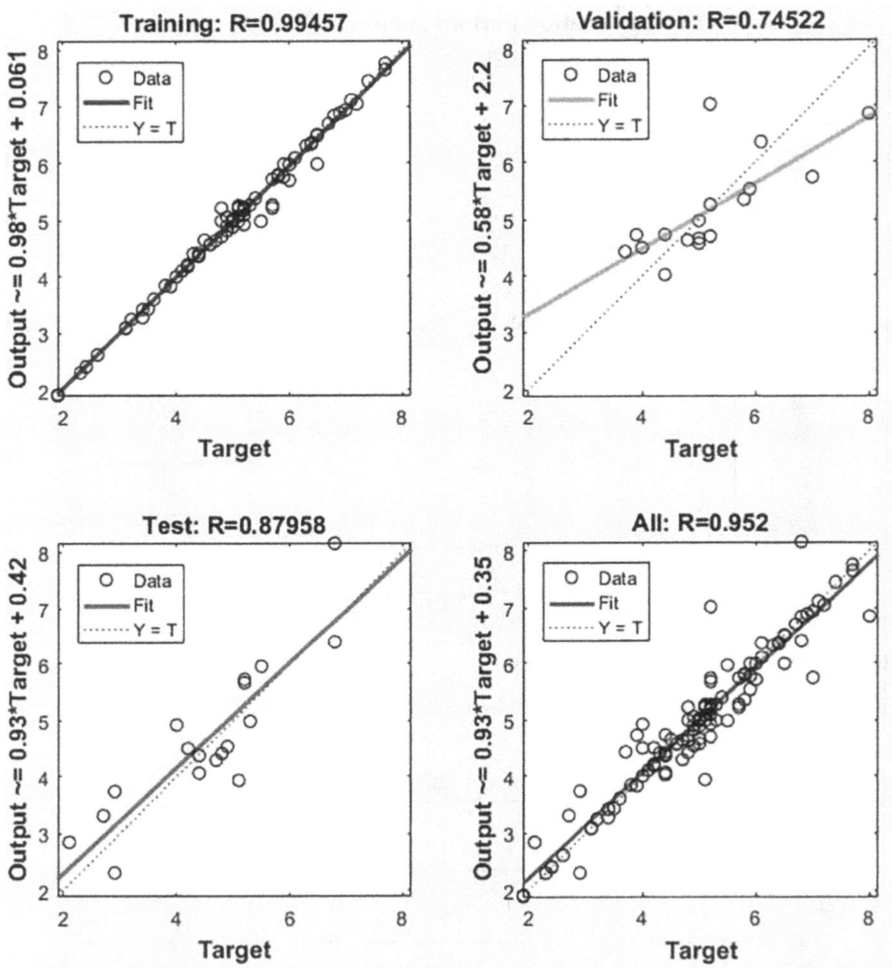

FIGURE 28.9 Plot of regression for training, validation, and test.

Figure 28.9 shows a neural network training regression that matches the test dataset on a slope of 0.952. It also demonstrates the correlation coefficients for training, validation, and test processes, which is the degree of similarity between the target and outcome. R-values serve as primary measures of how well a dataset resembles the best fit line. The "Measured" values are represented by the "Target" in the regression plots, whereas he "Predicted" values are represented by the "Output." Both during the phases of training and validation, the R-values demonstrated the satisfactory accuracy rate of model in the regression plot.

28.5 CONCLUSION

The purpose of this research was to contribute to the presently held information on the use of new technologies for assessing the strength of RAC. By making it easier to develop quick and economical material characteristic forecasting methodologies, this sort of study will be helpful to the construction industry. Further, by using these methods for promoting ecologically responsible construction, the use and adoption of RAC in the construction industry might be ramped up. As a result of urbanization and industrialization, real infrastructure renovation is essential, leading to huge levels of construction and demolition debris. Consequently, valuable places are being turned into rubbish dumps, land prices are rising, trash dumping expenses are rising, and landfill space becomes extremely scarce. Consequently, waste minimization has risen to prominence in developing countries and has evolved into a worldwide issue requires time consuming solutions. Additionally, the mining and processing of natural aggregates for use in concrete consumes a tremendous amount of energy and produces a significant quantity of CO_2 (Limbachiya et al., 2012). Consequently, adding RAC to concrete manufacturing could lead to decreased energy usage, cost savings, resource preservation, and a considerable drop in construction and demolition waste.

This study demonstrates how machine learning techniques may be utilized to foresee RAC flexural and compressive strength. To establish the best effective prediction, the study used the machine learning technique ANN. Unless there are few incorrect values in the model, it is more accurate. It might be challenging to choose and recommend the optimal machine learning model for foreseeing outputs across a number of industries since the validity of the model significantly depends on the input components and the dimension of the dataset utilized.

REFERENCES

Ahmad, A., Farooq, F., Niewiadomski, P., Ostrowski, K., Akbar, A., Aslam, F., & Alyousef, R. (2021). Prediction of compressive strength of fly ash based concrete using individual and ensemble algorithm. *Materials*, *14*(4), 1–21. https://doi.org/10.3390/ma14040794

Ahmad, W., Ahmad, A., Ostrowski, K. A., Aslam, F., & Joyklad, P. (2021). A scientometric review of waste material utilization in concrete for sustainable construction. *Case Studies in Construction Materials*, *15*, e00683. https://doi.org/10.1016/j.cscm.2021.e00683

Awoyera, P. O. (2016). Nonlinear finite element analysis of steel fibre-reinforced concrete beam under static loading. *Journal of Engineering Science and Technology*, *11*(12), 1669–1677.

BKA, M. A. R., Ngamkhanong, C., Wu, Y., & Kaewunruen, S. (2021). Recycled aggregates concrete compressive strength prediction using Artificial Neural Networks (ANNs). *Infrastructures*, *6*(2), 17. https://doi.org/10.3390/infrastructures6020017

Bui, N. K., Satomi, T., & Takahashi, H. (2017). Improvement of mechanical properties of recycled aggregate concrete basing on a new combination method between recycled aggregate and natural aggregate. *Construction and Building Materials*, *148*, 376–385. https://doi.org/10.1016/j.conbuildmat.2017.05.084

Butler, L., West, J. S., & Tighe, S. L. (2011). The effect of recycled concrete aggregate properties on the bond strength between RCA concrete and steel reinforcement. *Cement and Concrete Research, 41*(10), 1037–1049. https://doi.org/10.1016/j.cemconres.2011.06.004

Cao, M., Mao, Y., Khan, M., Si, W., & Shen, S. (2018). Different testing methods for assessing the synthetic fiber distribution in cement-based composites. *Construction and Building Materials, 184*, 128–142. https://doi.org/10.1016/j.conbuildmat.2018.06.207

Chou, J. S., Tsai, C. F., Pham, A. D., & Lu, Y. H. (2014). Machine learning in concrete strength simulations: Multi-nation data analytics. *Construction and Building Materials, 73*, 771–780. https://doi.org/10.1016/j.conbuildmat.2014.09.054

Deshpande, N., Kulkarni, S. S., & Patil, N. (2011). Effectiveness of using coarse recycled concrete aggregate in concrete. *International Journal of Earth Sciences and Engineering, 4*(SPL), 20410439.

Dutta, S., Samui, P., & Kim, D. (2018). Comparison of machine learning techniques to predict compressive strength of concrete. *Computers and Concrete, 21*(4), 463–470. https://doi.org/10.12989/cac.2018.21.4.463

Etxeberria, M., Vázquez, E., Marí, A., & Barra, M. (2007). Influence of amount of recycled coarse aggregates and production process on properties of recycled aggregate concrete. *Cement and Concrete Research, 37*(5), 735–742. https://doi.org/10.1016/j.cemconres.2007.02.002

Farooq, F., Ahmed, W., Akbar, A., Aslam, F., & Alyousef, R. (2021). Predictive modeling for sustainable high-performance concrete from industrial wastes: A comparison and optimization of models using ensemble learners. *Journal of Cleaner Production, 292*, 126032. https://doi.org/10.1016/j.jclepro.2021.126032

Gupta, S. (2013). Concrete mix design using artificial neural network. *Journal on Today's Ideas-Tomorrow's Technologies, 1*(1), 29–43. https://doi.org/10.15415/jotitt.2013.11003

Ilyas, I., Zafar, A., Javed, M. F., Farooq, F., Aslam, F., Musarat, M. A., & Vatin, N. I. (2021). Forecasting strength of CFRP confined concrete using multi expression programming. *Materials, 14*(23), 7134. https://doi.org/10.3390/ma14237134

Khan, S. U., Ayub, T., & Rafeeqi, F. A. S. (2013). Prediction of compressive strength of plain concrete confined with ferrocement using Artificial Neural Network (ANN) and comparison with existing mathematical models. *American Journal of Civil Engineering and Architecture, 1*(1), 7–14. https://doi.org/10.12691/ajcea-1-1-2

Kou, S. C., Poon, C. S., & Chan, D. (2007). Influence of fly ash as cement replacement on the properties of recycled aggregate concrete. *Journal of Materials in Civil Engineering, 19*(9), 709–717. https://doi.org/10.1061/(asce)0899-1561(2007)19:9(709)

Limbachiya, M., Meddah, M. S., & Ouchagour, Y. (2012). Use of recycled concrete aggregate in fly-ash concrete. *Construction and Building Materials, 27*(1), 439–449. https://doi.org/10.1016/j.conbuildmat.2011.07.023

Olalusi, O. B., & Awoyera, P. O. (2021). Shear capacity prediction of slender reinforced concrete structures with steel fibers using machine learning. *Engineering Structures, 227*, 111470. https://doi.org/10.1016/j.engstruct.2020.111470

Rahal, K. (2007). Mechanical properties of concrete with recycled coarse aggregate. *Building and Environment, 42*(1), 407–415. https://doi.org/10.1016/j.buildenv.2005.07.033

Saha, S., & Rajasekaran, C. (2016). Mechanical properties of recycled aggregate concrete produced with Portland Pozzolana Cement. *Advances in Concrete Construction, 4*(1), 27–35. https://doi.org/10.12989/acc.2016.4.1.027

Singh, N., Kumar, P., & Goyal, P. (2019). Reviewing the behaviour of high volume fly ash based self compacting concrete. *Journal of Building Engineering, 26*, 100882. https://doi.org/10.1016/j.jobe.2019.100882

Xiao, J., Li, W., Fan, Y., & Huang, X. (2012). An overview of study on recycled aggregate concrete in China (1996–2011). *Construction and Building Materials*, *31*, 364–383. https://doi.org/10.1016/j.conbuildmat.2011.12.074

Xiao, J., Li, W., & Poon, C. (2012). Recent studies on mechanical properties of recycled aggregate concrete in China: A review. *Science China Technological Sciences*, *55*(6), 1463–1480. https://doi.org/10.1007/s11431-012-4786-9

Yuan, X., Tian, Y., Ahmad, W., Ahmad, A., Usanova, K. I., Mohamed, A. M., & Khallaf, R. (2022). Machine learning prediction models to evaluate the strength of recycled aggregate concrete. *Materials*, *15*(8). https://doi.org/10.3390/ma15082823

29 Examining the Benefits of Lean Manufacturing

A Comprehensive Review

Ravish Arora, Shaman Gupta, Neeraj Sharma and Vijay Kumar Sharma

29.1 INTRODUCTION

The lean manufacturing (LM) process is an important manufacturing process and management technique that helps to mitigate waste and helps to be efficient. This literature review will discuss LM's general aspect regarding its philosophy, theories, importance, barriers to implementation, and competitive advantage it provides. Moreover, this review will discuss various steps of implementing LM processes with various tools used in this process. This aspect will help to understand the importance of the LM processes importance. Moreover, this aspect will help understand its implementation in various industries and leverage this aspect. It helps to produce and manufacture different products, reducing waste and irrelevant additives. This aspect creates value in these products and helps to attract consumers.

29.2 METHODOLOGY

This writing is based on a systematic literature review of LM. Information is collected through different kinds of articles and journals available on the internet. Thousands of journals and articles have been found worldwide regarding LM as the starting point of data collection. Those articles and journals have identified different kinds of relevant and effective information.

The initial process of finding information appears with a list of more than 10,000 research papers related to this particular topic. After that, the list is reduced its length slowly by using different types of important words about the main topic. Using different filters in the search engine, research papers have been sorted out, and relevant research papers have been selected for this literature review. This research list is reduced to about ten research papers on lean philosophy, 20 research papers on lean surveys, and 20 research papers on a case study. Figure 29.1 depicts the methodology flowchart, whereas Figure 29.2 represents the conceptual framework of LM. Figure 29.2 represents LM publications.

 DOI: 10.1201/9781003367154-29

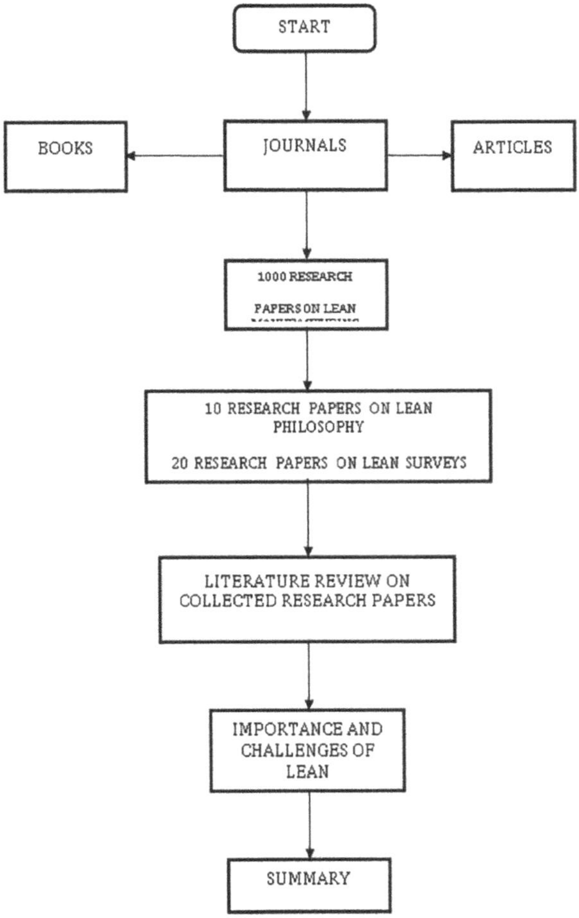

FIGURE 29.1 Methodology flowchart.

29.2.1 Conceptual Framework

Table 29.1 represents classification of literature.

29.3 LITERATURE REVIEW

29.3.1 Lean Philosophy

The focus of LM is on improving processes and eliminating waste. This particular technique can help reduce costs and deliver customers' demanded products and services they want to pay for [1]. The main concepts and main principles of lean come under lean philosophy. Lean refers to one process that includes five steps: defining customer value as the first stem and one of the most significant steps for LM [2].

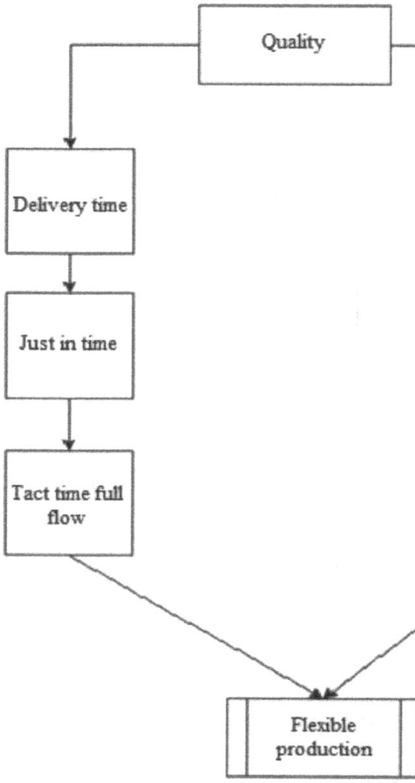

FIGURE 29.2 Conceptual framework of LM.

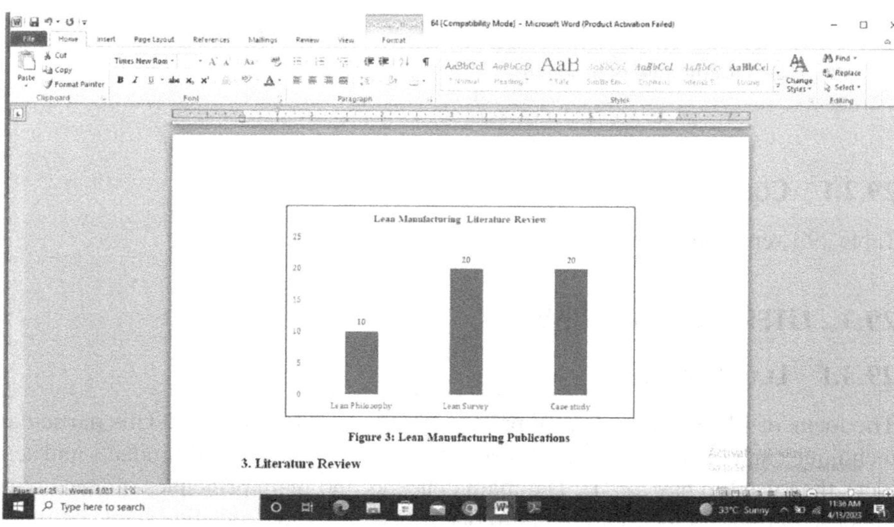

FIGURE 29.3 Lean manufacturing publications.

TABLE 29.1
Classification of Literature

Focus Area	Chronological List of Publications
Lean Philosophy	Malihi & Shee [1]
	Price, Pepper, & Stewart [2]
	Wang *et al.* [3]
	Rodgers *et al.* [4]
	Chauhan & Chauhan [5]
	Chiarini, Claudio, & Vittorio [6]
	Isack *et al.* [7]
	Badeeb, Abdulaal, & Bafail [8]
	Chetthamrongchai & Jermsittiparsert [9]
	De Vin & Jacobsson [10]
	1.
Lean Surveys	Orlowski, Shanaka, & Mendis [11]
	Böhme et al. [12]
	Ferry & Gengatharen [13]
	Bailey et al. [14]
	Prior et al. [15]
	Shaw & Luisa [16]
	Ferreira et al. [17]
	Wang [18]
	Xiao et al. [19]
	Andrew [20]
	Youness et al. [21]
	Sablok et al. [22]
	Nakandala et al. [23]
	Haifer et al. [24]
	Pap et al. [25]
	Reiner, Danese, & Gold [26]
	Nik Nazli & Dewan [27]
	Gilham, Hall, & Woods [28]
	Eva et al. [29]
	Hasan et al. [30]
Case Studies	Singh, Singh, & Singh [34]
	Uhrin, Bruque-Cámara, & Movano-Fuentes [35]
	Perera & Perera [36]
	Randhawa & Ahuja [37]
	Psomas, Antony, & Bouranta [38]
	Ghobakhloo & Azar [39]
	Leksic, Stefanie, & Veza [40]
	Rishi *et al.* [41]
	Goshime, Kitaw, & Jilcha [42]
	Zhu & Lin [43]
	Knol et al. [44]

After that, the second one is defining the value stream. The third is making it flow, the fourth is establishing pull, and the last one of the five-step of lean is striving for excellence [2].

One of the major parts of this lean philosophy is that it focuses on customer demand and manufacturing high-quality products and services by applying the most effective and financial manner. LM will be characterized by important factors and areas, such as alternative approaches, cultural diversities, and total quality management [3]. CSR (corporate social responsibility) and sustainable development are two major factors of LM [4].

LM processes help to eliminate waste and unnecessary processes and products from any manufacturing process [5, 6]. Thus, this aspect helps create a more valuable and efficient product in their manufacturing process. Moreover, these values are more towards consumer preferences, which help these products, attract consumers, and give them a competitive advantage. This aspect can be demonstrated using *lean principle theory*, which has a similar approach [7]. This theory is divided into five aspects, and it helps to create efficiency in the manufacturing process, as the primary process is to identify the value of the product regarding the consumers and then identify the essential aspect that helps to create value and mitigate the additional aspect, which is not important and irrelevant [8, 9]. After this, the product proceeds to value-creating steps more tightly; the approach will help to attract more consumers [10].

29.3.2 Lean Surveys

This approach needs to be done so that when the consumers need some improvement, they will get that in time. This aspect will help to align the product with consumers. Moreover, the product must strive for perfection, which will help them be efficient and have a competitive advantage in its market. This theory is essential for any manufacturing process as this implication can help most of the manufacturing processes to be efficient in their production standards and help them to save some expenses. Moreover, it helps to create products that can create a competitive advantage for any company in the market. However, as seen, this process is a continuous process, which helps improve products progressively aligned to consumer needs.

The LM process is the core of manufacturing operations in today's business world and this international market. On the other hand, this process has been identified as one of the most important things for the other individual sectors in this global competitive market. Therefore, it has been recognized that in this competitive market, the LM process has different benefits or advantages, such as reducing waste and enhancing manufacturing efficiency [11]. This process can help increase the productivity level, which can help increase the profit within the business [12].

One of the most significant aspects of this process is that it can help enhance the quality of the product by improving the level of efficiency among the employees and resources for innovating and maintaining quality control that would have previously been wasted [13]. Improving the lead times is one of the best parts of the LM process in a business. A manufacturing business can get help by adopting this process to give a better performance and give tough competition to other businesses in an individual industry. It can help to maintain lead times in the fluctuation of the market of a business. This process can help maintain business sustainability by reducing waste and increasing a better adaptability process that can help make a business perfect and sustainable for its future [14].

LM increases productivity within a business; however, it can help boost employee satisfaction. Previously, employees thought about the unnecessary packaging and productivity faults; however, after applying this process within a business, employees have no pressure regarding these, and they are now focusing on their performance [15]. One of the most significant benefits of the LM process is that it helps increase the profit level of a business by enhancing the productivity and quality of the products and boosting employee efficiency and satisfaction [16]. LM, also known as lean production, can help minimize the risk of waste of labour and material while maintaining the production level [17]. The result is that it can improve a business by minimizing environmental and social risks. This is one of the main parts of CSR (corporate social responsibility).

One of the most significant aspects of this LM process is improving customer service quality within a business. This is based on the first principle of lean philosophy: the identification of value. A company can be successful when its customers are satisfied with its services. This process can help a company to develop its customers' services by using this process within its organizational culture and value. This process helps a company maintain an easy workforce management system for its customers' understanding and reduces the chances of risk [18].

LM is one kind of philosophy and strategy that is developed to reduce the time of customers' requests for orders and delivery. This LM process sounds easy. However, its implementation is not so easy in SMEs (small- and medium-sized enterprises) [19]. LM's introduction reveals that it can make positive and effective changes to the work culture of a business. These changes sometimes become a barrier to implementing this process within a business. Therefore, one of the biggest challenges is to recognize the changes within the organizational culture [20].

Adopting and responding will become difficult for the employees when they observe that a change has occurred within the organizational workforce. Therefore, the shortage of management timings is one of the most significant barriers to implementing LM processes within an SME [21]. The knowledge gap in understanding the potential benefits of the LM process within a business becomes a barrier [22]. Employee attitude or resistance to change is one of the major challenges to this process's implementation within a business. Employees are habituated to the traditional work culture and need time to acknowledge the importance of the LM process.

The insufficient workforce culture of a company has become a barrier to implementing lean within the business [23]. Furthermore, backsliding to the old, inefficient ways of working is another barrier to the LM process within a business [24]. Three main barriers to implementing LM processes within a business are a poor understanding of the concept of LM, the gap of commitment between senior and middle management, and the attitudes of the workers of a business. Besides that, the inability of the production schedule is one of the most significant barriers to carrying out lean practices. However, these barriers will be overcome by adopting accurate and appropriate communication and training. These can help to understand the potential benefits of this process for the stakeholders of a business [25].

Implementing LM tries to create a value flow at the pull of customers' demand and removes waste in processes. Waste comprises seven components: inventory, transport, waiting, motion, overproducing, overprocessing, and defects. These components have a direct impact on the performance of LM. If LM can be utilized properly, it can lead to different kinds of positive and effective improvements. Besides these advantages, the LM process has a lot of hidden advantages that play an important role in business success stories [26]. Though these benefits have no direct importance in the success of business stories regarding LM processes, these advantages have a very important indirect role in the success stories of a business that cannot be neglected.

Reducing the time for traceability is one of the most significant competitive advantages of the LM process in today's competitive market [27]. A business can be developed by applying the principles of LM processes and store management techniques [28]. One of the most effective competitive advantages of the LM process is improving the quality and safety of the products and services. Customers always demand quality and safe products that cannot harm them; however, LM can help mitigate this demand. In this way, a company can increase customer engagement [29]. This manufacturing technology process can help minimize the risk of mistakes that would be committed by a company's staff [30].

The major purpose of the LM process is to reduce the risk of mistakes in the organizational work culture and increase productivity and quality. Therefore, it has been identified that LM processes are beneficial for a business to give tough competition to the competitors and make effective business sustainability for its future perspectives. It has many competitive advantages that can help a business to reduce the risk of waste and increase business sustainability. It can change the work culture within an organization, which would be an effective result for its future.

29.3.2.1 Steps of Lean Manufacturing Implementation

29.3.2.1.1 Identification of Wastes

Organizations have various kinds of waste that arise from the manufacturing process of work. Therefore, organizations must find the waste to maintain a clean environment.

29.3.2.2 Different Types of Wastes in the Organization

Organizations have different types of waste, including manufacturing disposals. It is important to identify the types of waste and its main causes. LM has found different tools and methods to reduce these wastes.

29.3.2.3 Find the Solution to Root Causes

One of the most significant steps of lean implementation is identifying the solutions to reduce waste.

29.3.2.4 Test the Solution

Lean implementation will be done when tested for a business's better result. There are various ways that Lean principles can be implemented in various manufacturing processes; however, all of these processes follow five common aspects: finding the value, mitigating unnecessary aspects, providing updates, providing updates according to consumer's needs and striving to improve. These aspects are discussed in the following sections. However, a clear goal must be set with a lean mind-set [31].

29.3.2.4.1 Finding Value

Any manufacturer must find the value they are willing to offer their consumers through their products. This aspect will help them create more desirable products for their consumers. Thus, this aspect helps in improving their sales. However, these manufacturers need to understand their consumers for this aspect and what they expect from their products. After they have found the necessary aspect, they need to evaluate the consumer group and manufacture their product for their target consumers [31].

29.3.2.4.2 Value Creation

Value is important for a product; however, mitigating various unnecessary factors for this aspect will help this product be more attractive to the consumer. Moreover, the manufacturer needs to be aware that this mitigation process only needs to follow factors that are not important to the product and do not hamper the product value or quality. This aspect is important to make the product lean and rich in value. Furthermore, this aspect will help the manufacturer save resources and time as an unnecessary product that is not used will affect the production time and expenses. Therefore, this implementation will help the manufacturing process. However, for implementation, a proper mapping process needs to be done to understand whether these removals impact the product [32].

29.3.2.4.3 Update

The product needs to be frequently updated, and these steps need to add additional value to the product. This aspect will help attract more consumers. However, this process is difficult to implement as it depends on various factors, such as consumer needs, trends, competitors, and innovation. These aspects are ever-changing aspects of the market and create complexities regarding updates. Moreover, consumer needs can be different from each other; trends can change in minimal time gaps. Therefore, proper research needs to be done on this aspect, and the updates need to target the long-term, which will help them be relevant. This aspect will help updates be relevant when implemented regarding their consumer needs [33].

29.3.2.4.4 Consumer Pull

Manufacturers need to understand that they need to implement these valuable updates when consumers need them, which will help them mitigate issues regarding consumer experiences and irrelevancy. Moreover, this aspect will help these implementation processes align with consumer preferences, which will help shape the product according to consumers' needs. However, implementing updates without concerning consumers will harm this aspect as it would not create any product value. Furthermore, this aspect will attract consumers as they expect their needs to be fulfilled when they need them. Therefore, this implementation helps create consumer pull towards the product.

29.3.2.4.5 Improvement

This process is self-explanatory, as implementing this process will help the quality of the product continuously. This aspect can be implemented by investing time and resources in the product. Moreover, the saved resources and time from the value creation process can be invested in this aspect, which will help the product to have a greater quality and help it have a competitive advantage in its market.

29.3.3 Case Study

There are various lean tools used to implement lean processes; however, some of them will be discussed here, which are 5S, Kanban, Kaizen, and PDCA (plan, do, check, and act). Table 29.2 represents the function of various lean tools [34, 35].

5S TOOL

5S is an important process that can be implemented in LM. This process comprises five factors: sort, set in order, shine, standardize, and sustain [36]. In the sorting process, things that are not needed are eliminated. Setting in order helps to organize the remaining factors that are valuable in the process. In the shining process, work areas are cleaned, and maintenance and inspection are done. The standardization process helps to create standards for the work, which helps to maintain the efficiency of the work [37]. The sustaining process helps to maintain the standardization process to sustain the work environment.

KANBAN

This process helps LM control the flow of goods inside the factory, from supplier to delivery to consumers. Moreover, efficiency is seen in every process; as in the production phase, it can be seen that the initial process is controlled by after processes, and they only supply what the next process requires. Moreover, products that have

TABLE 29.2
Functions of Various Lean Tools

Lean Tools	Functions
Kanban	This process helps LM control the flow of goods inside the factory, from supplier to delivery to consumers.
Kaizen	Kaizen is a tool that helps to improve manufacturing processes regularly. This tool follows three aspects: planning, implementation, and follow-up.
5S	The *sustain* process helps to maintain the standardization process to sustain the work environment.
PDCA cycle	These aspects help counter issues occurring in the production process and mitigate them in each cycle, which makes this process important as it accounts for each step. Proper planning is done initially to mitigate risks.

defects are not permitted in the manufacturing process [38]. Furthermore, all these processes follow Kanban guidelines. Therefore, this process helps the manufacturing process by implementing proper control of the manufacturing process. This aspect can be implemented in supply processes and product delivery processes [39].

KAIZEN

Kaizen is a tool that helps to improve manufacturing processes regularly. This tool follows three aspects: planning, implementation, and follow-up [40]. The planning process is important as this aspect helps prepare for risks and achieve goals properly. Then implementation is done. The follow-up process helps to sustain the improvement of the process regularly [41].

PDCA CYCLE

This process comprises four factors that work in a cycle in LM: *planning, doing, checking the results and acting on improvements* [42, 43]. These aspects help counter issues occurring in the production process and mitigate them in each cycle, which makes this process important as it accounts for each step. Proper planning is done initially to mitigate risks. This approach is primarily taken when a new manufacturing process is initiated, and there is a need to improve the design of products or services, a need to improve a product or implement new changes [44].

29.4 CONCLUSION

This literature review primarily focuses on the general idea of LM and implementing lean principles in this aspect. Moreover, this review discussed the philosophy of LM, then it proceeded to the general theory of lean management upon which all manufacturing processes can be efficient. However, these aspects were visible in all LM processes and followed the same conditions. Furthermore, this reviewed the importance and barriers of LM with the inclusion of its competitive advantage in the manufacturing-goods market. Moreover, the general implementation process was discussed and focused on primordial aspects. Nevertheless, some tools of LM are discussed in this review following its completion.

REFERENCES

1. Malihi, K., & Shee, H. (2017). Strategic vehicles import supply chain: A paradigm shift in Australian automotive industry. *Asian Academy of Management Journal*, *22*(1), 103–130. http://doi.org/10.21315/aamj2017.22.L5
2. Price, O. M., Pepper, M., & Stewart, M. (2018). Lean six sigma and the Australian business excellence framework. *International Journal of Lean Six Sigma*, *9*(2), 185–198. http://doi.org/10.1108/IJLSS-01-2017-0010
3. Wang, P., Wu, P., Wang, X., Chen, X., & Zhou, T. (2020). Developing optimal scaffolding erection through the integration of lean and work posture analysis. *Engineering, Construction and Architectural Management*, *27*(9), 2109–2133. http://doi.org/10.1108/ECAM-04-2019-0193

4. Rodgers, B. A., Antony, J., He, Z., Cudney, E. A., & Laux, C. (2019). A directed content analysis of viewpoints on the changing patterns of lean six sigma research. *TQM Journal*, *31*(4), 641–654. http://doi.org/10.1108/TQM-03-2019-0089
5. Chauhan, G., & Chauhan, V. (2019). A phase-wise approach to implement LM. *International Journal of Lean Six Sigma*, *10*(1), 106–122. http://doi.org/10.1108/IJLSS-09-2017-0110
6. Chiarini, A., Claudio, B., & Vittorio, M. (2018). Lean production, Toyota production system and kaizen philosophy. *TQM Journal*, *30*(4), 425–438. http://doi.org/10.1108/TQM-12-2017-0178
7. Isack, H. D., Mutingi, M., Kandjeke, H., Vashishth, A., & Chakraborty, A. (2018). Exploring the adoption of lean principles in medical laboratory industry. *International Journal of Lean Six Sigma*, *9*(1), 133–155. http://doi.org/10.1108/IJLSS-02-2017-0017
8. Badeeb, A. M., Abdulaal, R. M. S., & Bafail, A. O. (2017). An application of LM techniques in paint manufacturing company: A case study. *Journal of King Abdulaziz University*, *28*(2), 51–73. http://doi.org/10.4197/Eng.28-2.5
9. Chetthamrongchai, P., & Jermsittiparsert, K. (2019). Impact of LM practices on financial performance of pharmaceutical sector in Thailand. *Systematic Reviews in Pharmacy*, *10*(2), 208–217. http://doi.org/10.5530/srp.2019.2.29
10. De Vin, L. J., & Jacobsson, L. (2017). Karlstad lean factory: An instructional factory for game-based LM training. *Production & Manufacturing Research*, *5*(1). http://doi.org/10.1080/21693277.2017.1374886
11. Orlowski, K., Shanaka, K., & Mendis, P. (2018). Design and development of weatherproof seals for prefabricated construction: A methodological approach. *Buildings*, *8*(9). http://doi.org/10.3390/buildings8090117
12. Böhme, T., Escribano, A., Heffernan, E. E., & Beazley, S. (2018). Causes and mitigation for declining productivity in the Australian mid-rise residential construction sector. *Built Environment Project and Asset Management*, *8*(3), 253–266. http://doi.org/10.1108/BEPAM-10-2017-0097
13. Ferry, J., & Gengatharen, D. (2019). Australian food retail supply chain analysis. *Business Process Management Journal*, *26*(2), 271–287. http://doi.org/10.1108/BPMJ-03-2017-0065
14. Bailey, G., Huynh, L., Govenlock, L., Jordan, D., & Jenson, I. (2017). Low prevalence of Salmonella and Shiga toxin–producing Escherichia coli in lymph nodes of Australian beef cattle. *Journal of Food Protection*, *80*(12), 2105–2111. http://doi.org/10.4315/0362-028X.JFP-17-180
15. Prior, S. J., Mather, C., Miller, A., & Campbell, S. (2019). An academic perspective of participation in healthcare redesign. *Health Research Policy and Systems*, *17*, 1–6. http://doi.org/10.1186/s12961-019-0486-2
16. Shaw, N. E., & Luisa, H. H. (2018). Editorial. *International Journal of Productivity and Performance Management*, *67*(2), 222–223. http://doi.org/10.1108/IJPPM-12-2017-0332
17. Ferreira, L. A. F., dos Santos, I. L., dos Santos, Ana Carla de, Souza Gomes, & Reis, A. D. C. (2020). Discrete event simulation for problem solving in the context of an emergency department. *Independent Journal of Management & Production*, *11*(5), 1515–1531. http://doi.org/10.14807/ijmp.v11i5.1286
18. Wang, M. (2018). Impacts of supply chain uncertainty and risk on the logistics performance. *Asia Pacific Journal of Marketing and Logistics*, *30*(3), 689–704. http://doi.org/10.1108/APJML-04-2017-0065
19. Xiao, X., Skitmore, M., Li, H., & Xia, B. (2019). Mapping knowledge in the economic areas of green building using scientometric analysis. *Energies*, *12*(15), 3011. http://doi.org/10.3390/en12153011

20. Andrew, A. A. (2020). Revisiting the housing market dynamics and its fundamentals: New evidence from Cyprus. *Journal of Economic Studies, 48*(1), 200–216. http://doi.org/10.1108/JES-07-2018-0237
21. Youness, E., Amir, A., Manoochehr, N., & Arab, M. A. (2018). Holistic performance management of virtual teams in third-party logistics environments. *Team Performance Management, 24*(3), 186–202. http://doi.org/10.1108/TPM-05-2017-0020
22. Sablok, G., Stanton, P., Bartram, T., Burgess, J., & Boyle, B. (2017). Human resource development practices, managers and multinational enterprises in Australia. *Education & Training, 59*(5), 483–501. http://doi.org/10.1108/ET-02-2016-0023
23. Nakandala, D., Samaranayake, P., Lau, H., & Ramanathan, K. (2017). Modelling information flow and sharing matrix for fresh food supply chains. *Business Process Management Journal, 23*(1), 108–129. http://doi.org/10.1108/BPMJ-09-2015-0130
24. Haifer, C., Kelly, C. R., Paramsothy, S., Andresen, D., Papanicolas, L. E., McKew, G. L., & Leong, R. W. (2020). Australian consensus statements for the regulation, production and use of faecalmicrobiota transplantation in clinical practice. *Gut, 69*(5), 801. http://doi.org/10.1136/gutjnl-2019-320260
25. Pap, R., Shabella, L., Morrison, A. J., Simpson, P. M., & Williams, D. M. (2018). Teaching improvement science to paramedicine students: Protocol for a systematic scoping review. *Systematic Reviews, 7.* http://doi.org/10.1186/s13643-018-0910-7
26. Reiner, G., Danese, P., & Gold, S. (2017). The 22nd international EurOMA conference. *International Journal of Operations & Production Management, 37*(11), 1582–1584. http://doi.org/10.1108/IJOPM-09-2017-0574
27. Nik Nazli, N. A., & Dewan, M. H. (2019). Exploring the meaning of climate change discourses: An impression management exercise? *Accounting Research Journal, 32*(2), 113–128. http://doi.org/10.1108/ARJ-07-2016-0085
28. Gilham, B., Hall, R., & Woods, J. L. (2018). Vegetables and legumes in new Australasian food launches: How are they being used and are they a healthy choice? *Nutrition Journal, 17,* 1–9. http://doi.org/10.1186/s12937-018-0414-2
29. Eva, N., Sendjaya, S., Prajogo, D., Cavanagh, A., & Robin, M. (2018). Creating strategic fit: Aligning servant leadership with organisational structure and strategy. *Personnel Review, 47*(1), 166–186. http://doi.org/10.1108/PR-03-2016-0064
30. Hasan, A., Baroudi, B., Elmualim, A., & Rameezdeen, R. (2018). Factors affecting construction productivity: A 30 year systematic review. *Engineering, Construction and Architectural Management, 25*(7), 916–937. http://doi.org/10.1108/ECAM-02-2017-0035
31. Centobelli, P., Cerchione, R., Esposito, E., & Passaro, R. (2021). Determinants of the transition towards circular economy in SMEs: A sustainable supply chain management perspective. *International Journal of Production Economics, 242*, 108297. https://doi.org/10.1016/j.ijpe.2021.108297
32. Rani, S., Koundal, D., Kavita, F., Ijaz, M. F., Elhoseny, M., & Alghamdi, M. I. (2021). An optimized framework for WSN routing in the context of industry 4.0. *Sensors, 21*(19), 6474. https://doi.org/10.3390/s21196474
33. Hoque, N., Biswas, W., Mazhar, I., & Howard, I. (2020). Life cycle sustainability assessment of alternative energy sources for the western Australian transport sector. *Sustainability, 12*(14), 5565. http://doi.org/10.3390/su12145565
34. Singh, J., Singh, H., & Singh, G. (2018). Productivity improvement using LM in manufacturing industry of northern India. *International Journal of Productivity and Performance Management, 67*(8), 1394–1415. http://doi.org/10.1108/IJPPM-02-2017-0037
35. Uhrin, Ã., Bruque-Cámara, S., & Moyano-Fuentes, J. (2017). Lean production, workforce development and operational performance. *Management Decision, 55*(1), 103–118. http://doi.org/10.1108/MD-05-2016-0281

36. Perera, S., & Perera, C. (2019). Performance measurement system for a LM setting. *Measuring Business Excellence*, *23*(3), 240–252. http://doi.org/10.1108/MBE-11-2018-0087
37. Randhawa, J. S., & Ahuja, I. S. (2017). 5S—A quality improvement tool for sustainable performance: Literature review and directions. *The International Journal of Quality & Reliability Management*, *34*(3), 334–361. http://doi.org/10.1108/IJQRM-03-2015-0045
38. Psomas, E., Antony, J., & Bouranta, N. (2018). Assessing lean adoption in food SMEs: Evidence from Greece. *The International Journal of Quality & Reliability Management*, *35*(1), 64–81. http://doi.org/10.1108/IJQRM-05-2016-0061
39. Ghobakhloo, M., & Azar, A. (2018). Business excellence via advanced manufacturing technology and lean-agile manufacturing: IMS. *Journal of Manufacturing Technology Management*, *29*(1), 2–24. http://doi.org/10.1108/JMTM-03-2017-0049
40. Leksic, I., Stefanie, N., & Veza, I. (2020). The impact of using different LM tools on waste reduction. *Advances in Production Engineering & Management*, *15*(1), 81–92. http://doi.org/10.14743/apem2020.1.351
41. Rishi, J. P., Srinivas, T. R., Ramachandra, C. G., & Abhishek. (2019). LM to green manufacturing: Practices and its implementation in SME's. *Applied Mechanics and Materials*, *895*, 21–25. http://doi.org/10.4028/www.scientific.net/AMM.895.21
42. Goshime, Y., Kitaw, D., & Jilcha, K. (2019). LM as a vehicle for improving productivity and customer satisfaction. *International Journal of Lean Six Sigma*, *10*(2), 691–714. http://doi.org/10.1108/IJLSS-06-2017-0063
43. Zhu, X., & Lin, Y. (2017). Does LM improve firm value?: IMS. *Journal of Manufacturing Technology Management*, *28*(4), 422–437. http://doi.org/10.1108/JMTM-05-2016-0071
44. Knol, W. H., Slomp, J., Schouteten, R. L. J., & Lauche, K. (2019). The relative importance of improvement routines for implementing lean practices. *International Journal of Operations & Production Management*, *39*(2), 214–237. http://doi.org/10.1108/IJOPM-01-2018-0010

30 Sentiment Analysis for Promoting the Manufacturing Sector

Shaveta, Neeru Mago, Rajeev Kumar Dang

30.1 INTRODUCTION

Sentiment analysis is the latest and hot topic in today's research environment for analysing the opinions expressed by people on social media about a product or service. Opinions are user evaluations of any good or service [1]. Manufacturing businesses can use these reviews to enhance themselves and their products. Here, sentiment analysis enters the picture. Sentiment analysis is a technique used to assess the polarity of evaluations based on people's attitudes, opinions, and feelings [2]. By examining user reviews of any product, manufacturers may improve their goods and services. It is a type of text analytics that makes use of natural language processing [3]. Other names for sentiment analysis include "opinion mining" and "emotion artificial intelligence." Finding opinions through social media is crucial. These product reviews might be used by manufacturers to enhance the product's quality or design. Sentiment analysis seeks to automatically identify the subjectivity or polarity of the text [4]. A manufacturing company has a very difficult time achieving a respectable place in the market in today's cut-throat economy. Because technology advances so quickly, manufacturing companies must adapt themselves to keep pace with global competition. Sentiment analysis can be used to determine whether customers have positive, negative, or neutral opinions of the product. Time is money or more valuable than money; thus, we can utilize automated algorithms for sentimental analysis rather than spending time reading and determining the positivity or negativity of content [5]. A manufacturing company can categorize client feedback as favourable, bad, or neutral by employing sentiment analysis tools. Sentiment analysis enables a business to ascertain early on what customers like and dislike about a product. The world's largest and most competitive economic sector, the automotive industry, can use sentiment analysis to better its marketing strategy and goals by analysing customer sentiments [6]. Knowing this information enables firms to concentrate their efforts on the few things that affect people's happiness (whether they are employees, customers, patients, etc.) the most. Organizations can track their success by making changes and comparing recent ratings and sentiments to earlier data. Who is going to go through all of this content and utilize its insight when you are reaching consumers in the hundreds of thousands? A machine alone! It is a huge work that, quite honestly, no human being can complete since it's so tedious, time-consuming,

DOI: 10.1201/9781003367154-30

and subjective. To automatically and accurately evaluate massive amounts of internet data from many sources, sentiment analysis is required. Teams searching for problem areas, trends, and triumphs can gain valuable information by reading the actual words people have used to describe a product or service [7]. This text analysis reveals particular service sectors that are struggling even as other areas are flourishing. For instance, text sentiment analysis can identify customers who are regularly satisfied in nine of their interactions with a business and can also draw attention to one consistently unpleasant interaction [8].

In this chapter, Section 2 presents a literature review of different studies. Various techniques for sentiment analysis are discussed in Section 3. The proposed methodology is elaborated in Section 4. The experimental results and observations are presented in Section 5. Finally, the conclusions and future scope are discussed in Section 6.

30.2 LITERATURE REVIEW

Sentiment analysis continues to be the subject of active and extensive study, which is still ongoing. A review of various studies reads as follows: sentiment analysis is useful for a variety of application domains, including product performance improvement, understanding user attitudes toward a product, and outcome prediction. To forecast the outcome of India's general elections in 2019, sentiment analysis was applied. Between January and March 2019, an experiment was conducted using data from Twitter. The actual results of the general elections held in May 2019 matched those of the experiment [9]. Various sentiment analysis methods have been discussed for analysis of data [10]. In comparison to other options like methodologies or algorithms utilized independently, they discovered that sentiment analysis APIs are used extremely infrequently. As the largest databases of electronic journals, the digital scientific databases used for the research were ACM Digital Library, IEEE Xplore Digital Library, ScienceDirect (Elsevier), SpringerLink, Wiley Library, and Taylor & Francis. Among other social media, they discovered that Twitter is the most often used platform for sentiment analysis. They reviewed numerous machine learning and tools based on lexicons, and after discussing about a variety of technologies based on lexicons and machine learning, they came to the conclusion that SVM and Naive Bayes are the most popular techniques for sentiment analysis.

Convolutional neural networks (CNNs) and SVM text sentiment analysis were proposed as a combined model [11]. CNN model with SVM text sentiment analysis for experimental data combination. The NLPCC2014 emotional analysis evaluation task dataset based on deep learning technology was utilized as the experimental data in this study to assess the effectiveness of sentiment analysis technology based on deep learning. They used 5,000 positive and 5,000 negative emotional polarity data from the training set (10,000 data points) and 1,250 positive and 1,250 negative emotional polarity data from the test set (2,500 data points). They conducted a comparative experiment using the classic CNN-based text sentiment analysis approach and the NLPCC-SCDL-best method (the top system in the NLPCC-SCDL evaluation job), to assess the efficacy of the suggested method. Next, a method to predict the level of teaching performance by automatically analysing the text input from students is proposed [12]. To determine the polarity of words, a database of English sentiment words was established. Of the 745 words in the database, 448 are positive, 263 are negative, and the rest words are neutral. The range of the sentiment score is –3 to +3. A technique based on the lexicon is used to analyse sentiments. Scores between 1 and 3 are seen as positive, whereas scores between –1 and –3 are regarded as bad. The neutral category is defined as having a sentiment score of 0. The level of the opinion of any teacher is determined by the feedback comments of the students. Various strategies are used to categorize product reviews in the next [13]. They compared various classifiers to categorize a sizable number of English tweets about certain products. They talked about and contrasted several machine learning methods for analysing products. Sentiment analysis of educational data is conducted in this research [14]. There are too many internet resources for education. This can aid in enhancing teaching abilities because student reviews are subjected to sentiment analysis. They employed the Kalboard 360 dataset for their research, and Waikato Environment for Knowledge Analysis (WEKA) was used for data mining and data pre-processing. SVM, NLP, decision tree, K-star, Bayes net, Simple Logistics, multi-class classifier, and random forest are just a few of the classifiers that receive the data. The dataset is properly trained for each classifier, and a model is created. The model is then tested against test data to get the results. According to the results, the SVM and NLP–deep learning approach generally outperformed the other classifiers in terms of classification accuracy. Research work related to these has been summarized in Table 30.1.

30.3 TECHNIQUES FOR SENTIMENT ANALYSIS

Essentially, there are two methods for sentiment analysis: lexicon-based methods and machine learning methods [25] as shown in Figure 30.1.

30.3.1 Lexicon-Based Methods

Compare sentiment phrases from a sentiment dictionary to the available data to ascertain polarity. It is further divided into two subcategories that use statistical or semantic methods to find polarity: dictionary-based approaches and corpus-based approaches [27].

- **Dictionary-based approach:** Using well-known vectors, a small number of concepts are manually gathered. After that, the rehearsals and echoes of these terms are looked up and added to corpora like WordNet or thesaurus. Up until there are no more words to add, this collection slowly expands [28]. The weakness of the emotional classification depends on the size of the lexicon, which is a drawback to this method. When the dictionary size expands, this strategy is equally flawed.
- **Corpus-based approach:** They are based on the semantics and word vocabulary of big polysaccharides. A big-named database is needed since words generated could be reliant on the environment [29].

TABLE 30.1
Comparative Analysis of Some Research Papers

Author	Title	Objectivity of the Paper	Methodology	Experimental Results		
S. Sohangir et al. [15]	Big Data: Deep Learning for Financial Sentiment Analysis	To determine if deep learning models can be adapted to improve the performance of sentiment analysis for Stock Twits	RNN CNN	CNN gives better accuracy.		
Q. You et al. [16]	Joint Visual-Textual Sentiment Analysis with Deep Neural Networks	Used both the state-of-the-art visual and textual sentiment analysis techniques for joint visual-textual sentiment analysis	CNN	Algorithm	Precision	Accuracy
				Textual	0.806	0.696
				Visual	0.747	0.732
				Early fusion	0.778	0.763
				Late fusion	0.785	0.769
[17]	Sentiment Analysis Using Convolutional Neural Network	Proposed a framework called Word2vec + convolutional neural network	CNN Used 3 pairs of convolutional layers and pooling layers	Model	Fine-grained (%)	
				NB	41.0	
				SVM	40.7	
				BiNB	41.9	
				VecAvg	32.7	
				RNN	43.2	
				MV-RNN	44.4	
				Proposed framework	45	
C. Li et al. [18]	Recursive Deep Learning for Sentiment Analysis over Social Data	Introduced a novel recursive neural deep model (RNDM) to predict sentiment label based on recursive deep learning	Recursive deep learning	Model	Accuracy (%)	
				NB	78.65	
				ME	87.46	
				SVM	84.9	
				RNDM	90.8	

L. C. Chen et al. [19]	Exploration of Social Media for Sentiment Analysis Using Deep Learning	To construct a sentiment analysis framework and processes for social media in order to propose a self-developed military sentiment dictionary for improving sentiment classification	LSTM Bi-LSTM	Model LSTM Bi-LSTM	Accuracy (%) 84.08 92.68
A. Onaciu et al. [20]	Ensemble of Artificial Neural Networks for Aspect Based Sentiment Analysis	Presented a system consisting of an ensemble of classifiers built using deep learning strategies for aspect-based sentiment analysis	CNN RNN	System Proposed system XRCE IIT-T	F1-Score 65.31 68.70 63.05
A. Shah et al. [21]	Sentiment Analysis of Product Reviews Using Deep Learning	Used Word2vec to learn word embedding and convolution neural networks to train and classify the sentiment classes of the product reviews	Combined Word2vec-CNN model	Model Bigram+NaiveBayes BOW+NaiveBayes NormalizedBigram+ NB Word2Vec+CNN	Accuracy (%) 0.441 0.735 0.587 0.91323
S. Chen et al. [22]	A Deep Neural Network Model for Target-Based Sentiment Analysis	Proposed a deep neural network model combining (CNN-RLSTM) for the task of target-based sentiment analysis	CNN-RLSTM	Model CNN LSTM ATT-CNN ATT-LSTM MATT-CNN ATT-RLSTM CNN-RLSTM	Time(S) 11 126 28 386 82 136 153

TABLE 30.1 *(Continued)*
Comparative Analysis of Some Research Papers

Author	Title	Objectivity of the Paper	Methodology	Experimental Results	
M. Pota et al. [23]	A Subword-Based Deep Learning Approach for Sentiment Analysis of Political Tweets	Presented a neural-network-based approach to analyse the sentiment expressed on political tweets	CNN	Sentiment	Similarity of the CNN among the baseline models
				Positive	29%
				Negative	30%
				Neutral	93%
				Mean of value	61%
T. Wang et al. [24]	COVID-19 Sensing: Negative Sentiment Analysis on Social Media in China via BERT	Demonstrated how public sentiment on social media evolves as COVID-19 spreads	BERT	Method	Accuracy (%)
				Fine-tuned BERT	75.65

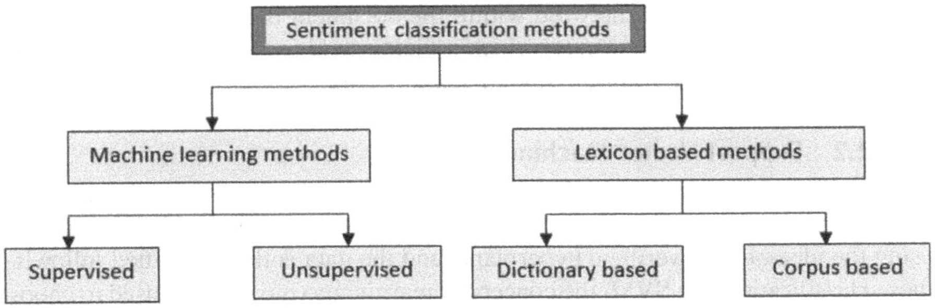

FIGURE 30.1 Types of machine learning techniques by S. Hamed et al. [26].

30.3.2 Machine Learning Methods

Through the use of machine learning techniques, computers are given the ability to implicitly learn from their programming. It analyses the sentences and provides feedback based on whether they express positive, negative, or neutral ideas. Supervised, unsupervised, and reinforcement learning are the three subcategories that can be used to categorize machine learning [30]. Machine learning algorithms mostly make use of sentiment classification techniques. Supervised learning, unsupervised learning, semi-supervised learning, and reinforcement learning are the general divisions. The model learns from the trained dataset through supervised learning, as the name suggests. It applies to classification and regression issues and trains on labelled data. Clustering and association rule mining use unsupervised learning to train models on unlabelled data. Unlabelled data is used in a semi-supervised strategy to train models with scant labelled data [31]. Iterative reinforcement learning continuously absorbs information from the surroundings. It is primarily utilized in the artificial intelligence industry. Supervised learning techniques are preferred because classification is done on sentiment analysis to determine the direction of sentiments.

- **Supervised learning:** Models for supervised learning gain knowledge from labelled data. It uses patterns to identify which label should be applied to fresh data and links the patterns to the new unlabelled data [32]. These models utilize an algorithm to learn the mapping function from input to output and have input and output variables. Various supervised learning algorithms exist; a few are covered in the following sections.

30.3.2.1 Naïve Bayes Classifier

The Bayes theorem with conditional probabilities forms the foundation of the Naive Bayes classifier. The independent assumption between its aspects is how it typically operates. The Nave Bayes algorithm is demonstrated to be a straightforward, ideal, and effective model for classification in machine learning.

To determine the appropriate class label for a feature, the posterior probability of a class is computed. The likelihood is calculated using the word distribution in the document; the position of the word within the document is not taken into account. When the input variables are categorical, it yields good results [33]. It requires less processing memory and training time for classification.

30.3.2.2 Support Vector Machine

The supervised learning method known as SVM is used to solve classification and regression issues in sentiment analysis. The training dataset is divided into classes using the idea of discovering a hyperplane, and the data is then classified following those classifications. In SVM, the concept of margin maximization is used to choose the hyperplane with the biggest feasible margin to distinguish classes clearly [34]. It does not make any conclusive assumptions about the data in the field. The data are not overfitted. It delivers effective outcomes for high-dimensional datasets.

30.3.2.3 Maximum Entropy Method

A conditional exponential classifier called a maximum entropy (ME) classifier is used to integrate the joint features produced from a set of features by the encoding method. Based on the weight of each feature, the encoded vector is used to determine the most probable label for the feature set. It is a feature-based model that uses the features to determine how these attributes are distributed among different classes. It does not make any unsupported feature assumptions [35].

- **Unsupervised learning:** Unsupervised machine learning involves training the computer using an unlabelled dataset, and the computer then makes output predictions without any human intervention. Unsupervised learning involves training models on data that is neither classified nor labelled, and then letting the models behave autonomously on that data. The unsupervised learning algorithm's primary goal is to classify or group the unsorted dataset based on commonalities, patterns, and differences. The hidden patterns in the input dataset are to be found by the machines. The categories for unsupervised learning are clustering and association [36].

30.3.3 Hybrid Approach

There is a long history of ensemble learning techniques outperforming other machine learning methods. Classification and regression issues are among the application domains of these programmes. Well-known ensemble models that combine weak learners to form an ensemble include the random forest model and the gradient boosting model. These models have a homogeneous collection of weak learners, which means that weak learners of the same kind are put together to demonstrate their combined strength.

30.4 METHODOLOGY

1. **Dataset**

   The Product Sentiment Analysis data collection from Kaggle, which includes descriptions of user reviews of a variety of distinctive products, was employed for this study. Dataset is in the form of a comma separated value (CSV) file. Based on the user-provided raw text review, this dataset is utilized to create machine learning models that accurately categorize diverse items into 4 different emotion groups. By examining these opinions, we can better serve our consumers and learn about several customer characteristics that are either obvious or covert in the evaluations.

2. **Proposed hybrid model**
   Machine learning techniques are used as the major classification method for tweets. Term and spelling errors could be present in the tweets. You must therefore investigate the consequences of these tweets. Three steps are taken in total. Pre-processing is carried out in the initial stage. The majority of words, errors, and other mistakes will be eliminated. A feature is produced in the second stage using features related to vectors. Using several classifiers, tweets can be categorized as favourable, negative, or neutral. The whole process can be divided into six stages, as shown in Figure 30.2.

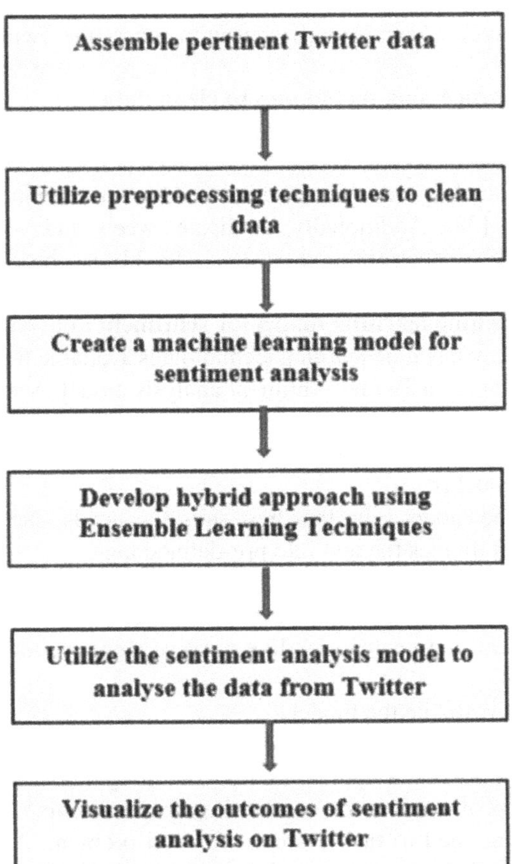

FIGURE 30.2 Sentiment analysis procedure.

a. **Assemble pertinent Twitter data**
The gathering and sorting of data is the first step in sentiment analysis. On Twitter, there is a sea of data; it is crucial to choose the data that is most pertinent to the issue we are trying to solve or the information we are seeking. Only pertinent data should be used to train the sentiment analysis model and determine whether it successfully analyses Twitter data. What kind of tweets, historical or current, has to be examined is another crucial point to cover [37]. Data must be collected from Twitter before sorting these issues. Some of the platforms that can be employed are as follows:
- Use Zapier, for instance, to set up an automatic workflow between Google Sheets and Twitter.
- IFTTT uses no-code Twitter data collection.
- Follow hashtags, keywords, and so on in real time or search through previous tweets and mentions, then export the tweet.
- Download tweets, including mentions and responses, from one's account.
- Access and analyse public tweets concerning keywords, brand mentions, hashtags, or tweets from specific people using the Twitter API.
- The Python package Tweepy is used to access the Twitter API and collect data.

b. **Utilize pre-processing techniques to clean data**
Before the data can be utilized to train the Twitter sentiment analysis model, it must first be collected, sorted, and cleaned. Emojis, special characters, and unused blank spaces must be eliminated because Twitter data is largely unstructured [38]. Additionally, duplicate tweets and very brief tweets, those with fewer than three characters, should also be eliminated. More accurate results can be produced with clean data.

c. **Create a machine learning model for sentiment analysis**
There are many machine learning technologies available that can be used to develop and apply a Twitter sentiment analysis model. Access to trained or ready-for-training models may be available through these platforms. Train these models for utilizing the data from Twitter [39]. The procedure for building a model is:
- Choose the model type that best suits the needs, such as a classifier model that divides the text into pre-defined tags.
- Choose the categorization type like sentiment analysis.
- Import the relevant Twitter data to the model for training.
- Sort the data into categories like positive, negative, or neutral to train the model.
- Finally, investigate the model.

d. **Develop hybrid approach using ensemble learning techniques**
In this research work, a hybrid ensemble learning model is created using a diverse group of weak learners. For this, many machine-learning-method types are combined to tackle a classification problem. A machine learning concept called ensemble learning uses the combined power of machine learning models to solve learning problems like classification or regression.

This method classifies a number of homogeneous machine learning models as weak learners and groups them together. Each of the weak learners exhibits its unique result when applied to the problem, whether it be on the complete training set or only a portion of it. The ultimate result is obtained by combining the findings of each poor learner.

e. **Utilize the sentiment analysis model to analyse the data from Twitter**
The model is prepared for deployment once it has undergone training and produced positive test results [40]. Establish a link between the sentiment analysis model and Twitter data at this point. There are numerous ways to carry it out. One approach is to classify and analyse a specific file of recent or unused tweets. Another approach is to use the model to evaluate Twitter data by integrating it with Zapier and Google Sheets.

f. **Visualize the outcomes of sentiment analysis on Twitter**
Tools are available to aid in the visualization of the data results, making them simple to understand and assimilate [41]. These appealing visualization tools, like Google Data Studio, Looker, and Tableau, produce visually appealing reports with charts, graphs, and data tables that are simple for more people to understand.

30.5 EXPERIMENTAL RESULTS

This research effort first applies a few machine learning techniques to a dataset that is a collection of various user-generated tweets regarding various products. On a given dataset, several machine learning techniques are applied, including KNN, logistic regression, Naive Bayes, SVM, and random forest classifier. Different values for precision, recall, and F1-score are shown by the experiment. The tabular form of each classifier's accuracy is shown in Table 30.2.

After this, a *hybrid ensemble learning model* is created using the various types of machine learning models for weak learners. These models include the Naive Bayes model, K nearest neighbour model, support vector machine, random forest classifier, and logistic regression model. In previous ensemble models, a homogenous group of weak learners is utilized; however, in this assignment, a heterogeneous group of weak learners is used, hence the word hybrid.

TABLE 30.2
Precision, Recall, F1-Score, and Accuracy by Individual Machine Learning Techniques

Classifier	Precision	Recall	F1-Score	Accuracy
SVM	0.68	0.69	0.69	69%
Logistic regression	0.69	0.69	0.61	69%
Random forest	0.64	0.67	0.60	67%
Naïve Bayes	0.43	0.65	0.52	65%
KNN	0.58	0.64	0.59	64%

TABLE 30.3
Precision, Recall, F1-Score, and Accuracy by Ensemble Learning Techniques

Classifier	Precision	Recall	F1-Score	Accuracy
KNN+LR	0.96	0.97	0.96	96%
KNN+SVM	0.93	0.92	0.92	92%
KNN+NB	0.88	0.90	0.89	90%
KNN+RF	0.87	0.83	0.85	83%
KNN+LR+NB+SVM+RF	0.95	0.95	0.94	95%

30.5.1 Observations

As shown in Table 30.3, the hybrid ensemble learning model has outperformed every individual learning model, as we can see from these comparisons. The confusion matrices and cross-validation results of each individual model and the ensemble model can be used to confirm this. Using the combined power of several machine learning models to solve a learning problem, such as a classification or regression problem, is known as ensemble learning. This method groups together a number of homogeneous machine learning models that are considered weak learners. Each weak learner exhibits its own result when applied to the problem, either on the entire training set or on a subset of the complete training set. In order to arrive at the ultimate result, each weak learner's results are merged. Each of the five machine learning models was defined five times in this study, creating a total of 25 weak learners. Finally, the max-voting classifier approach is employed, and the ensemble model's final class prediction will be the one that has been primarily predicted by the weak learners. By incorporating a wider variety of weak learners, we can create more hybrid ensemble models in the future. A collection of related models with various architectures is another option.

30.6 CONCLUSIONS AND FUTURE SCOPE

Consumer behaviour metrics that have changed can be used to assist the manufacturer in improving innovation. One of the most intriguing study fields now being used by modern networks and telecommunication technology is social media organizing and analysis. Through the measurement of insight, modern social media analytics can be utilized to widen how businesses track changes in consumer behaviour. Organizations and sectors like the manufacturing industry can benefit from these insights. To empower customers and track changes in consumer behaviour, manufacturers can use social media analysis. The manufacturer can attempt to improve innovation by using changes in consumer behaviour measures. In this chapter, we have deployed social media as a potent tool for communication. Customer opinion toward the industrial sector has been well-measured through social media posts and the proposed sentiment analysis model has been applied on the product sentiment analysis dataset to achieve high performance using ensemble learning methods. The hybrid ensemble learning model showed significant improvement over all other individual learning models, experiments show 96% accuracy using the ensemble learning model.

In the future, manufacturers can generalize customer behaviour as a whole by using statistical grouping analysis. The topic of big data analysis of social networking is still open despite the successful use of mathematical statistics employed to cluster and generalize user behaviour. This occurs as a result of the need to tailor user activity models and comprehend certain aspects of human behaviour. The analyst's capacity to use social media analytics in a way that can boost innovation is increased by human behaviour. By using sentiment analysis, this research's primary goal is to enhance user-machine interaction. Incorporating predictive analysis into this technology allows manufacturers to improve quality control, production control, and maintenance. Manufacturers can use this study to get more rewards and increase their revenues, which helps them become more competent and ensures their survival in the market.

REFERENCES

[1] K. L. Santhosh Kumar, J. Desai, and J. Majumdar, "Opinion mining and sentiment analysis on online customer review," in *2016 IEEE Int. Conf. Comput. Intell. Comput. Res. ICCIC 2016*, no. December, 2017, doi:10.1109/ICCIC.2016.7919584.

[2] C. Kaur and A. Sharma, "Social issues sentiment analysis using python," in *Proc. 2020 Int. Conf. Comput. Commun. Secur. ICCCS 2020*, no. October, 2020, doi:10.1109/ICCCS49678.2020.9277251.

[3] Q. Tul *et al.*, "Sentiment analysis using deep learning techniques: A review," *Int. J. Adv. Comput. Sci. Appl.*, vol. 8, no. 6, 2017, doi:10.14569/ijacsa.2017.080657.

[4] F. Xianghua, L. Guo, G. Yanyan, and W. Zhiqiang, "Multi-aspect sentiment analysis for Chinese online social reviews based on topic modeling and HowNet lexicon," *Knowledge-Based Syst.*, vol. 37, pp. 186–195, 2013, doi:10.1016/j.knosys.2012.08.003.

[5] Z. Drus and H. Khalid, "Sentiment analysis in social media and its application: Systematic literature review," *Procedia Comput. Sci.*, vol. 161, pp. 707–714, 2019, doi:10.1016/j.procs.2019.11.174.

[6] S. E. Shukri, R. I. Yaghi, I. Aljarah, and H. Alsawalqah, "Twitter sentiment analysis: A case study in the automotive industry," in *2015 IEEE Jordan Conf. Appl. Electr. Eng. Comput. Technol. AEECT 2015*, no. November, 2015, doi:10.1109/AEECT.2015.7360594.

[7] M. D. Devika, C. Sunitha, and A. Ganesh, "Sentiment analysis: A comparative study on different approaches," *Procedia Comput. Sci.*, vol. 87, pp. 44–49, 2016, doi:10.1016/j.procs.2016.05.124.

[8] C. Baecchi, T. Uricchio, M. Bertini, and A. Del Bimbo, "A multimodal feature learning approach for sentiment analysis of social network multimedia," *Multimed. Tools Appl.*, vol. 75, no. 5, pp. 2507–2525, 2016, doi:10.1007/s11042-015-2646-x.

[9] A. Sharma and U. Ghose, "Sentimental analysis of Twitter data with respect to general elections in India," *Procedia Comput. Sci.*, vol. 173, no. 2019, pp. 325–334, 2020, doi:10.1016/j.procs.2020.06.038.

[10] Ramírez-Tinoco, Francisco Javier, Giner Alor-Hernández, José Luis Sánchez-Cervantes, Beatriz Alejandra Olivares-Zepahua, and Lisbeth Rodríguez-Mazahua, "A brief review on the use of sentiment analysis approaches in social networks," in *Trends*

and Applications in Software Engineering: Proceedings of the 6th International Conference on Software Process Improvement (CIMPS 2017) 6, pp. 263–273, Springer International Publishing, 2018.

[11] Y. Chen and Z. Zhang, "Research on text sentiment analysis based on CNNs and SVM," in *2018 13th IEEE Conf. Ind. Electron. Appl. (ICIEA)*, pp. 2731–2734, 2018. doi:10.1109/ICIEA.2018.8398173.

[12] K. Z. Aung and N. N. Myo, "Sentiment analysis of students' comment using lexicon based approach," in *Proc.—16th IEEE/ACIS Int. Conf. Comput. Inf. Sci. ICIS 2017*, pp. 149–154, 2017, doi:10.1109/ICIS.2017.7959985.

[13] M. Dagar, A. Kajal, and P. Bhatia, "Twitter sentiment analysis using supervised machine learning techniques," in *2021 5th Int. Conf. Inf. Syst. Comput. Networks, ISCON 2021*, no. January, 2021, doi:10.1109/ISCON52037.2021.9702333.

[14] J. Sultana, N. Sultana, K. Yadav, and F. AlFayez, "Prediction of sentiment analysis on educational data based on deep learning approach," in *2018 21st Saudi Comp. Soc. Nat. Comp. Conf. (NCC)*, pp. 1–5, 2018. doi:10.1109/NCG.2018.8593108.

[15] S. Sohangir, D. Wang, A. Pomeranets, and T. M. Khoshgoftaar, "Big data: Deep learning for financial sentiment analysis," *J. Big Data*, vol. 5, no. 1, 2018, doi:10.1186/s40537-017-0111-6.

[16] Q. You, J. Luo, H. Jin, and J. Yang, "Joint visual-Textual sentiment analysis with deep neural networks," in *MM 2015—Proc. 2015 ACM Multimed. Conf.*, pp. 1071–1074, 2015, doi:10.1145/2733373.2806284.

[17] X. Ouyang, P. Zhou, C. H. Li, and L. Liu, "Sentiment analysis using convolutional neural network," in *Proc.—15th IEEE Int. Conf. Comput. Inf. Technol. CIT 2015, 14th IEEE Int. Conf. Ubiquitous Comput. Commun. IUCC 2015, 13th IEEE Int. Conf. Dependable, Auton. Se*, pp. 2359–2364, 2015, doi:10.1109/CIT/IUCC/DASC/PICOM.2015.349.

[18] C. Li, B. Xu, G. Wu, S. He, G. Tian, and H. Hao, "Recursive deep learning for sentiment analysis over social data," *Proc.—2014 IEEE/WIC/ACM Int. Jt. Conf. Web Intell. Intell. Agent Technol.—Work. WI-IAT 2014*, vol. 2, pp. 180–185, 2014, doi:10.1109/WI-IAT.2014.96.

[19] L. C. Chen, C. M. Lee, and M. Y. Chen, "Exploration of social media for sentiment analysis using deep learning," *Soft Comput.*, vol. 24, no. 11, pp. 8187–8197, 2020, doi:10.1007/s00500-019-04402-8.

[20] A. Onaciu and A. Nicoleta Marginean, "Ensemble of artificial neural networks for aspect based sentiment analysis," in *Proc.—2018 IEEE 14th Int. Conf. Intell. Comput. Commun. Process. ICCP 2018*, pp. 13–19, 2018, doi:10.1109/ICCP.2018.8516637.

[21] A. Shah, "Sentiment analysis of product reviews using supervised learning," *Reliab. Theory Appl.*, vol. 16, pp. 243–253, 2021, doi:10.1145/3447568.3448513.

[22] S. Chen, C. Peng, L. Cai, and L. Guo, "A deep neural network model for target-based sentiment analysis," *Proc. Int. Jt. Conf. Neural Netw.*, vol. 2018, pp. 1–7, 2018, doi:10.1109/IJCNN.2018.8489180.

[23] M. Pota, M. Esposito, M. A. Palomino, and G. L. Masala, "A subword-based deep learning approach for sentiment analysis of political tweets," in *Proc.—32nd IEEE Int. Conf. Adv. Inf. Netw. Appl. Work. WAINA 2018*, vol. 2018, pp. 651–656, 2018, doi:10.1109/WAINA.2018.00162.

[24] T. Wang, K. Lu, K. P. Chow, and Q. Zhu, "COVID-19 sensing: Negative sentiment analysis on social media in China via BERT model," *IEEE Access*, vol. 8, pp. 138162–138169, 2020, doi:10.1109/ACCESS.2020.3012595.

[25] L. Yue, W. Chen, X. Li, W. Zuo, and M. Yin, "A survey of sentiment analysis in social media," *Knowl. Inf. Syst.*, vol. 60, no. 2, pp. 617–663, 2019, doi:10.1007/s10115-018-1236-4.

[26] S. Hamed, M. Ezzat, and H. Hefny, "A review of sentiment analysis techniques," *Int. J. Comput. Appl.*, vol. 176, no. 37, pp. 20–24, 2020, doi:10.5120/ijca2020920480.
[27] W. Medhat, A. Hassan, and H. Korashy, "Sentiment analysis algorithms and applications: A survey," *Ain Shams Eng. J.*, vol. 5, no. 4, pp. 1093–1113, 2014, doi:10.1016/j.asej.2014.04.011.
[28] P. Tyagi, S. Chakraborty, R. C. Tripathi, and T. Choudhury, "Literature review of sentiment analysis techniques for microblogging site," *SSRN Electron. J.*, 2019, doi:10.2139/ssrn.3403968.
[29] M. Wankhade, A. C. S. Rao, and C. Kulkarni, "A survey on sentiment analysis methods, applications, and challenges," *Artif. Intell. Rev.*, vol. 55, no. 7, 2022. doi:10.1007/s10462-022-10144-1.
[30] R. Hu, L. Rui, P. Zeng, L. Chen, and X. Fan, "Text sentiment analysis: A review," in *2018 IEEE 4th Int. Conf. Comput. Commun. ICCC 2018*, pp. 2283–2288, 2018, doi:10.1109/CompComm.2018.8780909.
[31] V. A. Rohani and S. Shayaa, "Utilizing machine learning in sentiment analysis: SentiRobo approach," in *2015 Int. Symp. Technol. Manag. Emerg. Technol. (ISTMET)*, 2015, pp. 263–267. doi:10.1109/ISTMET.2015.7359041.
[32] S. Poria, I. Chaturvedi, E. Cambria, and A. Hussain, "Convolutional MKL based multimodal emotion recognition and sentiment analysis," in *2016 IEEE 16th International Conference on Data Mining (ICDM)*, 2016, pp. 439–448, doi:10.1109/ICDM.2016.0055.
[33] T. U. Haque, N. N. Saber, and F. M. Shah, "Sentiment analysis on large scale Amazon product reviews," in *2018 IEEE Int. Conf. Innov. Res. Dev. ICIRD 2018*, no. June 2019, pp. 1–6, 2018, doi:10.1109/ICIRD.2018.8376299.
[34] M. V. Mäntylä, D. Graziotin, and M. Kuutila, "The evolution of sentiment analysis—A review of research topics, venues, and top cited papers," *Comput. Sci. Rev.*, vol. 27, pp. 16–32, 2018, doi:10.1016/j.cosrev.2017.10.002.
[35] A. Giachanou and F. Crestani, "Like it or not: A survey of Twitter sentiment analysis methods," *ACM Comput. Surv.*, vol. 49, no. 2, 2016, doi:10.1145/2938640.
[36] S. Jardim and C. Mora, "Customer reviews sentiment-based analysis and clustering for market-oriented tourism services and products development or positioning," *Procedia Comput. Sci.*, vol. 196, no. 2021, pp. 199–206, 2021, doi:10.1016/j.procs.2021.12.006.
[37] L. Jiang, M. Yu, M. Zhou, X. Liu, and T. Zhao, "Target-dependent Twitter sentiment classification," *ACL-HLT 2011—Proc. 49th Annu. Meet. Assoc. Comput. Linguist. Hum. Lang. Technol.*, vol. 1, pp. 151–160, 2011.
[38] D. Zimbra, A. Abbasi, D. Zeng, and H. Chen, "The state-of-the-art in Twitter sentiment analysis," *ACM Trans. Manag. Inf. Syst.*, vol. 9, no. 2, pp. 1–29, 2018, doi:10.1145/3185045.
[39] J. F. Raisa, M. Ulfat, A. Al Mueed, and S. M. S. Reza, "A review on Twitter sentiment analysis approaches," in *2021 Int. Conf. Inf. Commun. Technol. Sustain. Dev. ICICT4SD 2021—Proc.*, pp. 375–379, 2021, doi:10.1109/ICICT4SD50815.2021.9396915.
[40] S. Yi and X. Liu, "Machine learning based customer sentiment analysis for recommending shoppers, shops based on customers' review," *Complex Intell. Syst.*, vol. 6, no. 3, pp. 621–634, 2020, doi:10.1007/s40747-020-00155-2.
[41] S. Ahuja and G. Dubey, "Clustering and sentiment analysis on Twitter data," *2nd Int. Conf. Telecommun. Networks, TEL-NET 2017*, vol. 2018, pp. 1–5, 2018, doi:10.1109/TEL-NET.2017.8343568.

31 Design and Comparative Analysis of Modified Multilevel Inverter for Harmonic Minimization

Mamatha Sandhu, Tilak Thakur

31.1 INTRODUCTION

Multilevel inverters have developed very vastly and are in high demand for high-power applications in the field of power electronics. Various topologies have evolved to improve the efficiency with respect to dv/dt stress on switches, losses, electromagnetic inference, and so on. The switches are in series with DC sources to obtain a staircase voltage waveform. To achieve higher voltage at inverter output stage, the switches are operated in on and off modes. Multilevel inverter operates on fundamental switching frequency and high switching frequency pulse width modulation [1]. The proposed inverter in this work generates 3^m—voltage levels (e.g. if m = 2, DC sources produce nine levels, or 3^2). To produce a smooth sinusoidal waveform, much research was carried out, but the obtained waveform was having high harmonics, to overcome this, the voltage levels were increased to a higher level with a good resolution [2].

The researcher has designed multilevel inverter with minimum switches to improve the sinewave [4], [21–25]. Problems like electromagnetic compatibility and common mode voltage, absorb a large current that causes a voltage swing having harmonics. In this chapter, the research work is carried out where a modified cascaded multilevel inverter [13–15] is designed using single-phase full-bridge inverter in series to reduce the harmonics [3], [16–20]. The DC sources connected here produce different output voltages, $+E_{dc}$, 0, and $-E_{dc}$. Various voltage levels, such as seven levels and nine levels of modified MLI, are designed and modelled. The FFT spectrum analysis of THD is carried out using MATLAB/Simulink for both the levels.

31.2 PROPOSED SYSTEM

The proposed method is illustrated in Figure 31.1. The modified multilevel inverter is grid connected. The independent DC sources from multijunction solar cell, wind, other renewable energy sources, and batteries are supplied to the grid through the multilevel inverter to provide voltage continuously. A boost converter is used to generate a constant DC to the modified inverter using a boost converter. For maximum

 DOI: 10.1201/9781003367154-31

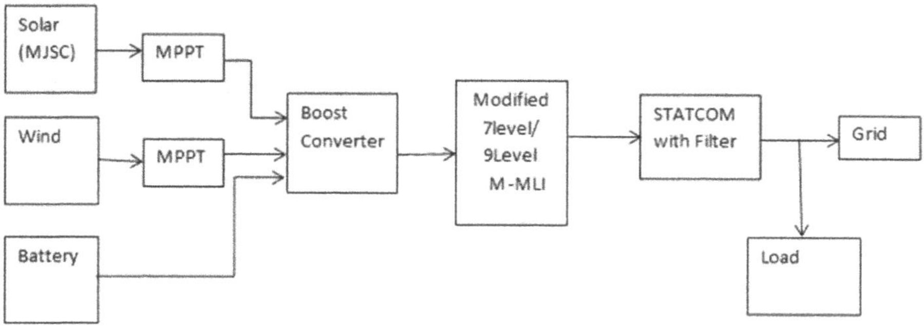

FIGURE 31.1 Modified multilevel inverter with grid connected system using renewable sources.

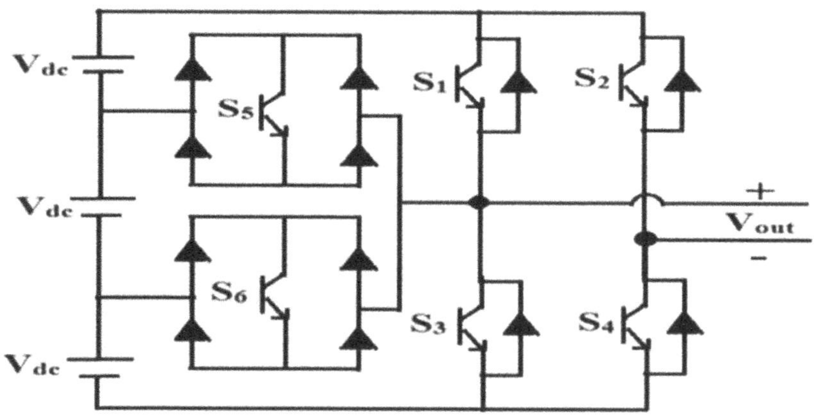

FIGURE 31.2 Seven-level –6 switches.

extraction of energy from renewable sources, incremental conductance algorithm is implemented. STATCOM and an LCL filter are used towards grid side [12].

31.3 MODELLING MODIFIED MULTILEVEL INVERTER

The MMLI is explained in [6], which is used in this work for designing the model. As seen in literature work, switches are reduced, but the requirement of diodes has increased with levels as discussed in [5]. Here, in our work, voltage building source is not required for increasing levels but cascading of different unit stage is necessary.

As shown in Figure 31.2, the symmetric input stages are constructed, which are individual cascaded modules having gate triggering switches [11]. The proposed unit stage topology for new modified MLI topology, consisting of five switches and one diode, to increase the voltage levels, which is symmetric in nature. By cascading individual units, the modified MLI is obtained [6]. The obtained output is sum of output voltage of multiple unit stages. If k is multiple unit stages, which are cascaded, the output voltage is as shown in equation (31.1).

$$V_{out} = V_{01} + V_{02} + V_{03} + V_{0p} \quad (31.1)$$

Methodology used for nine levels is that the value of two voltages in multiple unit stage as shown in Figure 31.3. The values are equal for the voltage source of cascaded

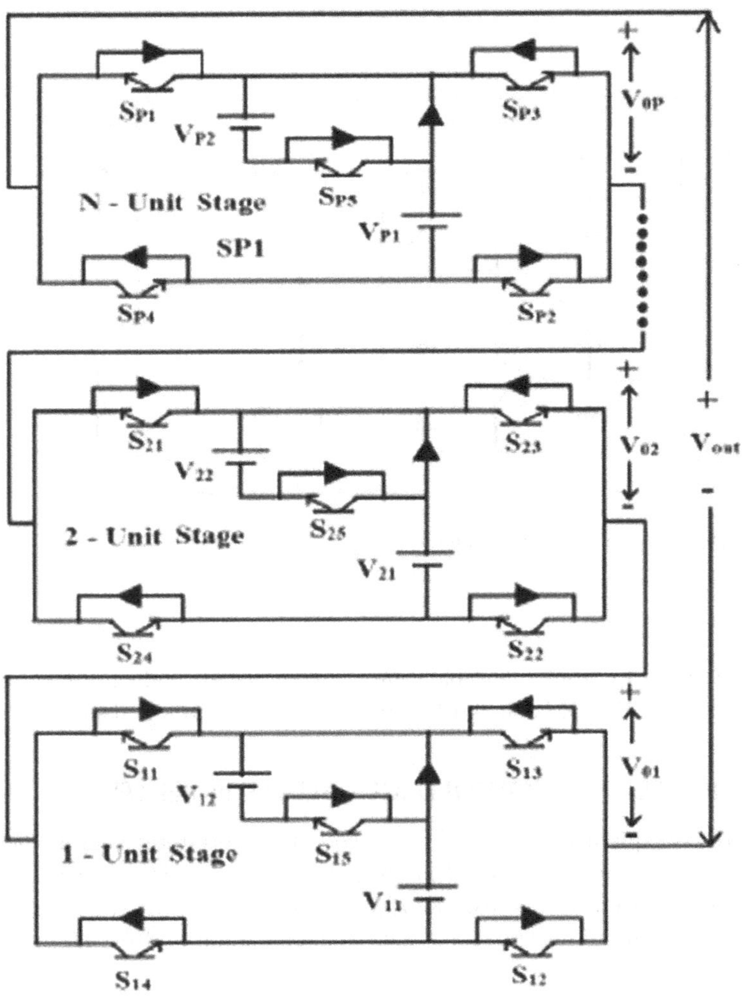

FIGURE 31.3 Proposed N-level modified MMLI.

unit stages based on the literature [6]. Output level obtained for proposed seven-level uses five switches with 2 DC sources and nine-level uses only ten switches with four DC sources instead of 12 switches for seven-level and 16 switches for nine-level as seen in conventional MLI. The new model has improved diodes and switches, which drastically help reducing gate triggering and switching losses [7–8]. The switching pattern to generate voltage levels (N) is obtained by cascading the k stages in 4k+1 form. The higher and lower positive and negative peaks are $2kV_{dc}$ and $-2kV_{dc}$, respectively. To generate nine-level, the required stages are two as discussed in [6]. To achieve higher levels, the cascading continues for harmonics reduction.

31.4 ARTIFICIAL NEURAL NETWORK SWITCHING TECHNIQUE

The ANN architecture work on input, hidden and output layers, where inputs are stored to process further then multiplied with weights to store the results. Artificial neural network calculates the data in parallel action; with sequential computations it is quite fast. In ANN numerous architectures, rules for learning can be implemented. Any one technique leads to generate voltage and current reference signals. For power electronic applications, the forward error back propagation is best suited.

The links' weights are changed, and error is computed if outputs found are not of desired targets. It is carried out from back to the front where the new weights process data of the given outputs that helps in convergence to occur fast when performance function is used with mean square error [9]. Here, MATLAB/Simulink is used to design ANN controller. The tool creates and trains a network, to compute the performance with the help of mean square error and regression analysis. To define a problem, an input and target-based data are given. By default, the tool accepts the samples per the errors received [10]. Back propagation is used for training and stops once the generalization stops improving. At the end of training, the file can be generated [15]. The ANN using PWM signals are given to the switches as proposed model shown in Figures 31.8 and 31.9.

31.5 SIMULATION USING MATLAB/SIMULINK OF PROPOSED SYSTEM

The simulated results are shown in Figure 31.4. and Figure 31.5.

31.6 DESIGN OF SOLAR (MJSC) AND WIND PARAMETERS

1. **Solar Input (MJSC) [3]**

   Temperature = 25°C; irradiance = 400; solar output = 36 V, as shown with and without MPPT in Figure 31.6 and Figure 31.7

2. **Wind Input**

   Wind output = 18 V, as shown in Figure 31.8. After converter: 46 V.

3. Battery and Filter

Battery voltage = 44 V; total DC voltage = 146 V; filter value: L= 5e-3H; C = 30e-6μF

4. STATCOM

STATCOM is used at the point of common coupling (PCC) to obtain neutralized harmonic and to control three-phase alternating current for the proper regulation of reactive current to obtain controllable reactive power shown in Figure 31.4 and Figure 31.5.

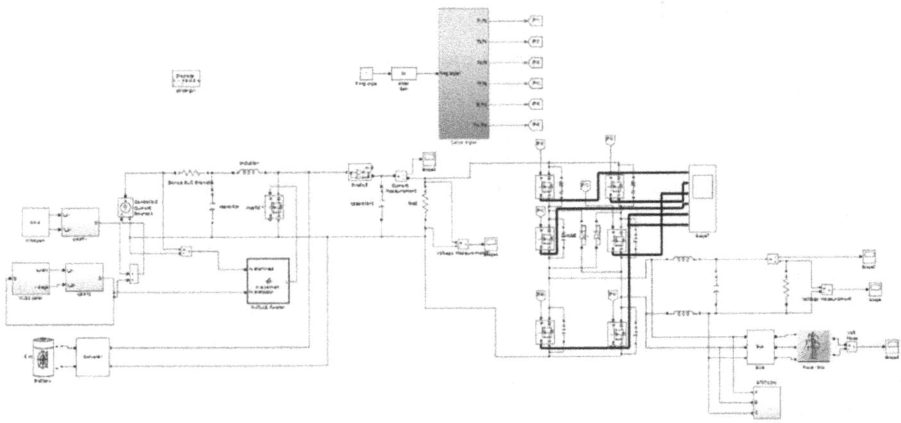

FIGURE 31.4 Seven-level grid-connected MLI system using MATLAB/SIMULINK.

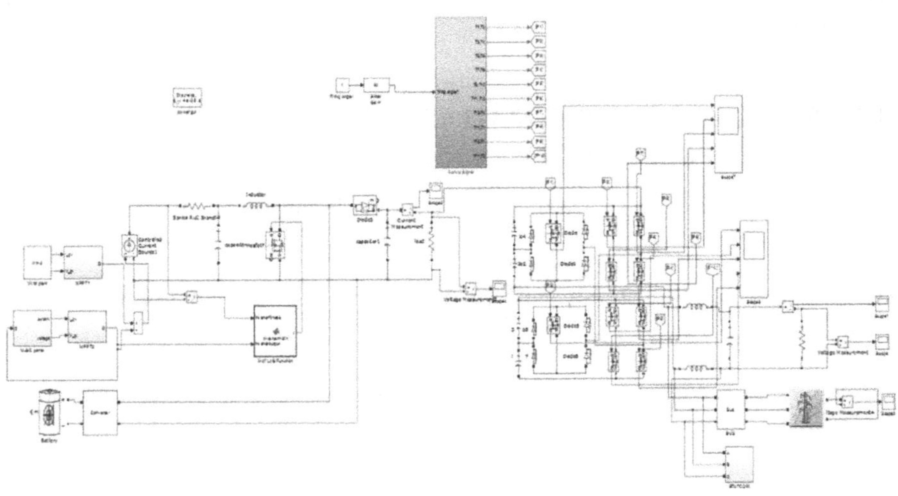

FIGURE 31.5 Nine-level grid-connected MLI using MATLAB/SIMULINK.

31.7 SIMULATION RESULTS

FFT Spectrum for THD analysis of CHBMLI with hybrid PV/wind source is shown graphically in Figures 31.9 and 31.10, respectively. The harmonics obtained is 4.38% and 4.18% per IEEE 519 standards for the seven-level and nine-level cascaded multilevel inverter for proposed systems. Harmonic analysis in the proposed system is improved.

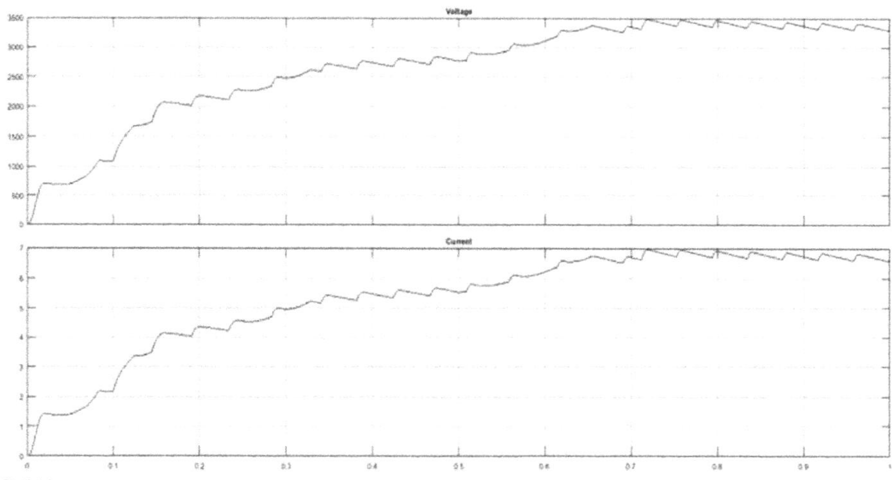

FIGURE 31.6 Current and voltage waveforms of the MJSC cell.

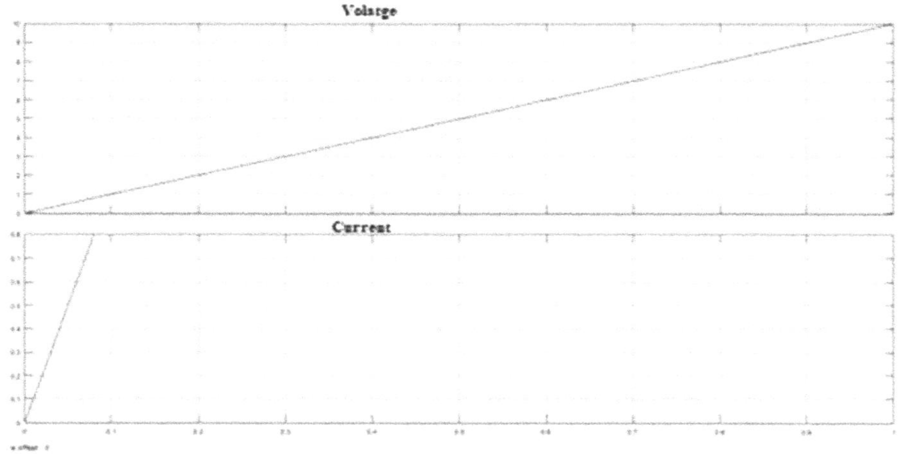

FIGURE 31.7 Incremental conductance—MPPT for MJSC.

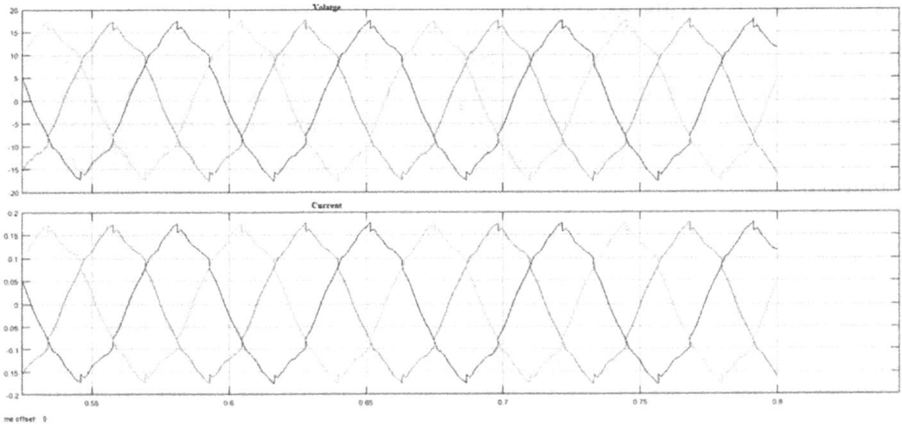

FIGURE 31.8 Voltage and current waveforms of a wind energy system.

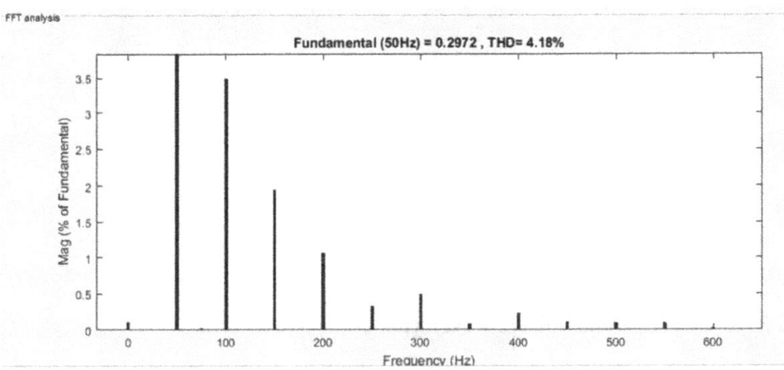

FIGURE 31.9 Nine-level THD of MMLI.

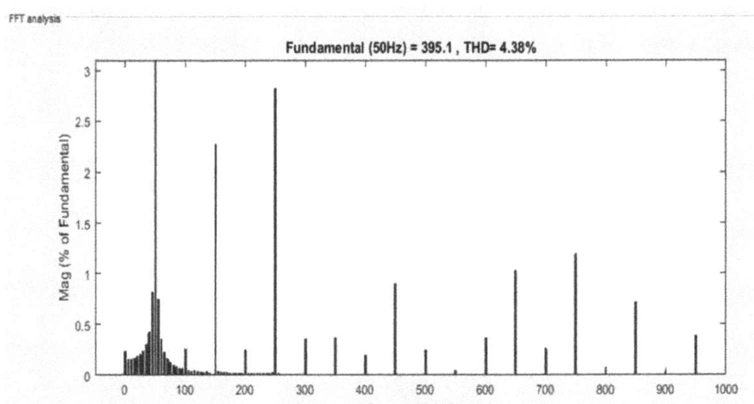

FIGURE 31.10 Seven-level THD analysis of MMLI.

31.8 COMPARATIVE ANALYSIS

The system has improved the harmonics as compared with various levels of conventional multilevel inverters and has reduced switching losses. Since reduced number of switches were used in the system the cost is effective. For the applications of power-based inverter; this is a very good solution as discussed in [11]. Various comparisons are carried out with respect to literature work of multilevel inverters as shown in Table 31.1. Comparative analysis of different multilevel inverters in the literature is carried out.

In Table 31.2, the cost of various topologies and the switching losses are shown. Comparative analysis of total harmonic distortion is as shown in Table 31.2. In Table 31.3, the comparison of THD and voltage stress of various topologies is shown, and the comparison of THD for the seven-level and nine-level modified multilevel inverter is shown in Table 31.4.

TABLE 31.1
Comparison of Different Multilevel Inverters

Type of MLI	Switching Losses	Cost
Modified MLI (2018) [25]	High	Low
Modified MLI (2021) [22]	Medium	High
Hybrid MLI (2020) [23]	Medium	Low
Proposed Modified MLI	**Low**	**Low**

TABLE 31.2
Comparative Analysis of Losses and Cost of MMLI

Name	Switches	Diodes	Capacitors	Source
Diode clamped	2(N−1)	$\frac{N^2-1}{2}$	$\frac{N-1}{2}$	1
Flying capacitor	2(N−1)	2(N−1)	$\frac{(N-1)^2}{4}$	1
Cascaded H Bridge	2(N−1)	2(N−1)	$\frac{N-1}{2}$	$\frac{N-1}{2}$
Proposed MLI	$\frac{5(N-1)}{4}$	$\frac{3(N-1)}{2}$	$\frac{N-1}{2}$	$\frac{N-1}{2}$

TABLE 31.3
Comparative Analysis of Components, THD, and Voltage Stress of MMLI for Seven Levels

Sl. No.	MMLI Levels	THD
1	7 levels	4.38%
2	9 levels	4.18%

TABLE 31.4
Comparison of THD for Seven and Nine Levels

Topology	DC Source	Switches	THD (%)	Voltage Stress (%)
Anil Kumar Yarlagadda (2018) [25]	—	6	4.60%	24.12
Gopal Y (2021) [22]	3	7	3.0%	34.71
Arshad MH (2020) [23]	2	6	21.41%	25.64
Proposed Topology	**2**	**6**	**4.38%**	**22.14**

31.9 CONCLUSION

As the comparative analysis clearly shows that the symmetrical MLI for hybrid PV/wind with modified topology is helpful in increasing voltage levels with reduced switches. The method is effective in the reduction of the switching losses and cost of the converter. The modified MLI with reduced harmonic distortion has reduced switching losses and stress due to the reduced switches and STACOM used. Thus, using the ANN technique system efficiency has raised with modified MLI compared with other modified MLIs. The remote areas will be provided with a continuous power supply with the help of these hybrid renewable sources in grid integration. The proposed system of various levels is compared, and the analysis is carried out through simulation using MATLAB/SIMULINK. The ANN technique is used as an intelligent technique for improved performance. The future scope is to improve by increasing voltage levels to reduce harmonics and use of various renewable sources, which will be an added advantage.

REFERENCES

[1]. S. Selvakumar, Dr S. Thiruvenkadam, "A New Nine Level Hybrid H-Bridge Inverter for Photovoltaic-Wind Energy System", *IEEE Sponsored 9th International Conference on Intelligent Systems and Control (ISCO)*. IEEE Xplore, Year: 2015, Pages: 1–7.

[2]. G. Buticchi, E. Lorenzani, and G. Franceschini, "A Five-Level Single-Phase Grid-Connected Converter for Renewable Distributed Systems", *IEEE Trans. Ind. Electron.*, Year: 2013, Volume: 60, Issue: 3, Pages: 906–918.

[3]. Giampaolo Buticchi, "A Nine Level Grid Connected Converter Topology for Single Phase Transformerless PV System", *IEEE Trans. Indus. Electron.*, Year: 2014, Volume: 61, Issue: 8, Pages: 3951–3960.
[4]. Krishna Kumar Gupta and Shailendra Jain, "A Novel Multilevel Inverter Based on Switched DC Sources", *IEEE Trans. Ind. Electron.*, Year: 2014, Volume: 61, Issue: 7, Pages: 3269–3278.
[5]. Nurul Aisyah Yusof, Norazliani Md Sapari, Hazlie Mokhlis, and Jeyraj Selvaraj, "A Comparative Study of 5-Level and 7-Level Multilevel Inverter Connected to the Grid", *International Conference on Power and Energy*, IEEE, PEcon, Year: 2012, Pages: 542–547.
[6]. Mahajan Sagar Bhaskar Ranjana, Prasad Wankhade, and Nira Gondhalekar, "A Modified Cascaded H-Bridge Multilevel Inverter For Solar Applications", *International Conference on Green Computing Communication and Electrical Engineering (ICGCCEE).* IEEE Xplore, Year: 2014, Pages: 1–7.
[7]. Ataollah Mokhberdoran and Ali Ajami, "Symmetric and Asymmetric Design and Implementation of New Cascaded Multilevel Inverter Topology", *IEEE Trans. Power Electron.*, Year: 2014, Volume: 29, Issue: 12, Pages: 6712–6724.
[8]. Marcelo A. Perez, Steffen Bernet, Jose Rodriguez, Samir Kouro, and Ricardo Lizana, "Circuit Topologies, Modelling, Control Schemes and Applications of Modular Multilevel Converters", *IEEE Trans. Power Electron.*, Year: 2015, Volume: 30, Issue: 1, Pages: 4–17.
[9]. Payam Farhadi, Mohammad Navidi, Milad Gheydi, Mehdi Pazhoohesh, and Hassan Bevrani, "Online Selective Harmonic Minimization for Cascaded Half-Bridge Multilevel Inverter Using Artificial Neural Network", *Intl Aegean Conference on Electrical Machines & Power Electronics (ACEMP), 2015 Intl Conference on Optimization of Electrical & Electronic Equipment (OPTIM) & 2015 Intl Symposium on Advanced Electromechanical Motion Systems (ELECTROMOTION).* IEEE Xplore, Year: 2015, Pages: 331–335.
[10]. T. Ilakkia and G. Vijayagowri, "Hybrid PV/Wind System for Reduction of Harmonics Using Artificial Intelligence Technique", *International Conference on Advances in Engineering, Science and Management (Icaesm-2012).* IEEE Xplore, Year: 2012, pages: 305–308.
[11]. A. Kumar, A. Dasgupta, and D. Chatterjee, "Comparative Analysis Different Optimization Technique: Harmonic Minimization in Multilevel Inverter", *7th International Conference on Power Electronics.* IEEE Xplore, Year: 2016, Pages: 1–5.
[12]. S. Vinnakoti and V.R. Kota, "Implementation of Artificial Neural Network Based Controller for a Five-Level Converter Based UPQC. *Alex. Eng. J.*, Year: 2018, Volume: 57, Issue: 3, Pages: 1475–1488.
[13]. M.S. Ranjana, P.S. Wankhade, and N.D. Gondhalekar, "A Modified Cascaded H-Bridge Multilevel Inverter for Solar Applications", *2014 International Conference on Green Computing Communication and Electrical Engineering (ICGCCEE).* IEEE, Year: 2014, Pages: 1–7.
[14]. R.M. Pachagade, M.S. Ranjana, P.K. Maroti, and R. Maheshwari, "A New Multilevel Inverter with Fewer Number of Control Switches", *2015 Conference on Power, Control, Communication and Computational Technologies for Sustainable Growth (PCCCTSG).* IEEE, Year: 2015, Pages: 246251.
[15]. M. Sandhu and T. Thakur, "Design and Modeling of Hybrid MPPT MJSC Photovoltaic and Wind Based Microgrid Using Multilevel Inverter", *2016 5th International Conference on Wireless Networks and Embedded Systems (WECON).* IEEE, Year: 2016, Pages: 1–5.

[16]. S. Selvakumar and S. Thiruvenkadam, "A New Nine Level Hybrid h-Bridge Inverter for Photovoltaic-Wind Energy System", *2015 IEEE 9th International Conference on Intelligent Systems and Control (ISCO).* IEEE, Year: 2015, Pages: 1–7.

[17]. B. Wang and M. Illindala, "Operation and Control of a Dynamic Voltage Restorer Using Transformer Coupled H-Bridge Converters", *IEEE Trans. Power Electron.*, Year: 2006, Volume: 21, Issue: 4, Pages: 1053–1061.

[18]. N.K. Dewangan, V. Gurjar, S.U. Ullah, and S. Zafar, "A Level-Doubling Network (LDN) for Cross-Connected Sources Based Multilevel Inverter (CCS-MLI)", *2014 IEEE International Conference on Power Electronics, Drives and Energy Systems (PEDES).* IEEE, Year: 2014, Pages: 1–4.

[19]. A. Bughneda, M. Salem, D. Ishak, S. Alatai, M. Kamarol, and K.B. Hamad, "Modified Five-level Inverter for PV Energy System with Reduced Switch Count", *2021 IEEE Industrial Electronics and Applications Conference (IEACon).* IEEE Xplore, Year: 2022, Pages: 103–107.

[20]. M.A. Memon, M.D. Siddique, S. Mekhilef, and M. Mubin, "Asynchronous Particle Swarm Optimization-Genetic Algorithm (APSO-GA) Based Selective Harmonic Elimination in a Cascaded H-Bridge Multilevel Inverter", *IEEE Trans. Ind. Electron.*, Year: 2021, Volume: 69, Issue: 2, Pages: 1477–1487.

[21]. M.A. Bimazlim, B. Ismail, M.Z. Aihsan, S.K. Mazalan, M.S. Walter, and M.N. Rohani, "Comparative Study of Optimization Algorithms for SHEPWM Five-Phase Multilevel Inverter", *IEEE International Conference on Power and Energy (PECon).* IEEE, Year: 2020, Pages: 95–100.

[22]. Y. Gopal, K.P. Panda, D. Birla, and M. Lalwani, "Swarm Optimization-Based Modified Selective Harmonic Elimination PWM Technique Application in Symmetrical H-Bridge Type Multilevel Inverters", *Eng. Technol. Appl. Sci. Res.*, Year: 2019, Volume: 9, Issue: 1, Pages: 3836–3845.

[23]. M.H. Arshad, S. Khodijah, N.A. Kajaan, N.M. Nayan, Z.M. Isa, B. Ismail, and N.I. Abdullah, "Multiple Switching Pattern for a Modified Reduce Switch Multilevel Inverter: A Comparison Analysis", *J. Phys. Conf. Series*, Year: 2020, Volume: 1432, Issue: 1, Pages: 012027. IOP Publishing.

[24]. S. Ramavath and N. Yadaiah, "A Comparative Analysis of Modified and Updated Modified Cascaded H-Bridge Multilevel Inverter", *IEEE International Conference on Smart Generation Computing, Communication and Networking (SMART GENCON).* IEEE Xplore, Year: 2021, Pages: 1–6.

[25]. Anil Kumar Yarlagadda, Vargil kumar Eate, Y.S. Kishore Babu, and Abanishwar Chakraborti, "A Modified Seven Level Cascaded H Bridge Inverter", *5th IEEE Uttar Pradesh Section International Conference on Electrical, Electronics and Computer Engineering (UPCON).* IEEE Xplore, Year: 2018, Pages: 1–7.

32 Applications of Big Data in the Healthcare Sector
SLR through Network Analysis

Pankaj, Dr Payal Bassi, Dr Cheenu Goel

32.1 INTRODUCTION

Big data is the current buzzword in all economic sectors, and people everywhere are trying to figure out how to use the vast amounts of data to make judgements that will not only increase earnings but will also help them to better understand their customers. Big-data-enabled techniques are used by everyone from academics to researchers to politicians to businesses to merchants to even the planning of political social connections (Goyal and Saini, 2022). Huge amounts of data are produced almost daily, and businesses can effectively trace the history of their clients and provide customized solutions for them. Big data is proven to be particularly useful in the healthcare industry, where all patient records are kept electronically, making it possible to retain data about a patient's ailment, allergies, treatments, onset of illness, and progress toward recovery (Imran et al., 2021). Experts in the medical community may be able to diagnose the issue and identify the causes of disease by using big data techniques on the information currently accessible. The use of artificial-intelligence-enabled techniques that are more precise and less prone to error has supersede the traditional method of manually diagnosing the issue, according to Kaur et al. (2020), especially with diseases affecting the heart, diabetes, nervous system, kidney, liver, and other important body organs. The use of big data in healthcare and hospital administration has thereby transformed how diagnosis and treatment were previously carried out.

32.2 REVIEW OF LITERATURE

Big data, according to Borckardt et al. (2008), is very effective for historical approaches in diagnosing the symptoms of any disease because it makes it possible to analyse every last detail about the patient, including how their age, hormone levels, dietary habits, and the effects of any prescribed medications, in minute detail. By uncovering buried information in the patient's new location, climatic disruption, and other factors, it is possible to detect the interconnection of many diseases and their causes with simplicity (Celi et al., 2013). With the use of big data

DOI: 10.1201/9781003367154-32

approaches, a vast amount of organized and unstructured data can be filtered, thus making the process of understanding and treating the condition easier and clearer (Belle et al., 2015).

The use of big data analytics tools, according to De Silva et al. (2015), has made it possible for medical professionals to identify the underlying causes of infections, their scope, the causes of extended illnesses, their symptoms, and the efficacy of treatments. The healthcare industry, according to Kazançoğlu et al. (2021), is extremely fragile and is continually evolving, which presents new difficulties for healthcare professionals in terms of making wise decisions. Big data uses analytical techniques on the gathered data to offer a solution to these problems. According to Mehta and Pandit (2018), using big data analytics makes it possible to recognize patterns in previous data that has been digitally preserved, improving patient care.

The use of big data analytics has increased as a result of the e-recording of data in the healthcare industry, according to Hiba et al. (2015). This is because every piece of information makes it possible to access, process, and comprehend the data that has been stored, allowing for the drawing of conclusions. According to Mohammad et al. (2015), careful handling of this data can facilitate effective service delivery design, thereby reducing gaps between customer expectation and service delivery. The presence of various amounts of information related to a single patient and a single disease result in heaps of information in the digital repositories (Khalifa and Zabani, 2016). Instead of treating all patients with a single treatment that produces similar outcomes for all of them, the single solution is to apply big data analytics tools and techniques to transform electronic data into personalized individual solutions.

Electronic records of patients who visit a medical professional and the problems they describe on a regular basis can help the medical representative to plan for the staff, facility, and medical equipment needed to treat the patients (Sadineni, 2020). By using big data analytics effectively, big data offers amazing answers for the healthcare sector's problems and results in cost reduction (Roski et al., 2014). Companies must effectively manage the vast amounts of data they have at their disposal, according to Bassi et al. (2022), because only then will they be able to develop competitive strategies in this fierce market. The present study deals with the research questions that were developed to attain answers after thoroughly gazing in the selected papers chosen as a sample in the study: Which journals and countries have published the most articles, especially in the field of big data analytics in healthcare? Who are the authors who have collaborated with other nations for the purpose of research in the selected field? What is the trend of year-wise publication? Which keywords are the most commonly occurring words across different countries and documents?

32.3 LITERATURE SEARCH

Prisma model, as shown in Figure 32.1, was designed after application of related keywords and hence a final string for both the databases (ProQuest and Scopus) was attained thus resulting in 1,707 papers in phase 1. In the second phase, full-text, peer-reviewed, and scholarly journals published in the said domain from 2012 to 2022

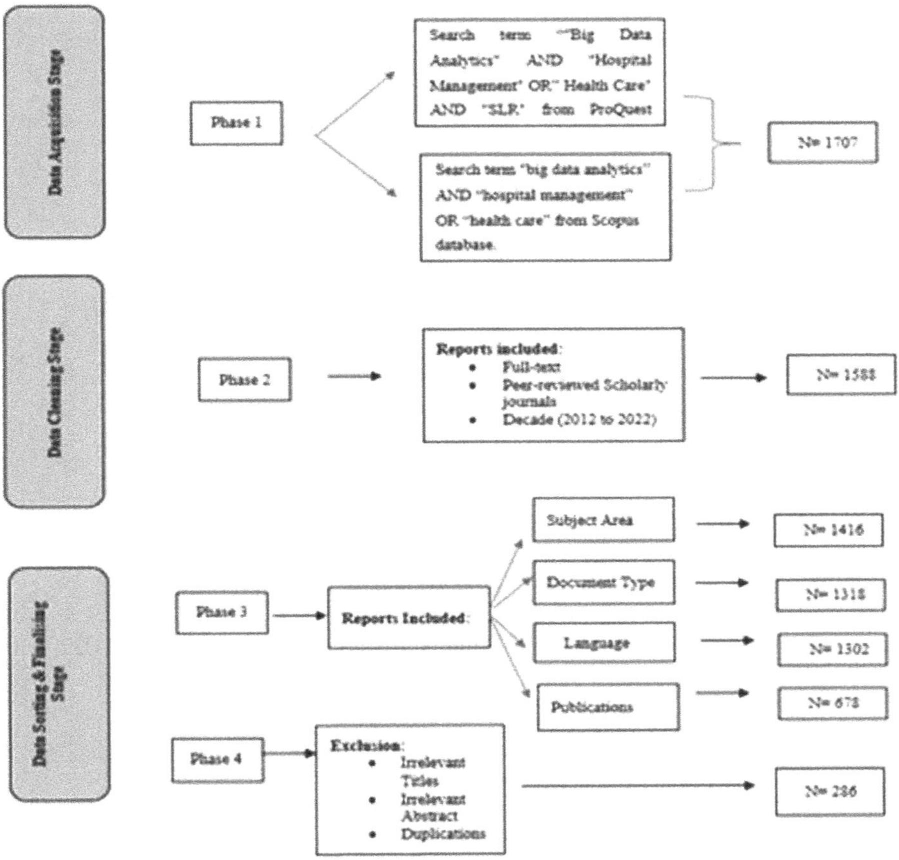

FIGURE 32.1 Prisma model.

were included, which resulted in repository of 1,588 papers. During the third phase, subject area, document type, language restriction, and publication stage in both the databases were applied, and hence, 678 papers were attained. Further, all the articles were thoroughly inspected in terms of titles, abstracts, and full-text papers and duplications were removed and thus reducing the sample size to 286 in phase 4.

32.4 NETWORK ANALYSIS

32.4.1 Year-Wise Publication Details

Figure 32.2 demonstrates the annual publication detail in terms of numbers and percentage in the field of big data and healthcare sector. In year 2012, no paper was recorded in the selected databases; however, from 2013 onwards, the number of publications increased steadily, reaching out to the highest publication score of 62 papers

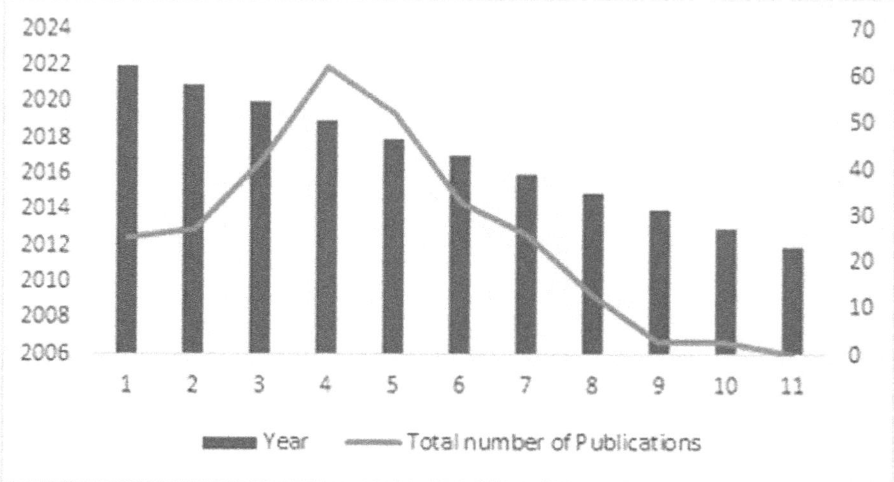

FIGURE 32.2 Publications by year.

in year 2019. The publications felled in number from year 2020 to August 2022, with 25 documents to its credit in 2022.

32.4.2 Publication by Document Type

The document types that are used in the current study for the aim of systematic literature are shown in Figure 32.3. The sample indicated that the whole sample consists of 286 articles, 139 conference papers, and 6 academic journals.

32.4.3 Top Ten Journals

Table 32.1 demonstrates the information related to the most influential journals in the field of big data and healthcare. It is further evident that the ACM International Conference Proceeding Series, Procedia Computer Science, and IEEE Access Journals may all be easily seen as the top three journals in the category based on the number of articles they have published in the aforementioned topic. However, Technological Forecasting and Social Change has the highest impact factor of 10.88, but only 4 papers are able to get published in the journal during the selected period of study.

32.4.4 Keyword and Title Clouds

Wordart.com, an open-source word cloud generator, is used to produce a word cloud of the terms used by authors. According to the search results in Figure 32.4 (a), authors often use the phrases "big data," "data analytics," "healthcare," "algorithm," "diagnosis," "data mining," "machine learning," and many more. The keywords

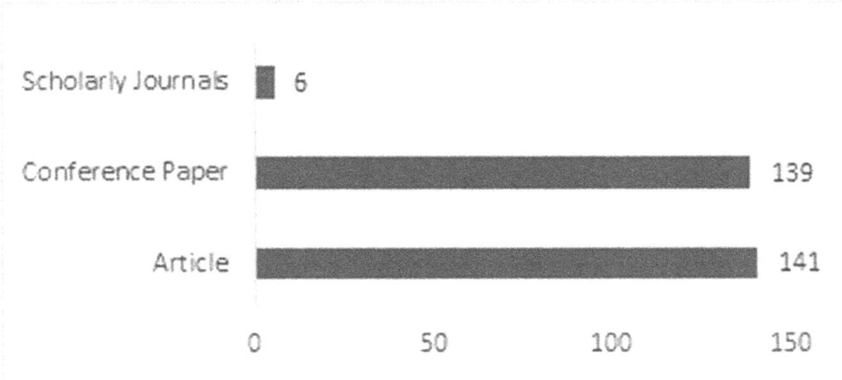

FIGURE 32.3 Document type.

TABLE 32.1
Influential Journals on Big Data and Healthcare*

Source	Documents	Impact Factor
ACM International Conference Proceeding Series	8	0.55
Procedia Computer Science	7	2.27
IEEE Access Journals	6	3.476
International Journal of Environmental Research and Public Health	5	3.39
International Journal of Recent Technology and Engineering	5	0.675
Computers and Electrical Engineering	4	4.89
Future Generation Computer Systems	4	7.187
IEEE Journal of Biomedical and Health Informatics	4	5.772
International Journal of Scientific and Technology Research	4	0.987
Technological Forecasting and Social Change	4	10.88

*The table shows the top ten journals with the column headings being the source, document, and impact factor. ACM International Conference Proceeding Series is the top journal with eight articles and an impact factor of .55.

written in small fonts has less representation in the chosen articles taken from both databases, the most prominent keywords with large font sizes are the most often searched key phrases. However, this does not mean that the words that appear in small fonts are not or less important for the chosen subject. The word cloud in Figure 32.4 (b) shows the titles' specific word cloud, which includes terms like "Internet of Things," "Cloud Computing," "Healthcare System," "Challenge," and "Artificial Intelligence," in addition to "Big Data," "Big Data Analytics," and "Healthcare."

FIGURE 32.4 (a) Word cloud of authors' keywords (b) Word cloud of titles.

32.4.5 Network Association of Co-authorship with Countries

Figure 32.5 indicates that authors from India published the most number of papers, 119 papers, with a citation score of 1,883 and total link strength of 33, followed by the United States and the United Kingdom, with 59 and 24 documents, respectively. However, Turkey, Morocco, and Romania are the least co-authored countries, with the publication of 5 documents each. Thus, authors from United States are considered the most influential researchers in the field of big data and the healthcare sector as their work is most cited.

32.4.6 Co-occurrence of All Keywords

Figure 32.6 presents the insights based on co-occurrence of all keywords. The data objectively revealed that 140 out of 2,283 keywords meet the threshold with

FIGURE 32.5 Co-authorship with countries.

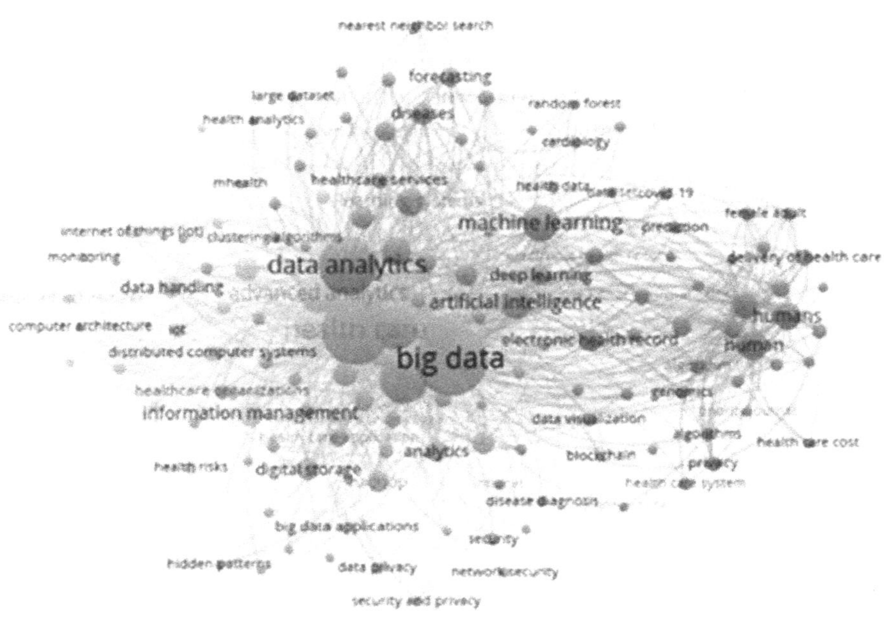

FIGURE 32. 6 Network diagram of co-occurrence of all keywords.

the conditional threshold of minimum 5 occurrences. It was discovered that there appeared 3 clusters with the 41 links and 15 items where "big data" appeared 234 times as a keyword, followed by "healthcare" and "data analytics," which appeared 202 and 147 times as keywords in all the selected papers, thus being gauged as second and third most preferred keywords, respectively. The results were obtained with the help of the VOS viewer software.

32.4.7 Network Association of Co-occurrence of Index Keywords

The network association of indexed terms among the chosen articles under examination from the ProQuest and Scopus databases is shown in Figure 32.7 This network connection was created in August 2022. The network uses a comprehensive counting approach to disclose information about co-occurrence and indexed keywords. "Big data," "healthcare," "data analytics," "advance analytics," and "data mining" are the five most often occurring keywords out of all those stated, with total link strength of 1601, 1426, 1091, 531, and 398, respectively. The least frequently used terms were big data analytics, healthcare sectors, and diseases with total occurrences of 25, 24, and 21, respectively.

32.4.8 Citation of Documents

Figure 32.8 represents the information related to citation of documents, and it comprises the information of 5 clusters and 206 documents, which are cited in the work of other authors. Wang et al. (2018) is the most cited author among the field of big data in health sector. His work has been cited 688 times, followed by Chen et al. (2017) and Zhang et al. (2015), whose work is the second and third most cited work in the selected domain of study.

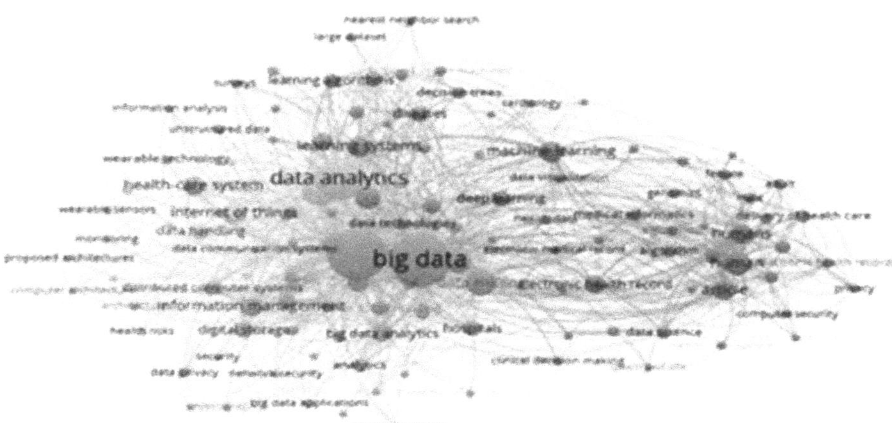

FIGURE 32.7 Network analysis of co-occurrence of index keywords.

32.4.9 Citation of Countries

The fields of big data and healthcare are being studied by researchers all around the world. Figure 32.9 shows the detail of citations and countries among research articles obtained from the two selected databases, ProQuest and Scopus, during the last ten years for the period of 2012–2022. It is evidently clear that the research work conducted by the authors in the United States, China, and the United Kingdom received the highest number of citations, with total scores of 3105, 2059, and 1883, respectively.

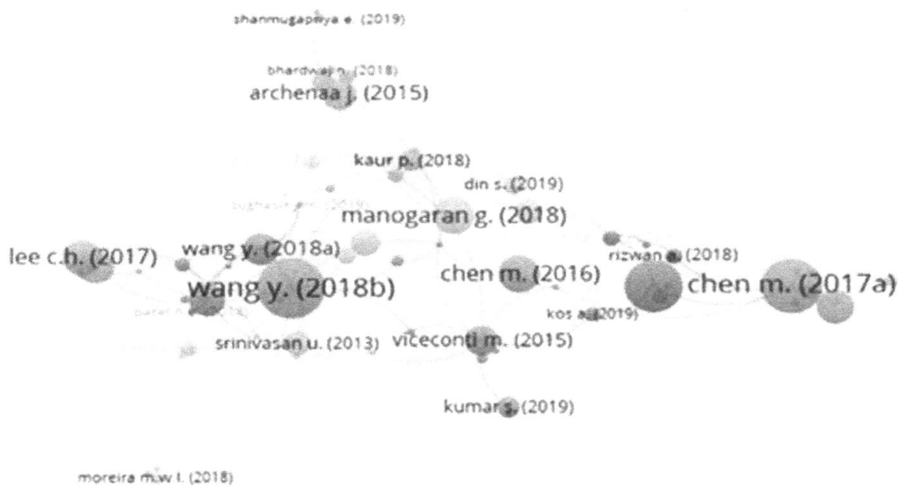

FIGURE 32.8 Network analysis of citation of documents.

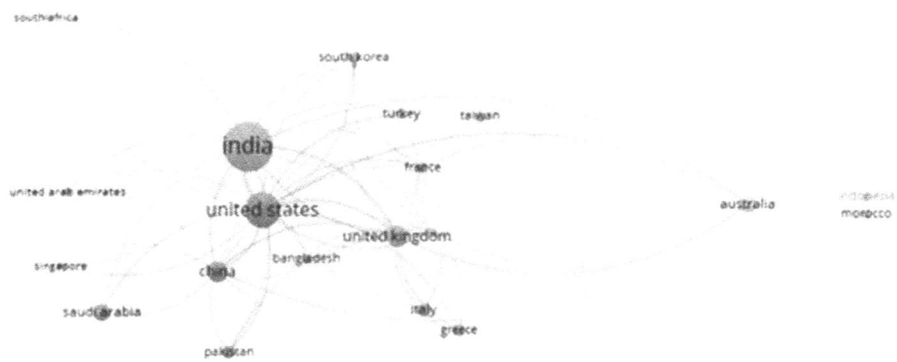

FIGURE 32.9 Network analysis diagram of the citation of countries.

32.4.10 Bibliographic Coupling of Documents

Figure 32.10 depicts the network retrieved via bibliographic coupling of the same set of referenced publications. The network was created with the help of the VOS viewer software with the minimum number of 4 citations per document, and 286 documents met the threshold. Articles published by Wang and Hajli (2017) and Wang et al.

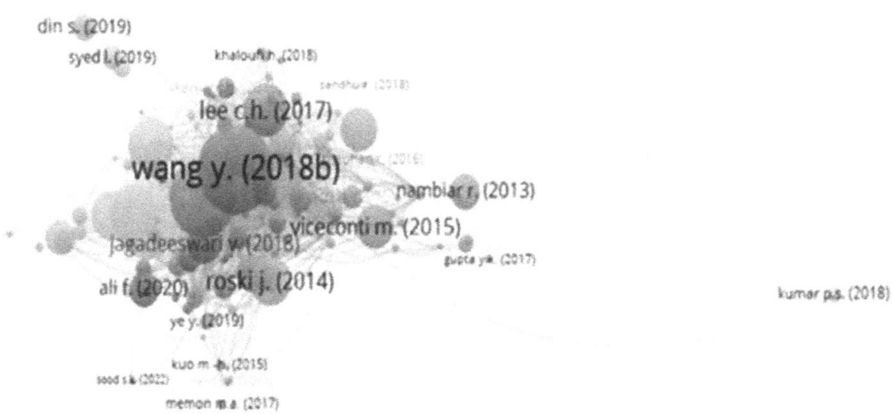

FIGURE 32.10 Bibliographic coupling of documents.

FIGURE 32.11 Bibliographic coupling of countries.

(2018), have shown maximum bibliographic coupling of documents as in his research his previous works are also well cited, followed by Ta et al. (2016) and Kumar et al. (2019) as the second and third most cited bibliographic coupling of documents.

32.4.11 Bibliographic Coupling of Countries

Bibliographic country-specific contributions in the field of big data and healthcare can be visualized in Figure 32.11. The association network was created in VOS viewer software. The restriction of minimum 5 documents was applied, and thus, 20 clusters with 170 items emerged with links of 1,552 and total link strength of 2,165. India, the United States, and the United Kingdom have shown the maximum bibliographic coupling of countries in their research work during the chosen period of study.

32.5 CONCLUSION

Big data analytics has emerged as a potential topic which has provided much gainful insights to the people in business and have enabled them to upscale their business to the new heights of success, and the field of healthcare is not an exception. Practitioners of healthcare are able to regulate and record the data of their patients in a much effective manner, thus keeping a better track of their patients and the diseases that affect them. This chapter has provided with a sound literature where the transfusion of technology in the form of big data has permitted healthcare representatives to gather knowledge from their clinical and other data repositories and make wise judgements for the treatment of patients. The study comprehends the publications undertaken in the said domain during the last decade, and the data shows that maximum number of papers published in 2019. The authors of related research often collaborate with each other to examine how a particular concept could strengthen the theoretical and practical foundations of a particular field, and thus, in the present study, authors from India, the United States, and the United Kingdom are teaming up to explore the aspects of big data and healthcare with the keywords such as "big data," "machine learning," "healthcare," "healthcare system," "internet of things," "advance analytics," "decision-making," "future research," "data technology," and many more. Apart from various journals, articles, conference papers, conference proceedings, and book chapters, the most cited literature has been published by the sources such as ACM International Conference Proceeding Series, Procedia Computer Science, IEEE Access Journals, International Journal of Environmental Research and Public Health, International Journal of Recent Technology and Engineering, and other cited Scopus and ProQuest sources. Research work of authors such as Wang Y., Ta V. D., and Kumar S. have emerged as the most influential work for the bibliographic studies taken by emerging researchers. The study has acted as a novelty in already-published literature and has developed the key insights that are being studied as of now. It has successfully established the gap between the application of big data in the field of healthcare, thus acting as a layout for the future researchers to develop further understanding in the simulation of AI, big data, and machine learning in the domain of healthcare.

REFERENCES

Alkhatib, M. A., Talaei-Khoei, A., & Ghapanchi, A. H. (2015). Analysis of research in healthcare data analytics. *Australasian Conference on Information Systems*, 1–16.

Asri, H., Mousannif, H., Al Moatassime, H., & Noel, T. (2015). Big data in healthcare: Challenges and opportunities. *International Conference on Cloud Technologies and Applications (CloudTech)*, 1–7.

Bassi, P., & Kaur, J. (2022). Comparative predictive performance of BPNN and SVM for Indian insurance companies. In Sood, K., Balusamy, B., Grima, S., & Marano, P. (Eds.), *Big Data Analytics in the Insurance Market (Emerald Studies in Finance, Insurance, and Risk Management)*, Emerald Publishing Limited, Bingley, pp. 21–30. https://doi.org/10.1108/978-1-80262-637-720221002.

Belle, A., Thiagarajan, R., Reza Soroushmehr, S. M., Navidi, F., Beard, D. A., & Najarian, K. (2015). Big data analytics in healthcare. *BioMed Research International*, 2015(Article ID 370194), 16. https://doi.org/10.1155/2015/370194.

Borckardt, J. J., Nash, M. R., Murphy, M. D., Moore, M., Shaw, D., & O'Neil, P. (2008). Clinical practice as natural laboratory for psychotherapy research: A guide to case-based time-series analysis. *The American Psychologist*, 63(2), 77–95.

Celi, L. A., Mark, R. G., Stone, D. J., & Montgomery, R. A. (2013). "Big data" in the intensive care unit: Closing the data loop. *American Journal of Respiratory and Critical Care Medicine*, 187(11), 1157–1160.

Chen, M., Hao, Y., Hwang, K., Wang, L., & Wang, L. (2017). Disease prediction by machine learning over big data from healthcare communities. *IEEE Access*, 5, 8869–8879.

De Silva, D., Burstein, F., Jelinek, H. F., & Starnieri, A. (2015). Addressing the complexities of big data analytics in healthcare: The diabetes screening case. *Australian Journal Information System*, 19.

Goyal, I., Singh, A., & Saini, J. K. (2022). Big data in healthcare: A review," *2022 1st International Conference on Informatics (ICI)*, pp. 232–234. https://doi.org/10.1109/ICI53355.2022.9786918.

Imran, S., Mahmood, T., Morshed, A., & Sellis, T. (2021). Big data analytics in healthcare a systematic literature review and roadmap for practical implementation. *IEEE/CAA Journal of Automatica Sinica*. http://doi.org/10.1109/JAS.2020.1003384.

Kaur, S., Singla, J., Nkenyereye, L., Jha, S., Prashar, D., Joshi, G. P., El-Sappagh, S., Islam, Md. S., Islam, S. M. R. (2020). Medical diagnostic systems using artificial intelligence (AI) algorithms: Principles and perspectives. *IEEE Access*, 8, 228049–228069. http://doi.org/10.1109/ACCESS.2020.3042273.

Kazançoğlu, Y., Sağnak, M., Lafcı, Ç., Luthra, S., Kumar, A., & Taçoğlu, C. (2021). Big data-enabled solutions framework to overcoming the barriers to circular economy initiatives in healthcare sector. *International Journal of Environmental Research and Public Health*, 18, 7513. https://doi.org/10.3390/ijerph18147513.

Khalifa, M., & Zabani, I. (2016). Utilizing health analytics in improving the performance of healthcare services: A case study on a tertiary care hospital. *Journal of Infection and Public Health*, 9. https://doi.org/10.1016/j.jiph.2016.08.016.

Kumar, S. R., Gayathri, N., Muthuramalingam, S., Balamurugan, B., Ramesh, C., & Nallakaruppan, M. K. (2019). Medical big data mining and processing in e-healthcare. In Mary Beth Privitera, (Ed.), *Internet of Things in Biomedical Engineering*, Cambridge, Massachusetts, US: Academic Press, pp. 323–339.

Mehta, N., & Pandit, A. (2018). Concurrence of big data analytics and healthcare: A systematic review. *International Journal of Medical Informatics*, 114, 57–65.

Roski, J., Bo-Linn, G. W., & Andrews, T. A. (2014). Creating value in health care through big data: Opportunities and policy implications. *Health Affairs*, 33(7), 1115–1122. http://doi.org/10.1377/hlthaff.2014.0147.

Sadineni, P. K. (2020). Developing a model to enhance the quality of health informatics using big data. *2020 Fourth International Conference on I-SMAC (IoT in Social, Mobile, Analytics and Cloud) (I-SMAC)*. http://doi.org/10.1109/I-SMAC49090.2020.9243395.

Ta, V. D., Liu, C. M., & Nkabinde, G. W. (2016, July). Big data stream computing in healthcare real-time analytics. *2016 IEEE International Conference on Cloud Computing and Big Data Analysis (ICCCBDA)* (pp. 37–42). IEEE.

Wang, Y., & Hajli, N. (2017). Exploring the path to big data analytics success in healthcare. *Journal of Business Research*, 70, 287–299.

Wang, Y., Kung, L., & Byrd, T. A. (2018). Big data analytics: Understanding its capabilities and potential benefits for healthcare organizations. *Technological Forecasting and Social Change*, 126, 3–13.

Zhang, Y., Qiu, M., Tsai, C. W., Hassan, M. M., & Alamri, A. (2015). Health-CPS: Healthcare cyber-physical system assisted by cloud and big data. *IEEE Systems Journal*, 11(1), 88–95.

Index